NATO ASI Series

Advanced Science Institutes Series

A series presenting the results of activities sponsored by the NATO Science Committee, which aims at the dissemination of advanced scientific and technological knowledge, with a view to strengthening links between scientific communities.

The Series is published by an international board of publishers in conjunction with the NATO Scientific Affairs Division

A Life Sciences B Physics	Plenum Publishing Corporation London and New York
C Mathematical and Physical Sciences D Behavioural and Social Sciences E Applied Sciences	Kluwer Academic Publishers Dordrecht, Boston and London
F Computer and Systems Sciences G Ecological Sciences H Cell Biology I Global Environmental Change	Springer-Verlag Berlin Heidelberg New York London Paris Tokyo Hong Kong Barcelona Budapest

PARTNERSHIP SUB-SERIES

1. Disarmament Technologies	Kluwer Academic Publishers
2. Environment	Springer-Verlag/Kluwer Academic Publishers
3. High Technology	Kluwer Academic Publishers
4. Science and Technology Policy	Kluwer Academic Publishers
5. Computer Networking	Kluwer Academic Publishers

The Partnership Sub-Series incorporates activities undertaken in collaboration with NATO's Cooperation Partners, the countries of the CIS and Central and Eastern Europe, in Priority Areas of concern to those countries.

NATO-PCO DATABASE

The electronic index to the NATO ASI Series provides full bibliographical references (with keywords and/or abstracts) to about 50000 contributions from international scientists published in all sections of the NATO ASI Series. Access to the NATO-PCO DATABASE is possible via a CD-ROM "NATO Science & Technology Disk" with user-friendly retrieval software in English, French and German (© WTV GmbH and DATAWARE Technologies Inc. 1992).

The CD-ROM can be ordered through any member of the Board of Publishers or through NATO-PCO, Overijse, Belgium.

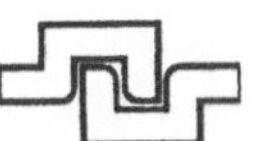

Series H: Cell Biology, Vol. 106

Springer
Berlin
Heidelberg
New York
Barcelona
Budapest
Hong Kong
London
Milan
Paris
Santa Clara
Singapore
Tokyo

Lipid and Protein Traffic

Pathways and Molecular Mechanisms

Edited by

Jos A. F. Op den Kamp

Centre for Biomembranes and Lipid Enzymology
Padualaan 8, P.O. Box 80054
3508 TB Utrecht, The Netherlands

With 91 Figures

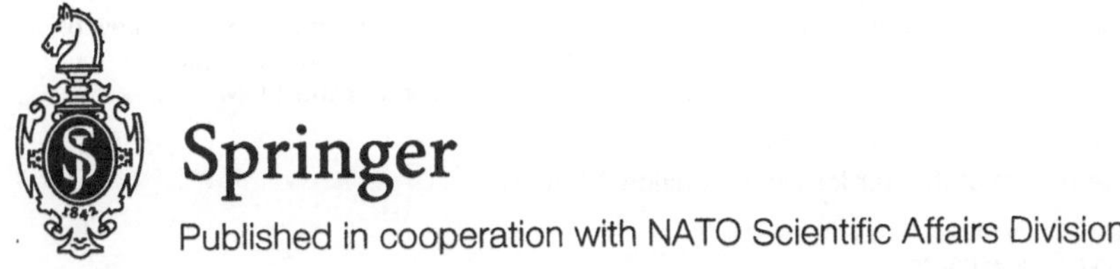

Proceedings of the NATO Study Institute "Molecular Mechanisms of Lipid and Protein Traffic", held at Cargèse, France, June 16–28, 1997.

Library of Congress Cataloging-in-Publication Data

NATO Study Institute "Molecular Mechanisms of Lipid and Protein Traffic" (1997: Cargèse, France)
Lipid and protein traffic: pathways and molecular mechanisms / edited by Jos A. F. Op den Kamp.
p. cm. – (NATO ASI series. Subseries H "Cell biology"; 106)
Includes index.

1. Membrane proteins–Physiological transport–Congresses.
2. Membrane lipids–Physiological transport–Congresses.
I. Kamp, Jos A. F. op den (Jos Arnoldus Franciscus), 1939– . II. Title. III. Series.
QP552.M44N38 1997 571.6'4–dc21 98-2792 CIP

ISSN 1010-8793

ISBN 978-3-642-51465-4 ISBN 978-3-642-51463-0 (eBook)
DOI 10.1007/978-3-642-51463-0

Softcover reprint of the hardcover 1st edition 1998

Typesetting: Camera ready by authors/editors
Printed on acid-free paper
SPIN 10553982 31/3137 - 5 4 3 2 1 0

PREFACE

The meeting on "Molecular Mechanisms of Lipid and Protein Traffic", sponsored by NATO Scientific Affairs Division as an Advanced Study Institute and by the Federation of European Biochemical Societies as a Lecture Course was held in Cargèse, France, in June 1997. The program included introductory lectures, specialized up-to-date contributions, poster sessions, tutorials and workshops. Emphasis was laid on the new developments in the field of membrane dynamics, in particular on the insertion and translocation of proteins; on intracellular protein, lipid and membrane traffic and sorting and on the mutual interactions between the various events occurring during membrane biogenesis. Mitochondrial biogenesis, chloroplast assembly and the insertion of proteins into peroxisomes were highlighted.

Much progress in these research areas has been made in recent years and the ASI provided an excellent opportunity to illustrate this progress in comparison with previous meetings on a similar topic. Not only graduate students and postdocs took advantage from this program but also experienced scientists were given the opportunity to obtain a complete overview of recent progress and the remodeling of ideas and concepts.

The present publication presents most of the important contributions by teachers as well as students in the meeting. Overviews of protein insertion and assembly in mitochondria, chloroplasts and peroxisomes clearly illustrate that our knowledge of membrane biogenesis is increasing rapidly. A similar conclusion can be drawn for the complicated processes of protein translocation and secretion and the accompanying sorting processes. Detailed information on the basic mechanistic characteristics of these events become available and some examples of the new developments are presented. Toxins, acting both from the intracellular as well as from the outside on (plasma)membranes are discussed in detail including the effects of these compounds on membrane pores, channels and conductance.

Lipid structure and function are discussed in relation to the above mentioned processes of protein insertion into and translocation through membranes and finally some insight is presented in the mechanisms underlying intracellular sorting and trafficking of specific phospholipids.

December, 1997

Jos A.F. Op den Kamp

CONTENTS

Import, Folding and Degradation of Proteins by Mitochondria

Gottfried Schatz, Biozentrum, University of Basel, CH-4056 Basel, Switzerland. phone: ++41 61 267 2150; fax: ++41 61 267 2175; schatz@ubaclu.unibas.ch

The problem

A typical eukaryotic cell contains in the order of 10^4 different proteins. About 10% of these are located in mitochondria. As mitochondria themselves synthesize only approximately one dozen proteins, >99% of the mitochondrial proteins are made in the cytosol and then imported (1). How do mitochondria recognize the proteins they must import? How do they transport them across one or even both of their membranes? And how do they degrade them? The answers to these questions bear not only on the mechanism of mitochondrial biogenesis, but also tell on the mechanism(s) by which proteins are transported across biological membranes (2).

The mitochondrial protein import machine

Most of the work on the import of proteins into mitochondria has been conducted with the yeast *Saccharomyces cerevisiae* and the mold *Neurospora crassa* (3,4). The results obtained from these systems agree well and the key findings seem to apply also to plants and mammals. In brief, a protein destined to be imported into the mitochondrial interior (the "matrix") is made with an N-terminal signal sequence that can form a basic amphiphilic helix. This "precursor" is bound by cytosolic chaperones that prevent its aggregation in the cytosol. The chaperone-borne precursor then binds to a hetero-oligomeric import receptor on the mitochondrial outer membrane, inserts “head first” into a protein transport channel (the Tom channel) of the outer membrane, and binds via its presequence to acidic domains of several outer membrane proteins at the inner face of the outer membrane. The positively charged presequence is then pulled across a protein transport channel (the Tim channel) in the inner membrane and is then captured by a mitochondrial 70 kDa heat-shock protein that is bound to the inner face of the inner membrane. Mhsp70, together with a mitochondrial GrpE homolog and the membrane anchor Tim44 functions as an ATP-driven protein import motor that uses the energy of ATP

NATO ASI Series, Vol. H 106
Lipid and Protein Traffic
Pathways and Molecular Mechanisms
Edited by Jos A. F. Op den Kamp

hydrolysis to pull the precursor chain completely across the inner membrane. As neither of the two protein transport channel allows the passage of native, folded proteins, the precursor emerges in the matrix in an unfolded conformation. Its presequence is removed by a specific processing protease (termed MMP) and the mature protein then refolds with the aid of one or more of several chaperones in the matrix (Fig. 1).

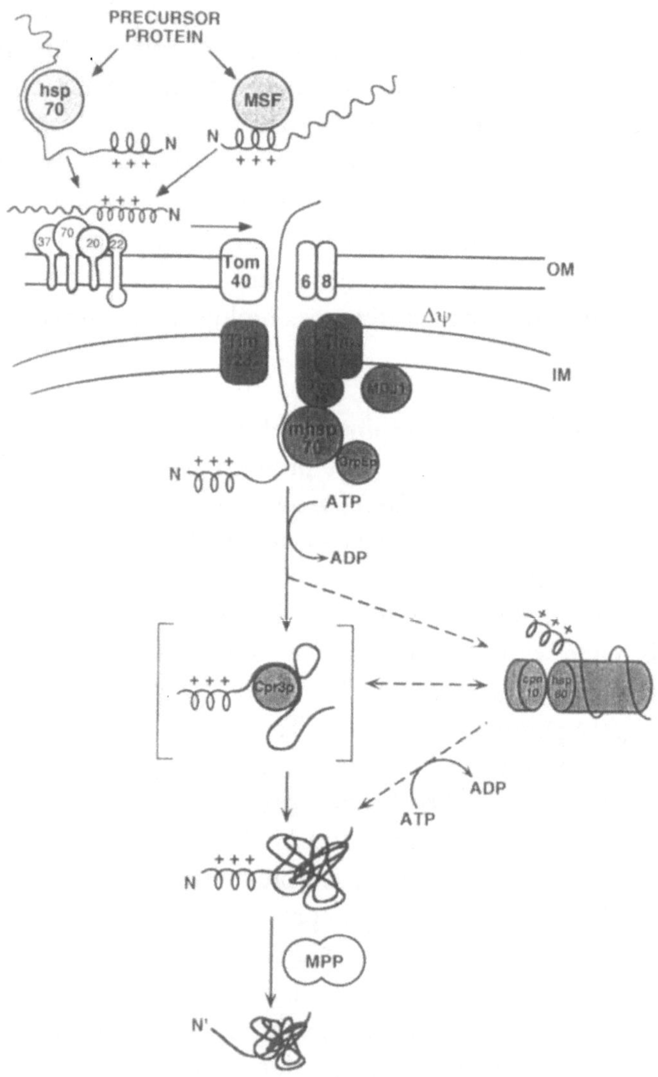

Fig. 1: The key components of the protein import machine of yeast mitochondria.

Molecular dissection of the receptor system

At least four integral outer membrane proteins function as import receptor (5): Tom20, Tom22, Tom37 and Tom70. Tom20/22 form a subcomplex which exposes acidic regions to the cytosol and seems to bind the basic mitochondrial presequences by a largely electrostatic interaction. This "acid bristle" receptor subcomplex appears to bind preferentially precursors that are delivered as a complex with cytosolic hsp70. Tom37/70 appears to prefer precursors bound to another class of cytosolic chaperones that, in mammals, are represented by mitochondrial import stimulating factor (MSF). Tom37/70 also bind the precursor's presequence, but not through electrostatic interactions. Once Tom37/70 has bound a precursor, it delivers it to Tom20/22. Only then is the precursor transported across the outer membrane (Fig. 2).

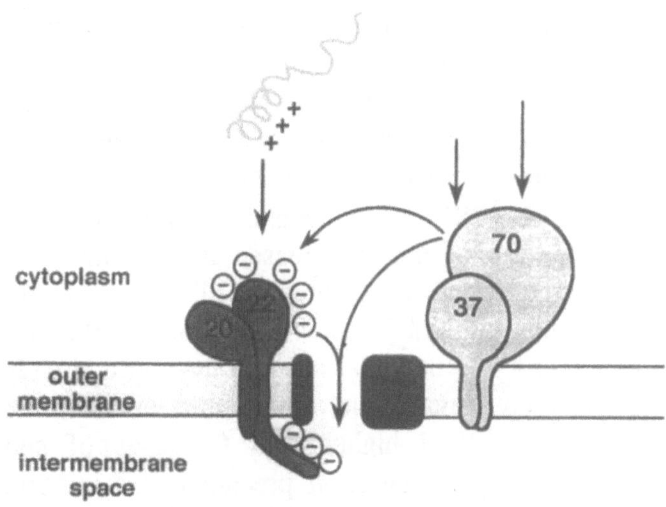

Fig. 2: Interaction of mitochondrial protein import receptor domains on both sides of the outer membrane: The "binding relay" model.

We have reconstituted several segments of this postulated "binding relay" in a completely soluble, defined system composed of a purified precursor protein, purified cytosolic chaperones, and recombinantly expressed and purified extra-membraneous receptor domains (6). These studies show that such a mechanism can work in principle.

The ATP-driven protein import motor in the inner membrane

Work by B. Glick in our laboratory has led to the suggestion (7) that mhsp70 binds the incoming precursor chain and then undergoes a large ATP-driven conformational change which pulls a segment of the precursor chain into the matrix. Release of the ADP from mhsp70 by the nucleotide-exchange factor GrpE would then reset the system for another round. One key feature of this model is that it proposes that the motor can actively unfold those precursors that fold in the cytosol before import into the mitochondria. One such precursor is that of cytochrome b_2, a soluble, heme-binding protein of the intermembrane space. In order to test this model, Andreas Matouschek has created two series of artificial precursors containing either mouse dihydrofolate reductase or bacterial barnase as "mature" regions (8). Each of these proteins was fused at its N-terminus to a series of matrix-targeting sequences whose length varied from 25 to 95 amino acids. When these constructs were synthesized in a reticulocate lysate in the presence of [^{35}S]-methionine and then presented to energized yeast mitochondria, the rate of import increased dramatically (40 to < 1000 fold) when the presequence was long enough to allow the folded precursor to span both membranes and bind to the import motor in the matrix. With the denatured precursors, the maximal import rates were already reached even with the "short" constructs. The import rates of the folded "long" precursors were up to 20fold higher than the rates of spontaneous unfolding. The results suggest that short precursors are imported slowly because they must first spontaneously unfold outside the mitochondria in order to bind to the ATP-driven import motor. In contrast, the long precursors bypass the slow spontaneous unfolding step because they are actively unfolded by the import motor which, by pulling at the precursor's N-terminal region, unfolds its on the mitochondrial surface (Fig. 3). We are currently purifying large amounts of the hetero-trimeric import motor (mhsp70, mGrpE, Tim44) in order to study its mode of action.

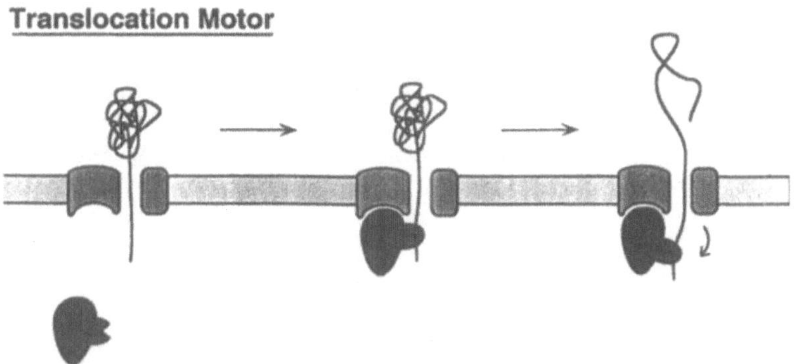

Fig. 3: Model for a force-generating ATP-driven mitochondrial import motor (adapted from ref. 7).

Protein insertion into the inner membrane

Many transmembrane carriers of the mitochondrial inner membrane are first transported across the outer membrane by the pathway just described and then insert directly into the inner membrane without passing through the matrix space. Accordingly, their insertion into the inner membrane does not require the action of the ATP-driven import motor in the matrix. We have reconstituted the insertion of the adenine nucleotide carrier into liposomes containing solubilized inner membrane proteins (9). Insertion requires an electrochemical potential across the liposome membrane and is inhibited by antibody against Tim23, a subunit of the Tim channel. We are currently using this functional assay to identify the components that participate in the insertion of multispanning inner membrane proteins. The results obtained so far suggest that protein insertion into the inner membrane may be a partial reaction of protein transport into the matrix.

Folding of imported proteins in the matrix

Initially it was assumed that all proteins imported into the matrix fold with the aid of the hsp60/hsp10 chaperonin system, a close homolog of the bacterial GroEL/GroES system. However, S. Rospert has shown that different imported proteins fold with the aid of different matrix-localized chaperones (10). She has also suggested the possibility that some proteins may fold without chaperone-assistance, in contrast to current dogma. The matrix-located proline rotamase Cpr3p can also accelerate folding of some proteins. The folding of each protein thus represents a separate case (Fig. 4)

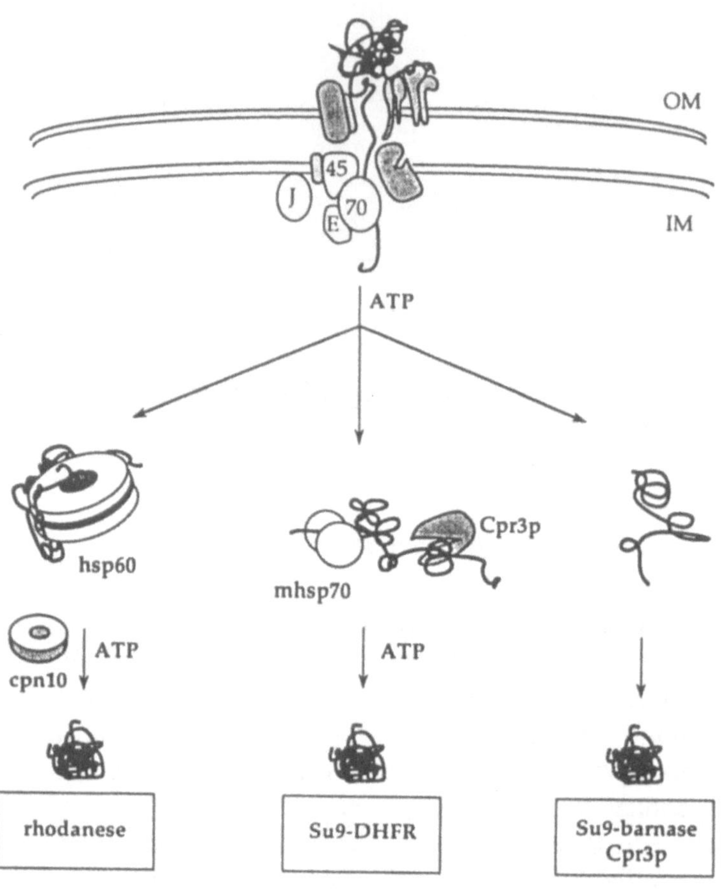

Fig. 4: The matrix-localized protein refolding system.

Turnover of mitochondrial proteins

As mitochondria have evolved from bacteria-like ancestors, K. Suzuki has speculated that they might have retained the ATP-dependent protein degradation systems known to exist in present-day bacteria. She has used the polymerase chain reaction to amplify and clone a nuclear yeast gene encoding a homolog of the ATP-dependent bacterial protease Lon (11). The predicted protein product of the yeast *LON* gene closely resembled Lon from *Escherichia coli* except for the presence of an N-terminal matrix-targeting signal and a highly charged region separating the ATP-binding region (characterized by typical ATP-binding motifs) from the proteolytic region (characterized by the diagnostic serine residue of bacterial Lon proteases). Indeed, the protein was located in the matrix. Yeast deleted for the *LON* gene lost the ability to degrade soluble matrix proteins, accumulated electron-dense inclusion bodies in the matrix, and also lost functional mitochondrial DNA. The same effects were seen when the essential serine residue in the proteolytic site was mutated to alanine. Surprisingly, however, the proteolytically inactive Lon could suppress the phenotypic defects of mutants lacking another type of protease which is located in the mitochondrial inner membrane (12). Further study of this phenomenon revealed that the ATP-binding region of Lon can function akin to an ATP-requiring chaperone. Lon can thus not only degrade proteins, but also promote their assembly. We propose that the ATP-dependent chaperone domain of Lon functions as an "antenna" that recognizes malfolded proteins, binds them, and then delivers them to the protease domain for degradation (13). We have shown that this function can be reconstituted *in vivo* from the genetically isolated Lon domains. We are currently isolating these domains in large amounts in order to learn how they cooperate in the selective degradation of mitochondrial proteins.

References

(1) G. Attardi and G. Schatz (1988). Biogenesis of Mitochondria. Annu. Rev. Cell Biol. 4, 289-333.

(2) G. Schatz and B. Dobberstein (1996). Common principles of protein translocation across membranes. Science 271, 1519-1526.

(3) G. Schatz (1996). The protein import system of mitochondria Biol. Chem. 271, 31763-31766.

(4) R. Lill, F. E. Nargang and W. Neupert (1996). Biogenesis of mitochondrial proteins. Curr. Opinion Cell Biol. 8, 505-512.

(5) T. Lithgow, B.S. Glick and G. Schatz (1995). The protein import receptor of mitochondria. Trends Biochem. Sci. 20, 98-101.

(6) T. Komiya, S. Rospert, G. Schatz and K. Mihara (1997). Binding of mitochondrial precursor proteins to the cytoplasmic domains of import receptors Tom70 and Tom20 is determined by cytoplasmic chaperones EMBO J., in press.

(7) B. S. Glick (1995). Can hsp70 proteins act as force-generating motors? Cell 80, 11-14.

(8) A. Matouschek, A. Azem, B. S. Glick, K. Schmid and G. Schatz (1997). Protein unfolding my mhsp70 suggests that it can act as an import motor. EMBO J., in press.

(9) V. Haucke and G. Schatz (1997). Reconstitution of the protein insertion machinery of the mitochondrial inner membrane EMBO J., in press.

(10) S. Rospert, R. Looser, Y. Dubaquié, A. Matouschek, B. S. Glick and G. Schatz (1996). Hsp60-independent protein folding in the matrix of yeast mitochondria. EMBO J. 15, 764-774.

(11) C. K. Suzuki, K. Suda, N. Wang and G. Schatz (1994). Requirement for the yeast gene LON in intramitochondrial proteolysis and maintenance of respiration Science, 264, 273-276.

(12) M. Rep, J. M. van Dijl, K. Suda, L. A. Grivell, G. Schatz and C. Suzuki (1996). A proteolytically inactive Yeast Lon promotes assembly of mitochondrial membrane complexes. Science 274, 103-106.

(13) C. K. Suzuki, M. Rep, J. M. van Dijl, K. Suda, L. A. Grivell and G. Schatz (1997). ATP-dependent proteases that also chaperone protein biogenesis. Trends Biochem. Sci. 22, 118-123.

Protein Complexes Involved in Import of Mitochondrial Preproteins

Peter J.T. Dekker

Institut für Biochemie und Molekularbiologie, Universität Freiburg, Hermann-Herder-Str. 7, D-79104 Freiburg, Germany.

Keywords: protein import, mitochondria, blue native electrophoresis, *Saccharomyces cerevisiae*, membrane proteins

1 Introduction

Transport of preproteins across the mitochondrial outer and inner membrane involves the participation of separate protein complexes in both membranes. In recent years many subunits of the translocation channels have been characterized by biochemical and genetic studies in the yeast *Saccharomyces cerevisiae* (reviewed by Kübrich *et al.*, 1995; Schatz and Dobberstein, 1996). In principle the different subunits can be functionally divided into four major groups: (i) the preprotein receptors, (ii) the translocation channel subunits of the outer membrane [together termed Tom proteins for "Translocase of the Outer Membrane"], (iii) the translocation channel(s) of the inner membrane and (iv) translocation motor subunits. The inner membrane proteins are termed Tim proteins for "Translocase of the Inner Membrane". Contacts between individual subunits of the Tom or Tim complexes have been studied extensively by co-immune precipitation experiments after solubilization of mitochondria under non-denaturing conditions (generally with the detergent digitonin), or by *in vivo* cross-linking experiments (reviewed by Pfanner and Meijer, 1997). From these experiments the observer might have the impression that all proteins bind to all other proteins in the individual membranes, and all of them bind to translocating polypeptide chains. It should be emphasized, however, that these experiments do not make a distinction between stable structural interactions and interactions that are more transient and sub-stoichiometric, but might still be functionally important. Furthermore, the relative abundance of many individual Tom and Tim proteins is unknown, making any prediction concerning the constitution of the membrane complexes unreliable.

In order to obtain a more realistic picture of the translocases of the mitochondrial outer and inner membrane, we recently decided to analyze the individual membrane protein complexes involved in mitochondrial protein import by blue-native electrophoresis (BN-PAGE; Dekker *et al.*, 1996; 1997). This method was originally developed by Schägger and von Jagow for the separation and analysis of the respiratory chain complexes, and is especially suited for the determination of the

NATO ASI Series, Vol. H 106
Lipid and Protein Traffic
Pathways and Molecular Mechanisms
Edited by Jos A. F. Op den Kamp

relative molecular weight of membrane protein complexes (Schägger and von Jagow, 1991; Schägger *et al.*, 1994). After solubilization of the membranes, the detergent is substituted for the negatively charged, non-denaturing dye Coomassie Brilliant Blue G250 that binds to hydrophobic patches on the proteins and keeps them soluble. The size of the protein complexes can then be assessed on a gradient polyacrylamide gel in comparison to a set of marker proteins of known size. This method is much more accurate in molecular weight determination in the range of 50-1000 kDa than more conventional methods like gel filtration and sucrose density centrifugation, since it circumvents problems with detergent micelles and most membrane protein complexes run as sharp bands in the gradient gels.

Here I will discuss recent progress we have made in the analysis of the translocases of the mitochondrial outer and inner membrane using BN-PAGE. These experiments do not only give us important structural information concerning the constitution of both translocation channels; the comparison of the abundance of individual subunits and the analysis of the translocation complexes in the presence of a preprotein on route to the mitochondrial matrix also give us new clues to the mechanism of mitochondrial protein import.

2 The Translocation Channels

2.1 The Tom Complex

Most mitochondrial proteins are synthesized on cytoplasmic ribosomes and post-translationally transported into mitochondria. Mitochondrial targeting sequences, generally located at the amino-terminus, are recognized by specialized receptors on the outer mitochondrial surface. The cytosolic, protease accessible domains of these receptors are thought to interact with the targeting sequences in the preproteins and thereby provide selectivity to the mitochondrial translocation apparatus. The receptors (Tom70, Tom37 or Tom20) are thought to interact loosely with a translocation pore in the mitochondrial outer membrane (reviewed by Lithgow *et al.*, 1995; Lill and Neupert, 1996). We recently found that the translocation pore is a stable assembly consisting of (at least) the integral membrane proteins Tom40, Tom22, Tom7, Tom6 and Tom5 (Dekker *et al.*, 1996; 1997). This relatively abundant protein complex (approximately 60 pmol/mg mitochondrial protein) has an apparent molecular weight of 400K, as determined by BN-PAGE. The 400K Tom complex functions as the translocation pore of the mitochondrial outer membrane and interacts with a preprotein in transit across the mitochondrial membranes. Therefore, this assembly will here be termed the Tom core complex.

2.2 The Tim Complex

The translocase of the inner mitochondrial membrane is less well characterized. Three Tim subunits have been identified in a genetic screen that also yielded matrix Hsp70 mutants (Pfanner *et al.*, 1994). Tim44 is a peripheral membrane protein that interacts with mtHsp70 in an ATP regulated fashion. Tim23 and Tim17 are integral membrane proteins and thought to form the translocation pore. Both proteins are assembled in a 90K complex that interacts with a preprotein in transit across the inner membrane, and is termed the Tim core complex (Dekker *et al.*, 1997). Newly imported Tim23 and Tim17 assemble into a pre-existing 90K complex, suggesting that assembled subunits readily exchange for exogenous subunits. The presence of a precursor protein spanning the Tim core complex, however, prevents the exchange of Tim subunits. This indicates that a preprotein in transit across the inner membrane has a direct effect on the constitution of the Tim core complex. This excludes that the Tim core complex is only a passive diffusion channel; it might interact with a translocating preprotein itself. Tim44 and mtHsp70 do not seem to have a stable stoichiometric interaction with the Tim core complex, although both proteins might bind transient. Indeed, Tim44 is less abundant than the channel subunits, and might have a more catalytic function during protein translocation (Blom *et al.*, 1995).

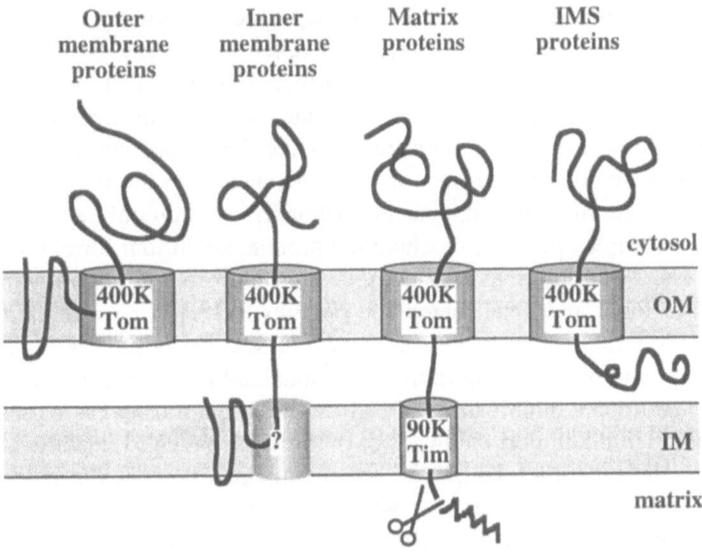

Figure 1: Translocation of different types of mitochondrial proteins into and across the outer (OM) and inner membrane (IM). While presequence containing preproteins (mainly matrix proteins) require the presence of both the 400K Tom and 90K Tim core complexes, other proteins only use the 400K Tom complex. (?) denotes a putative complex in the inner membrane required for insertion of inner membrane proteins lacking a cleavable presequence.

2.3 Specificity of the Translocation Channels

Tim11 has recently been identified by crosslinking to a cytochrome b_2 precursor and might be specialyzed in the recognition of sorting signals of intermembrane space proteins (Tokatlidis *et al.*, 1996). Tim11 is not a structural member of the 90K Tim core complex, but might also interact transiently (unpublished data). Interestingly, mitochondrial proteins that lack a cleavable presequence (mostly membrane proteins) do not seem to use the Tim core complex for correct membrane insertion. The inner membrane protein Tim22, that has sequence homology to the Tim core complex subunits Tim23 and Tim17, was shown to be required for the import of the ATP/ADP carrier and phosphate carrier (Sirrenberg *et al.*, 1996). Tim22 is apparently not a subunit of the Tim core complex, and therefore might be a component of a second import site specifically for mitochondrial membrane proteins lacking a cleavable presequence.

Additional evidence for a second import site in the mitochondrial inner membrane comes from the analysis of a Tim23 mutant, where the Tim core complex is destabilized (Dekker *et al.*, 1997). The import of all proteins containing a cleavable presequence is diminished in this *tim23-2* mutant, while membrane insertion of proteins lacking a cleavable presequence is normal (such as AAC, Tim17 and Tim23). Indeed, in contrast to preproteins containing a cleavable presequence, import of Tim23 and Tim17 themselves do not seem to require the function of mtHsp70, indicating a drastically different import route (Bömer *et al.*, 1997). Furthermore, when the Tim core complex was blocked by a preprotein accumulated in transit across both mitochondrial membranes, import of presequence containing preproteins was completely inhibited, while both outer and inner membrane proteins lacking a cleavable presequence could be readily inserted. Besides showing that inner membrane proteins lacking a cleavable presequence use a different import site for membrane insertion, this result indicates that the Tom core complex is not fully blocked under these conditions, since it can still function in the import of outer and inner membrane proteins. Indeed, yeast mitochondria contain only approximately 15 pmol Tim core complex per mg mitochondrial protein, about four times less than Tom core complexes. The additional Tom complexes, therefore, might be involved in the translocation of outer and inner membrane proteins lacking a cleavable presequence (Figure 1). This arrangement makes perfect biological sense since proteins like the carriers, Tom40 and porin belong to the most abundant proteins in mitochondria and might require additional import sites. Already at the receptor stage there seems to be a (partial) division in the import pathways of preproteins having or lacking a cleavable presequence. The receptors Tom70/37 are mainly involved in the recognition of preproteins lacking a presequence, while Tom20 seems to be specialized in presequence containing preproteins. Therefore, one might speculate that also the Tom core complexes have a division of labour, a special subclass for each group of preproteins; one subclass interacting with the Tom70/37 receptors and one for the Tom20 receptor. Direct evidence for such completely independent import pathways, however, is currently lacking.

3 The Role of MtHsp70 during Protein Translocation

3.1 The Brownian Ratchet Model

This model explaining the role of mtHsp70 during protein translocation is mainly based on the assumption that transport of preproteins through the import channels occurs by biased diffusion (Brownian motion). Preproteins diffuse freely in or out of the translocation channels and directionality to the movement is only supplied by the interaction of mtHsp70 to the translocating chain on the matrix side. MtHsp70 thereby functions as a ratchet to prevent reverse translocation of the polypeptide chain. The Brownian ratchet model was proposed by Neupert and co-workers to explain the fate of accumulated preproteins in ATP depleted mitochondria (Ungermann *et al.*, 1994).

When matrix ATP was depleted by treatment with apyrase, previously accumulated preproteins tended to slip back through the import channel and into solution. Under these conditions, mtHsp70 is not able to interact with the preprotein anymore. The conclusion of this study was that mtHsp70 functions as a ratchet (in the presence of ATP) and thereby prevents retrograde translocation. Obvious prerequisites for mtHsp70 to function as a molecular ratchet are that preproteins should diffuse freely through the translocation channels and that at least one mtHsp70 molecule should stably bind to every preprotein to prevent retrograde translocation. As will be explained hereafter, neither of these prerequisites could be verified by recent experiments.

3.2 The Import Motor Model

In this model the role of mtHsp70 is that of an ATP powered motor (Glick, 1995). MtHsp70 binds on one hand to the preprotein in transit, on the other hand it is linked to the mitochondrial inner membrane. By a conformational change in the chaperone, the preprotein is thought to be "pulled in". The import motor model explains the lack of import of some preproteins in a mtHsp70 mutant that bound stably to preproteins in transit, an observation that is difficult to explain by a sole ratchet function of mtHsp70 (Gambill *et al.*, 1993). Denaturation of the preprotein prior to import led again to efficient translocation, indicating that mtHsp70 has both a translocase and unfoldase function during preprotein import. The unfoldase function of mtHsp70 might be essential for efficient import of folded domains.

A possible scenario is that mtHsp70 is bound to a Tim protein in the inner membrane and awaits the incoming polypeptide chains. After binding to the preprotein, the chaperone changes conformation and thereby unfolds domains on the outside. After the power stroke, mtHsp70 releases the polypeptide and/or the membrane anchor to provide space for the next mtHsp70. Eventually the polypeptide is fully translocated across the mitochondrial membranes. It is essential to note that the "holding" and "pulling" models for mtHsp70 function do not necessarily contradict each other and might function co-ordinately in preprotein import.

3.3 Does MtHsp70 Act as a Gate Keeper?

Recently we described the accumulation across both mitochondrial membranes of a preprotein in chemical amounts (Dekker *et al.*, 1997). The 400K Tom core complex and the 90K Tim core complex were linked into a stable 600K supercomplex in this experiment. Most surprisingly, this stable complex did neither contain Tim44 nor mtHsp70. When this complex was isolated in the presence of EDTA (which stabilizes the interaction between mtHsp70 and preproteins) only approximately 5% of all preproteins spanning the membranes interacted with mtHsp70. The same low efficiency of mtHsp70 binding to preproteins spanning the mitochondrial membranes has been reported for co-immune precipitation experiments (Ungermann *et al.*, 1994). Apparently, both Tim44 and mtHsp70 are not required to hold a preprotein in position across the mitochondrial membranes. An essential molecular ratchet function of mtHsp70, therefore, is very unlikely. MtHsp70 might still function as an import motor, since here the chaperone has no structural function (as it does have in the ratchet model) but only has a catalytic function.

If not mtHsp70, what then holds the polypeptide chain in position and prevents retrograde translocation? I would like to propose here that the Tim core complex itself has this important function. Indeed, the experiment described above excludes that the 600K supercomplex is a passive diffusion channel; it apparently interacts actively with the translocating chain. The exchange of endogenous Tim subunits by newly imported ones is inhibited when a preprotein was accumulated first, indicating that the spanning polypeptide has a stabilizing function on the Tim core complex. A direct interaction between the Tim core complex and the translocating polypeptide, therefore, is very likely. A "scanning" of incoming polypeptide chains by the Tim core complex is not surprising if one imagines that some intermembrane space proteins contain targeting sequences that probably function as stop-transfer signals (Glick *et al.*, 1992) and, most likely, are recognized by the Tim complex and must leave the Tim complex laterally.

When mitochondria are depleted of ATP, preproteins that span the mitochondrial membranes are released from mitochondria (Ungermann *et al.*, 1994). Indeed, we observed dissociation of the 600K translocation complex when mitochondria were treated with apyrase at higher temperatures (unpublished data). Why is ATP required for the holding of polypeptide chains when mtHsp70 is only bound in substoichiometric amounts? One possibility is that another factor, besides mtHsp70, requires ATP to hold translocating polypeptides in the import channels. However, mtHsp70 is the only ATP requiring factor involved in import as yet identified in the mitochondrial matrix. Therefore, I would prefer the second possibility, that the Tim core complex binds to the polypeptide in transit only in the absence of mtHsp70. The chaperone dissociates from its membrane anchors by binding to ATP, and is not available at the import site under these conditions (von Ahsen *et al.*, 1995; Bömer *et al.*, 1997). In the absence of ATP (or presence of ADP), mtHsp70 is stably bound to its membrane anchors and might trigger the opening of the Tim core complex, thereby allowing forward but also retrograde translocation. This model might also explain the controversy about the function of ATP during protein import; on the one hand it was shown that mtHsp70 does not stably bind to preproteins in the presence of ATP (Voos *et al.*, 1994), on the other hand ATP is required to hold polypeptides in transit across the mitochondrial membranes (Ungermann *et al.*, 1994).

Indeed, the effect of mtHsp70 on the Tim core complex might be an important function of mtHsp70 during normal protein translocation. Here, mtHsp70 acts as a gate keeper, allowing import by opening up the translocation channel. Forward translocation might then occur by biased diffusion or by a pulling by the mtHsp70 import motor. Interestingly, a similar function for the Hsp70 of the endoplasmatic reticulum (ER), BiP, was recently proposed by Brodsky (1996). BiP is apparently required to release a translocating polypeptide chain from the ER translocon (Lyman and Schekman, 1997). Obviously, the verification of this and other models for the function of mtHsp70 will require additional experiments, and might only be convincingly studied in a reconstituted system, as done for ER translocation. Unfortunately, a reconstituted protein import system for the mitochondrial inner membrane is not yet available, rendering it impossible to study the role of the different factors in protein translocation directly.

4 Conclusions and Perspectives

The analysis of the membrane protein complexes (Tom and Tim) involved in mitochondrial protein import by blue native electrophoresis has led to several surprising and new findings (Dekker *et al.*, 1997). The results have both functional implications for the mechanism of protein import and provide us with a better understanding of the structure of the protein complexes involved in this process. The main conclusions of this study are: (i) The sizes of the major translocation complexes in the mitochondrial inner and outer membranes have been determined: the Tom core complex is 400K and the Tim core complex is 90K. Accumulation of a preprotein spanning the membranes links both complexes into a supercomplex of 600K. (ii) The Tom core complex is approximately 4-5 times more abundant than the Tim core complex. The Tim core complex, therefore, limits the number of translocation contact sites. (iii) Preproteins lacking a cleavable presequence do not require a functional Tim core complex for correct membrane insertion. Most likely an additional import site, especially for these proteins, is present in the mitochondrial inner membranes. (iv) MtHsp70 and Tim44 are not required to stably hold translocating chains across the mitochondrial membranes, rendering an essential molecular ratchet function for mtHsp70 unlikely. (v) The Tim core complex itself can hold preproteins spanning the inner membrane. The preprotein induces a structural change in the Tim core complex, preventing exchange of Tim subunits.

The results described here were obtained by an initial study of the Tom and Tim complexes by blue native electrophoresis. For the future this method promises to provide a major contribution to the functional and structural analysis of the membrane complexes involved in protein translocation. Currently we analyze the constitution of the Tom and Tim complexes in several mutants impaired in protein translocation. We hope to come to a description of the subunit composition and assembly pathway of these complexes. Furthermore, the accumulation across the membranes of a preprotein in chemical amounts, and its stable association with the translocation complexes after lysis of mitochondria, opens up the possibility to study the translocation process *in vitro* and to identify factors that are required to release and translocate the bound preproteins.

5 Acknowledgements

I would like to thank Dr Klaus Pfanner for critically reading the manuscript and the Human Frontiers Science Program for financial support for my stay in Freiburg.

6 References

Blom, J., Dekker, P.J.T. and Meijer, M. (1995) Functional and physical interactions of components of the yeast mitochondrial inner-membrane import machinery (MIM). *Eur. J. Biochem.* **232**, 309-314.

Bömer,U., Meijer,M., Maarse,A.C., Dekker,P.J.T., Pfanner,N. and Rassow,J. (1997) Multiple interactions of components mediating preprotein translocation across the inner mitochondrial membrane. *EMBO J.*, **16**, 2205-2216.
Brodsky, J.L. (1996) Post-translational protein translocation: not all hsc70s are created equal. *Trends Biochem. Sci.* **21**, 122-126.
Dekker, P.J.T., Müller, H., Rassow, J. and Pfanner, N. (1996) Characterization of the preprotein translocase of the outer mitochondrial membrane by blue native electrophoresis. *Biol. Chem.* **377**, 535-538.
Dekker, P.J.T., Martin, F., Maarse, A.C., Bömer, U., Müller, H., Guiard, B., Meijer, M., Rassow, J. and Pfanner, N. (1997) The Tim core complex defines the number of translocation contact sites and can hold arrested preproteins in the absence of matrix Hsp70/Tim44. *EMBO J.*, in press.
Gambill, B.D., Voos, W., Kang, P.J., Mao, B., Langer, T., Craig, E.A., Pfanner, N. (1993) A dual role for mitochondrial heat shock protein 70 in membrane translocation of preproteins. *J. Cell Biol.* **123**, 109-117.
Glick, B.S., Brandt, A., Cunningham, K., Müller, S., Hallberg, R.L. and Schatz, G. (1992) Cytochromes c_1 and b_2 are sorted to the intermembrane space of yeast mitochondria by a stop-transfer mechanism. *Cell* **69**, 809-822.
Glick, B.S. (1995) Can hsp70 proteins act as force-generating motors? *Cell* **80**, 11-14.
Kübrich, M., Dietmeier, K. and Pfanner, N. (1995) Genetic and biochemical dissection of the mitochondrial protein-import machinery. *Curr. Genet.* **27**, 393-403.
Lill, R. and Neupert, W. (1996) Mechanisms of protein import across the mitochondrial outer membrane. *Trends Cell Biol.* **6**, 56-60.
Lithgow, T., Glick, B.S. and Schatz, G. (1995) The protein import receptor of mitochondria. *Trends Biochem. Sci.* **20**, 98-101.
Lyman, S.K. and Schekman, R. (1997) Binding of secretory precursor polypeptides to a translocon subcomplex is regulated by BiP. *Cell* **88**, 85-96.
Pfanner, N., Craig, E.A. and Meijer, M. (1994) The protein import machinery of the mitochondrial inner membrane. *Trends Biochem. Sci.* **19**, 368-372.
Pfanner, N. and Meijer, M. (1997) Mitochondrial biogenesis: The Tom and Tim machine. *Curr. Biol.* **7**, R100-R103.
Schägger, H. and von Jagow, G. (1991) Blue Native Electrophoresis for isolation of membrane protein complexes in enzymatically active form. *Anal. Biochem.* **199**, 223-231.
Schägger, H., Cramer, W.A. and von Jagow, G. (1994) Analysis of molecular masses and oligomeric states of protein complexes by blue native electrophoresis and isolation of membrane protein complexes by two-dimensional native electrophoresis. *Anal. Biochem.* **217**, 220.
Schatz, G. and Dobberstein, B. (1996) Common principles of protein translocation across membranes. *Science* **271**, 1519-1526.
Sirrenberg, C., Bauer, M.F., Guiard, B., Neupert, W. and Brunner, M. (1996) Import of carrier proteins into the mitochondrial inner membrane mediated by Tim22. *Nature* **384**, 582-585.
Tokatlidis, K., Junne, T., Moes, S., Schatz, G., Glick, B.S. and Kronidou, N. (1996) Translocation arrest of an intramitochondrial sorting signal next to Tim11 at the inner-membrane import site. *Nature* **384**, 585-588.
Ungermann, C., Neupert, W. and Cyr, D.M. (1994) The role of Hsp70 in conferring unidirectionality on protein translocation into mitochondria. *Science* **266**, 1250-1253.
von Ahsen, O., Voos, W., Henninger, H. and Pfanner, N. (1995) The mitochondrial protein import machinery: role of ATP in dissociation of Hsp70-Mim44 complex. *J. Biol. Chem.* **270**, 29848-29853.
Voos, W., Gambill, B.D., Laloraya, S., Ang, D., Craig, E.A. and Pfanner, N. (1994) Mitochondrial GrpE is present in a complex with hsp70 and preproteins in transit across membranes. *Mol. Cell. Biol.* **14**, 6627-6634.

Protein Targeting into and within Chloroplasts

Steven M. Theg
Division of Biological Sciences
Section of Plant Biology
University of California - Davis
Davis, CA 95616

Introduction

Plastids, like mitochondria, synthesize a subset of their proteins from genes localized within the organelle. Most of their resident proteins, however, are encoded in the nucleus, translated in the cytoplasm and internalized posttranslationally. Many plastid proteins are additionally re-targeted internally to their final destinations and/or assembled into multimeric protein complexes. This chapter describes some aspects of the current state of knowledge regarding the import, targeting and assembly of proteins within chloroplasts. Where it has been investigated, distinct protein import properties have not been identified in plastids with differing morphologies and functions (Halpin et al., 1989b; Boyle et al., 1990; Dahlin, 1993). Consequently, the information presented has been obtained with chloroplasts, but is assumed to apply to all plastid types. Readers desiring more detailed information are referred to a number of excellent recent reviews on this topic (Cline and Henry, 1996; Kouranov and Schnell, 1996; Hauke and Schatz, 1997).

Chloroplast structure

Chloroplasts are structurally complex organelles. They are delineated from the cytoplasm by a pair of envelope membranes, with an aqueous intermembrane space between them. Together these three compartments contain only a small fraction of the total protein found in chloroplasts, a fact which greatly complicates attempts to study the envelope protein translocators. Bounded by the inner envelope membrane is the stromal space, which houses the plastid DNA, ribosomes and ploymerases, many of the cell's biosynthetic enzymes, and the enzymes involved in carbon fixation. Floating free within the plastid are

NATO ASI Series, Vol. H 106
Lipid and Protein Traffic
Pathways and Molecular Mechanisms
Edited by Jos A. F. Op den Kamp

the thylakoids, consisting of a sealed membrane system with an aqueous lumen. The thylakoid membranes contain the light-harvesting complexes as well as the four major protein complexes responsible for the photosynthetic light reactions. Interestingly, measurements of the change of membrane ion conductance in response to ionophores has led to the suggestion that, although they appear as stacked disks in electron micrographs, there is likely only a few or even just one thylakoid per chloroplast (Schoenknecht et al., 1990). The stroma and thylakoid membranes are extremely protein-rich environments, and together account for over 95 % of the total plastid protein, split approximately equally between them.

Chloroplast transit peptides

The topogenic sequences targeting cytosolic precursor proteins to chloroplasts are often referred to as transit peptides. Among the different types of topogenic sequences found in cells, transit peptides that direct proteins into the chloroplast stroma most closely resemble those which direct proteins to the mitochondria matrix. Both stromal and matrix targeting peptides are N-terminal, are cleaved after the targeting event, are relatively rich in hydroxylated amino acids and generally carry positive charges (Cline and Henry, 1996). However, here the resemblance ends, and compared to their mitochondria counterparts, stromal targeting peptides are considerably longer and possess a discernibly different amino acid composition (von Heijne et al., 1989). They are also not predicted to form amphipathic alpha-helices, which is thought to be the targeting determinant of mitochondrial transit peptides. Instead it has been suggested that they present a random coil configuration to the envelope membrane transport machinery (von Heijne and Nishikawa, 1991). These properties allow plant cells to differentiate between mitochondria and chloroplast precursors with relatively high efficiency, although a few instances have been reported in which proteins are imported into both organelles (Boutry et al., 1987; Whelan et al., 1990; Brink et al., 1994; Creissen et al., 1995; Filho et al., 1996).

Proteins destined to the thylakoid lumen possess bipartite transit peptides (Cline and Henry, 1996). The extreme N-terminal portion of these peptides are clearly recognizable as stromal targeting domains, which are removed by the stromal processing peptidase. Following this domain at the C-terminal portion of the peptides are targeting sequences that closely resemble the signal sequences responsible for protein export from bacteria or protein import into the endoplasmic reticulum. Some of these have been observed to function as such, causing the transport of passengers overexpressed in bacteria to the periplasm (Seidler and Michel, 1990; Meadows and Robinson, 1991; Betts et al., 1994). In line with this, the cleavage site for these signal peptides is the same as that utilized by bacterial leader peptidase, and is efficiently recognized by that enzyme (Halpin et al., 1989a). These observations presaged the identification of bacterial-like components involved in targeting of some proteins to thylakoids (see below).

Different authors have suggested that the initial interaction between precursors and their target organelles occurs via interactions between the amino acids in the targeting peptides and lipids presented on the external surface of the organelle (Briggs et al., 1986; Killian et al., 1990; Theg and Geske, 1992; de Kruijff et al., 1996). This may well be true for chloroplasts, as the transit peptides have been observed to interact strongly with lipid monolayers. Interestingly, van't Hof et al. (Van't Hof et al., 1991) reported that the interaction of the ferredoxin presequence was stronger with lipids derived from plastids than with phospholipids usually employed in such experiments, raising the possibility that the peptide-lipid interaction may play a part in the fidelity of chloroplast protein targeting.

It has recently been reported that chloroplast transit peptides possess phosphorylation sites that regulate their ability to interact productively with the translocation machinery (Waegemann and Soll, 1996). Since protein import was prevented by inhibitors of de-phosphorylation, these authors suggested that a phosphorylation/dephosphorylation cycle presents the possibility for in vivo regulation of chloroplast protein targeting.

Targeting of proteins to the envelope membranes

The initial information available for the targeting of proteins to the plastid outer envelope membrane suggested that, as a group, they are not synthesized as higher molecular weight precursors, and their targeting proceeded without the use of surface-exposed proteinaceous components or the input of metabolic energy (Salomon et al., 1990; Li, H. M. et al., 1991; Fischer et al., 1994). While a spontaneous insertion mechanism is attractive for these proteins, it presents a problem with respect to targeting fidelity. It is likely that additional factors yet to be described must be at work to keep these proteins from inserting opportunistically into all other, incorrect membranes.

Three outer membrane proteins have been described that do not conform to the traits described above, and interestingly, each is a component of the outer envelope membrane protein transport machinery. OEP34 is not synthesized with a cleavable transit peptide, nor is its insertion dependent on protease-digestible proteins in the membrane. However, its insertion is stimulated by GTP or GDP, and to a lesser extent, ATP (Chen and Schnell, 1997). This protein contains a GTP binding motif, modifications in which eliminate the stimulation of insertion by nucleotides. It appears likely that GTP induces a conformational change in OEP34 that promotes its spontaneous insertion across the outer membrane (Chen and Schnell, 1997).

In contrast to other outer envelope membrane proteins, OEP75 and OEP86 both contain cleavable transit peptides. OEP75 appears to engage the general import machinery used by stromal proteins (see below), but then is halted by an intraorganelle targeting signal present in the C-terminal portion of its transit peptide, perhaps via a stop transfer mechanism (Tranel and Keegstra, 1996). OEP86, on the other hand, is apparently inserted using a unique envelope machinery both sensitive to external proteases and utilizing energy from ATP hydrolysis (Hirsch et al., 1994). These latter proteins point to the dangers inherent in inferring a common mechanism from the study of only a few examples, and it is clear that more work must be done to determine whether they can be considered exceptions to a general mechanism of protein targeting to the outer envelope membrane, or represent larger classes of distinct pathways yet to be fully described.

Proteins destined to the inner envelope membrane appear to follow the general import pathway (Li, H. M. et al., 1992; Hirsch and Soll, 1995; Knight and Gray, 1995). Transit peptide swapping experiments have clearly demonstrated that the targeting peptides on inner envelope proteins can direct passengers into the stroma, and conversely, those from stromal proteins can serve to allow inner envelope proteins to take up residence in their correct location. An interesting question raised by such experiments concerns the manner in which these proteins become inserted into the membrane. One possibility is that the membrane anchor could serve as a stop-transfer signal, halting the progression of the protein across the membrane before it is fully transported into the stroma. The other possibility is that the protein is first directed to the stroma by its transit peptide, after which it is re-directed to the inner membrane. At the moment, there is no published information from which to distinguish these two mechanisms.

Protein targeting to the stroma

The import of proteins from the cytoplasm to the stroma is the best characterized protein targeting reaction in plastids. Using stromal resident proteins it was discovered early on that the import reaction could be experimentally dissected into separate binding and transport steps (Cline et al., 1985). Surprisingly, the binding reaction was found to be irreversible and dependent on the hydrolysis of NTPs (Theg et al., 1989; Olsen and Keegstra, 1992). This energy requirement was subsequently found to mediate the transfer of the preprotein from a putative receptor to the translocation machinery (Perry and Keegstra, 1994). The energy requirement for initial engagement of the protein translocation apparatus appears to be unique to the chloroplast envelope, and has no parallel in other protein transporting systems.

After the precursor is bound to the chloroplast surface, it is transported across both envelope membranes, perhaps most commonly at contact sites where the membranes are held in close appression (Cline and Henry, 1996; Kouranov and Schnell, 1996). Protein import is inhibited by pretreatment of plastids with exogenous protease, and is dependent on the hydrolysis of both ATP and GTP. It is now known that components of the outer envelope transport machinery are exposed on the chloroplast surface, and that two of them possess identifiable GTP binding motifs (see below).

All proteins entering the stroma are thought to do so via a single general import pathway. The strongest evidence for such a statement is the fact that of the many proteins tested, all proteins tested appear to compete for common sites on this one pathway (reviewed in Gray and Row, 1995). The number of transporters in the plastid envelope membranes have been independently determined a number of times, with estimates in the range of a few thousand per chloroplast (Friedman and Keegstra, 1989; Schnell and Blobel, 1993).

Although it appears that proteins cross the envelope membranes at contact sites, this is not a prerequisite for envelope transport. Evidence has accumulated from a number of directions suggesting that the protein translocation machineries in the inner and outer envelope membranes are separable, and can act independently and sequentially (Scott and Theg, 1996, reviewed in Kouranov and Schnell, 1996 and Cline and Henry, 1996). Accordingly, it may be that contact sites in chloroplast envelopes are transient structures formed by the protein import process, wherein the incoming preprotein engages the inner envelope machinery before it has cleared the outer transport apparatus. A similar view of contact site transport in mitochondria has also been proposed for mitochondria (Pfanner et al., 1992).

Identification of components involved in protein targeting to the stroma -- In the past three years there has been an explosion in our understanding of protein import into chloroplasts. This has resulted from the convergence of four independent research groups on the components of the envelope translocation machinery (Hirsch et al., 1994; Kessler et al., 1994; Schnell et al., 1994; Wu et al., 1994). The experimental techniques have been similar to those that yielded the first translocation components of the mitochondria import apparatus -- solubilization of the translocation machinery while it is in contact with a relatively stable membrane spanning translocation intermediate. In some instances, the translocation intermediates were chemically crosslinked to their membrane neighbors, but the association of incoming precursors with the translocation apparatus has often been found to be stable enough to allow for isolation of protein complexes without crosslinking.

There is now general agreement that both early and late-stage intermediates can be created en route across the envelope membrane, and that these intermediates interact with different, specific proteins. Early in the import process, translocating proteins are found to be associated with a complex consisting of four major proteins with molecular masses of 86, 75, 60 and 34 kDa (Figure 1, reviewed in (Cline and Henry, 1996; Kouranov and Schnell, 1996)). The 86 kDa protein, termed herein OEP86, is thought to be the receptor for stromally directed proteins, and is among the first proteins to associate with the precursor (Perry and Keegstra, 1994). Interestingly, this association precedes the

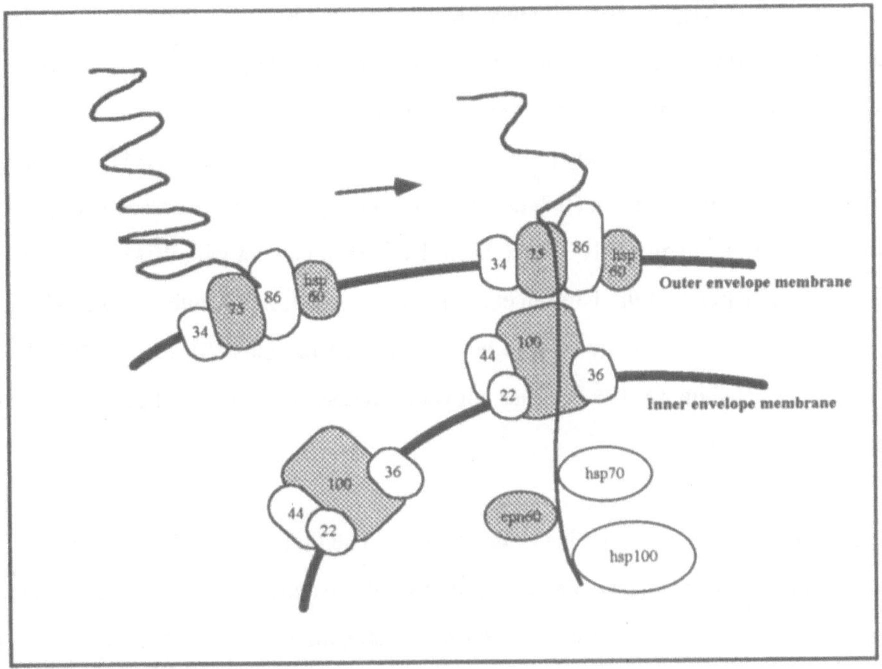

Figure 1. Envelope and stromal proteins involved in protein translocation across the envelope membranes.

ATP-dependent irreversible binding step (Perry and Keegstra, 1994). Direct evidence for its role in protein transport comes from experiments in which antibodies directed against OEP86 were seen to inhibit both binding and import of proteins into the stroma (Hirsch et al., 1994). Upon addition of low amounts of ATP, the precursor becomes newly associated with OEP75 (Perry and Keegstra, 1994), which may represent the translocation channel itself. Interestingly, OEP75, which is intrinsically embedded in the membrane, does not contain regions of membrane-spanning alpha-helices, but may instead cross the membrane in stretches of beta conformation (Schnell et al., 1994). The 70 kDa protein is an intrinsic hsp70 (Soll and Waegemann, 1992; Schnell et al., 1994; Wu et al., 1994), and as such, may provide the driving force for protein translocation via its ATPase activity. OEP34 shares sequence homology with OEP86, including a consensus GTP binding site, suggesting a common evolutionary origin of these two proteins (Hirsch et al., 1994; Kessler et al., 1994; Seedorf et al., 1995).

A later-stage translocation intermediate is found to be associated with additional and different proteins, some of which are in the inner envelope membrane and some of which are stromal. The inner envelope proteins IEP100, IEP36 and IEP44 are found to be associated with translocating proteins after addition of an amount of ATP sufficient to compete the import reaction (Schnell et al., 1994). In addition, two stromal chaperone proteins, cpn60 and an hsp100, have been detected in contact with these intermediates (Kessler and Blobel, 1996; Nielsen et al., 1997). Interestingly, while the stromal hsp70 has been seen to transiently bind to newly imported proteins (Madueno et al., 1993; Tsugeki and Nishimura, 1993), it has not yet been detected in association with the late-stage intermediate.

Energetics of protein import into the stroma -- The first paper to describe the in vitro import of proteins into intact isolated chloroplasts also examined the energy requirements for the process (Grossman et al., 1980). At that time it was determined that ATP hydrolysis was required to power the protein translocation reaction, and that a protonmotive force did not contribute to the energetics. This has been extensively re-examined by numerous investigators, and the original information has been found to be basically correct, if incomplete.

The study of the energetics of protein import is justified by the constraints this information places on the possible import mechanisms. In this light, the location of ATP utilization has been investigated by different research groups (Fluegge and Hinz, 1986; Schindler et al., 1986; Pain and Blobel, 1987; Theg et al., 1989; Scott and Theg, 1996). Although conflicting reports were originally published, examination of all the data leads to the conclusion that the ATP supporting precursor binding or translocation to the stroma is not hydrolyzed in the medium surrounding and external to the chloroplasts. This is based on experiments in which the internal and external levels of ATP are carefully controlled with external and internal ATP destroying or producing reactions. The most recent examination of this issue was made by Scott and Theg (1996), who found evidence that the import process was governed by three distinct ATPases operating at different locations and with different substrate affinities and NPT specificities (Figure 2). The first ATPase encountered provides energy for precursor binding to the external surface of the

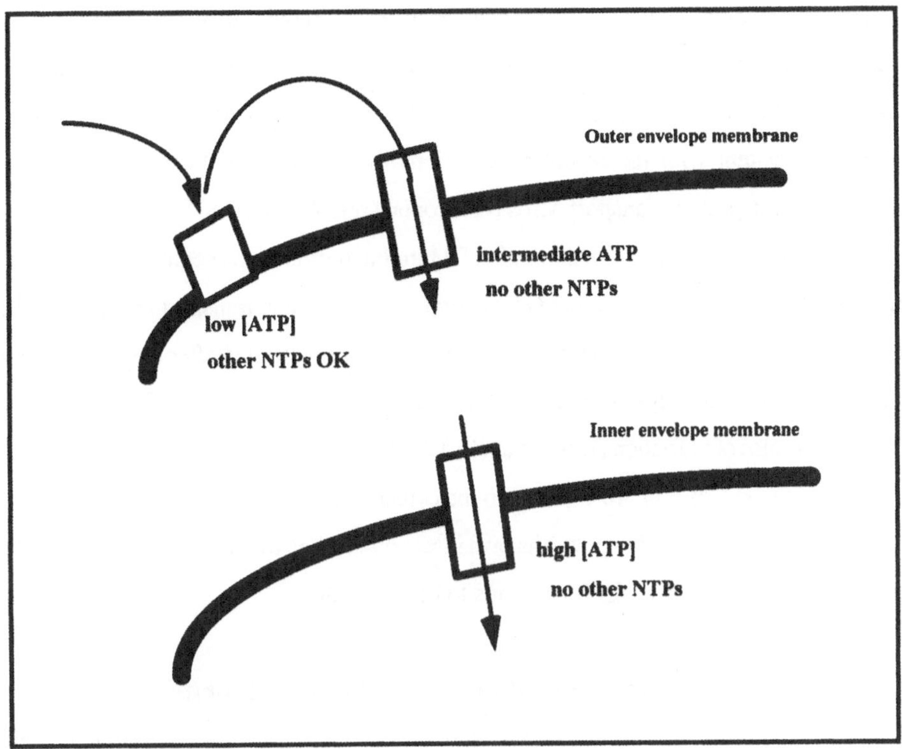

Figure 2. Sites of action of the three putative ATPases involved in protein transport across the envelope membranes.

outer envelope membrane by hydrolyzing low concentrations (approximately 0.05 mM) ATP or other NTPs in the intermembrane space (see also (Olsen and Keegstra, 1992). At ATP concentrations between 0.05 and 0.25 mM, the bound precursor is transported across the outer envelope membrane only and can be trapped in association with the external surface of the inner envelope membrane. This ATP is also hydrolyzed in the intermembrane space, but other NTPs do not support transport, suggesting that the ATPase involved is distinct from that utilized in the binding reaction. Even higher concentrations of ATP are required to transport the precursor across the inner envelope membrane to the stroma. This reaction appears to be powered by a stromal ATPase which is specific for ATP. The localization of each of these ATP requirements corresponds with the location of different hsp70s within the plastid, which is consistent with their involvement in protein import, although experimental evidence for their roles in this process is presently lacking. It is noteworthy that a requirement for external ATP hydrolysis has not been observed for chloroplast protein import, suggesting that cytosolic chaperones do not contribute to protein trafficking in this organelle as they do in others. This is consistent with the recent observations that protein unfolding is not a strict requirement for protein transport across chloroplast membranes (see below).

Following the identification of GTP binding motifs in the sequences of OEP86 and OEP34, it was discovered that GTP is required for protein transport into the stroma (Kessler et al., 1994). This GTP requirement was only manifested after treatment of the plastids with a non-hydrolyzable GTP analogue, thereby blocking access of required GTP to its normal sites of utilization (Kessler et al., 1994). Interestingly, the GTP binding sites of both OEP86 and OEP34 are exposed to the cytosol. While GTP hydrolysis by OEP34 may be specifically required to power its own insertion into the membrane (see above), the role of GTP hydrolysis by OEP86 remains to be elaborated.

Protein transport into or across the thylakoid membrane

Standing in contrast to the utilization of one general import machinery for protein translocation into the stroma, precursors follow one of four distinct pathways from the stroma into or across the thylakoid membrane (Figure 3). These pathways were first suggested on the basis of their different energy requirements, which has been followed by studies of their different substrate specificities and, in some cases, different interacting components.

The 54CP pathway -- In 1993, Franklin and Hoffman reported the coning of a stromal homologue to the 54 kDa subunit of the signal recognition particle involved in protein targeting to the endoplasmic reticulum, which they termed 54CP (Franklin and Hoffman, 1993). It was later shown that the integration of the major light harvesting chlorophyll a binding protein (LHCP) was driven not by ATP with the optional assistance of a protonmotive force as had been assumed, but by the protonmotive force and GTP (Hoffman and Franklin, 1994). This led these authors, in collaboration with Cline's group, to the realization that the soluble factor

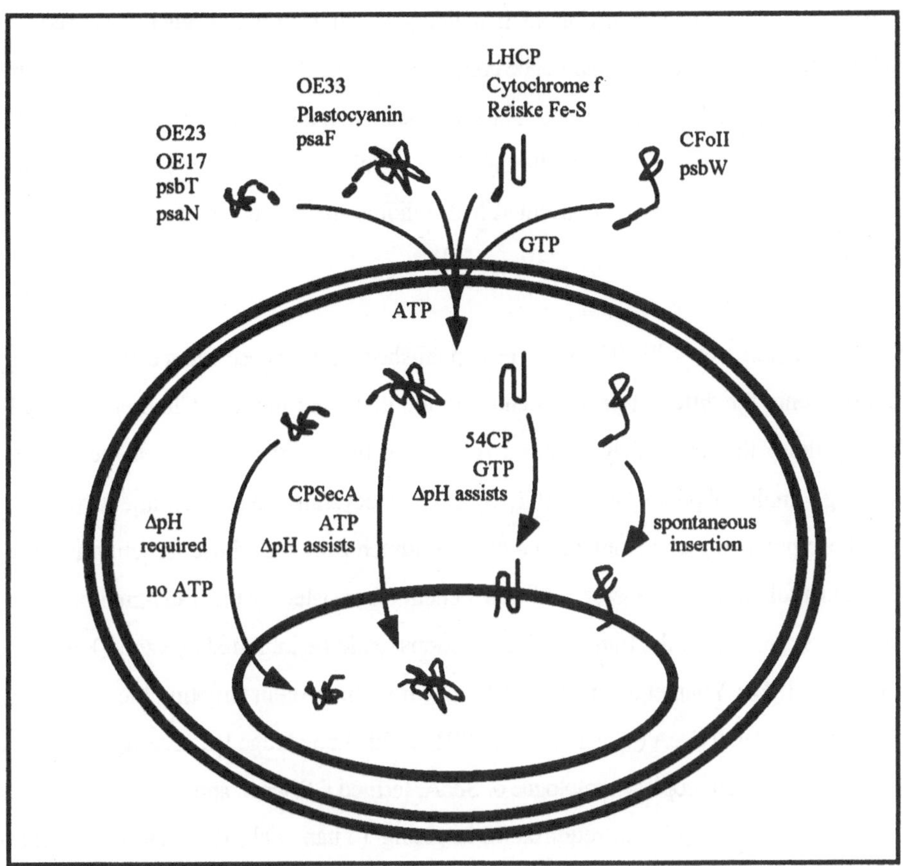

Figure 3. The four pathways for proteins into or across the thylakoid membrane.

long known to be required for LHCP integration was, in fact, 54CP (Li, X. et al., 1995). Recent experiments in which 54CP was crosslinked to nascent thylakoid membrane proteins revealed that it interacted with only a subset of thylakoid proteins with particularly hydrophobic transit peptides (Li, X. et al., 1995). Although LHCP is the only protein whose targeting has been specifically shown to depend on 54CP, it is presumed that these others which interact as nascent chains (cytochrome *f* and Rieske Fe-S protein) will be found to do so as well.

Spontaneous integration into the thylakoid membrane -- Two integral thylakoid membrane proteins, CFoII (Michl et al., 1994)and psbWp (Lorkovic et al., 1995), have been reported to take up residence in the membrane in the absence of soluble stromal factors, protease-sensitive thylakoid machinery or an identifiable energy source. In addition, their incorporation could not be competed with proteins known to be on the other three established thylakoid pathways. Thus it appears that a second pathway exists for the incorporation of proteins into the the thylakoid membrane in which the incorporation is driven spontaneously by the combined hydrophobic natures of the transit peptide and mature protein.

The CPSec pathway – In 1992, Cline et al. published a paper describing different energy requirements for different proteins entering the thylakoid lumen (Cline et al., 1992). Among those that absolutely required ATP were the 33 kDa subunit of the oxygen-evolving complex of photosystem II (OE33) and plastocyanin. Later experiments revealed that these polypeptides belonged to a class of lumen proteins whose targeting required ATP and a soluble stromal factor, and was sometimes assisted by the trans-membrane pH gradient. Significantly, the translocation reactions could be inhibited by azide (Knott and Robinson, 1994; Yuan et al., 1994), well known in the protein targeting literature as an inhibitor of bacterial SecA (Wickner et al., 1991). This knowledge led investigators to the identification of a chloroplast homologue of SecA, termed CPSecA, and determination that it is indeed the stromal factor required for targeting (Yuan et al., 1994; Berghofer et al., 1995; Nohara et al., 1995). A chloroplast homologue of SecY, another component of the bacterial secretion translocase, has also been identified (Laidler et al., 1995), and it is expected that other Sec homologues will be present in the complex that transports this set of proteins in the thylakoid lumen.

The delta-pH-dependent pathway -- The final pathway known to translocate proteins from the stroma to the thylakoid lumen is unique among all known targeting pathways known in its reliance solely on the energy contained in the trans-membrane pH gradient to power the transport reaction (Cline et al., 1992), making it the only energy-dependent translocation system yet described that does not hydrolyze NTPs during its reaction cycle. Nor does it require the addition of soluble components, i.e. stromal factors, for efficient targeting. Interestingly, the proteins on this pathway are those in the thylakoid lumen that do not have counterparts in cyanobacteria, the presumed progenitor endosymbiont that gave rise to modern plastids (Cline and Henry, 1996). It has been suggested, therefore, that the delta-pH-dependent pathway arose to handle those proteins that are newcomers to chloroplasts and which somehow could not be accommodated by the CPSec pathway.

The question of selection the CPSec or the delta-pH-dependent pathway has been addressed by transit peptide swapping and modification experiments (Robinson, C. et al., 1994; Henry et al., 1997). While the signals for the both pathways look similar, those for the delta-pH-dependent pathway possess a unique -RR- motif just before the signal's hydrophobic stretch (Chaddock et al., 1995). While proteins on the CPSec pathway will switch to the delta-pH-dependent pathway when cloned behind the latter's signal peptide, the reciprocal is not true, and delta-pH-dependent pathway proteins are not tolerated by the CPSec pathway (Robinson, C. et al., 1994; Henry et al., 1997). This lends further credence to the hypothesis that the delta-pH-dependent pathway arose late to allow transport of proteins that are difficult for the CPSec pathway. Further analysis of the different targeting motifs have led to the formation of a chimeric protein with a dual targeting signal peptide 6563(Henry et al., 1997). Still, further experiments in this direction are needed to understand why a particular protein prefers or requires one pathway over another.

Brock and Robinson (Brock et al., 1995) # made the only published effort so far to determine the magnitude of the pH gradient required to transport proteins on the delta-pH-dependent pathway. Using the permeable fluorescent amine 9-aminoacridine, they estimate the magnitude of the delta-pH developed in the light as a function of light intensity, and then correlated it with the rate of protein translocation at any given light intensity. They found that protein transport proceeded at high rates down to the lowest delta-pH that they could measure, which with this technique is greater than one pH unit. Consequently, it is still unknown whether, as in photophosphorylation, protein transport occurs only above a threshold level of protonmotive energy. Resolution of this question must await the development of a non-logarithmic measurement of the trans-membrane pH gradient developed in the light, or of an acid-base jump assay providing sufficient force to drive protein transport.

Thylakoid protein targeting pathways may not use common components -- The most convincing evidence that different proteins enter the thylakoid membrane or lumen via different pathways came from experiments in which bacterially expressed proteins on one pathway were seen to compete for common import sites with other members of their pathway family, but not with proteins on any of the different pathways (Cline et al., 1993). Such experiments suggest that the different pathways do not share components, although the possibility still exists that the regions not shared by pathways represent the major bottlenecks to protein transport, and that shared components between the pathways are passed relatively quickly by the incoming polypeptides.

The grouping of protein import pathways into different classes is also supported by analysis of two mutants in maize that are defective in thylakoid protein transport (Voelker and Barkan, 1995). One such mutant, *tha1*, does not accumulate mature forms of just those proteins thought from competition experiments to be on the CPSec pathway. Indeed, the mutation giving rise to the *tha1* phenotype has recently been identified as a lesion in CPSecA (Voelker et al., 1997). A different mutant, hcf106, is deficient in those proteins thought to be on the delta-pH-dependent pathway (Voelker and Barkan, 1995). Although the target gene giving rise to the hcf106 mutant has been identified, it is not obvious what role this protein plays in the disruption of the delta-pH-dependent pathway.

Directions for future research

The study of protein targeting in chloroplasts is far from complete. Although some of the components of the envelope translocation machinery are now known, it is likely that others will still be found and remain to be characterized. Many questions remain concerning the integration of proteins into the envelope membranes, some quite fundamental, such as whether inner envelope proteins are integrated by stop-transfer during translocation or whether they fully enter the stroma and then are re-exported. In addition, following earlier work suggesting that proteins must adopt unfolded or extended conformations to traverse the envelope membranes (della-Cioppa and Kishore, 1988; Guera et al., 1993; Schatz and Dobberstein, 1996), it has recently been shown that fully folded proteins can cross the envelopes (Ceccarelli et al., 1996; Clark and Theg, 1997), at least under certain conditions. The implications of this for the mechanism of chloroplast protein transport have not yet been fully explored.

Even more work remains to be done concerning the mechanism of protein translocation into or across the thylakoid membrane. Translocation intermediates have yet been positively identified in this membrane, and this has hindered the ability to study the translocation machinery biochemically. While it is thought that the CPSec pathway will use a molecular machinery very similar to the export apparatus of bacteria, this still remains to be demonstrated. Virtually nothing is known about the molecular components involved in transport on the delta-pH-dependent pathway, although a machinery is implicated by protease sensitivity (Robinson, D. et al., 1996) and saturable import kinetics (Cline et al., 1993). Interestingly, this pathway can also transport folded proteins (Clark and Theg, 1997), and does so without creating an ion leak in the membrane (Teter and Theg, unpublished observations), a condition necessary for uninterrupted energy transduction. Additionally, new experiments from the author's laboratory have revealed an unexpected azide sensitivity in a delta-pH-dependent pathway transport reaction, suggesting that a certain degree of crossover between this and the CPSec pathway can occur (Leheny et al., submitted). This suggestion is also supported by careful examination of the maize thylakoid targeting mutants described above. Clearly these and other questions not elaborated here will provide the impetus for research into chloroplast protein targeting for years to come.

Acknowledgments

Support by the USDA and the DOE for work in my laboratory is gratefully acknowledged. I am particularly grateful to Dr. Gottfried Schatz for the opportunity to spend a sabbatical stay in his laboratory, during which time this chapter was written.

References

Berghofer, J., Karnauchov, I., Herrmann, R. G., and Klosgen, R. B. (1995). Isolation and characterization of a cdna encoding the seca protein from spinach chloroplasts - evidence for azide resistance of sec-dependent protein translocation across thylakoid membranes in spinach. J. Biol. Chem. *270*, 18341-18346.

Betts, S. D., Hachigian, T. M., Pichersky, E., and Yocum, C. F. (1994). Reconstitution of the spinach oxygen-evolving complex with recombinant Arabidopsis manganese-stabilizing protein. Plant Mol Biol *26*, 117-130.

Boutry, M., Nagy, F., Poulssen, C., Aoyagi, K., and Chua, Nam-H. (1987). Targeting of bacterial chloramphenicol acetyltransferase to mitochondria in transgenic plants. Nature *328*, 340-342.

Boyle, S. A., Hemmingsen, S. M., and Dennis, D. T. (1990). Energy requirement for the import of protein into plastids from developing endosperm of Ricinus communis L. Plant Physiol. *92*, 151-154.

Briggs, M. S., Cornell, D. G., Dluhy, R. A., and Gierasch, L. M. (1986). Conformations of signal peptides induced by lipids suggest initial steps in protein export. Science *233*, 206-208.

Brink, S., Fluegge, U.-I., Chaumont, F., Boutry, M., Emmermann, M., Schmitz, U., Becker, K., and Pfanner, N. (1994). Preproteins of chloroplast envelope inner membrane contain targeting information for receptor-dependent import into fungal mitochondria. J Biol Chem *269*, 16478-16485.

Brock, I. W., Mills, J. D., Robinson, D., and Robinson, C. (1995). The delta pH-driven, ATP-independent protein translocation mechanism in the chloroplast thylakoid membrane. J. Biol. Chem. *270*, 1657-1662.

Ceccarelli, E. A., Krapp, A. R., Serra, E. C., and Carrillo, N. (1996). Conformational requirements of a recombinant ferredoxin-nadp(+) reductase precursor for efficient binding to and import into isolated chloroplasts. Eur. J. Biochem. *238*, 192-197.

Chaddock, A. M., Mant, A., Karnauchov, I., Brink, S., Herrmann, R. G., Klosgen, R. B., and Robinson, C. (1995). A new type of signal peptide: central role of a twin-arginine motif in transfer signals for the delta pH-dependent thylakoidal protein translocase. EMBO J. *14*, 2715-2722.

Chen, D. D., and Schnell, D. J. (1997). Insertion of the 34-kDa chloroplast protein import component, IAP34, into the chloroplast outer membrane is dependent on its intrinsic GTP-binding capacity. J. Biol. Chem. *272*, 6614-6620.

Clark, S. A., and Theg, S. M. (1997). A folded protein can be transorted across the chloroplast envelope and thylakoid membranes. Mol. Biol. Cell *in press*.

Cline, K., and Henry, R. (1996). Import and routing of nucleus-encoded chloroplast proteins. Annu. Rev. Cell. Dev. Biol. *12*, 1-26.

Cline, K., Werner-Washburne, M., Lubben, T. H., and Keegstra, K. (1985). Precursors to two nuclear-encoded chloroplast proteins bind to the outer envelope membrane before being imported into chloroplasts. J. Biol. Chem. *260*, 3691-3696.

Cline, K., Ettinger, W. F., and Theg, S. M. (1992). Protein-specific energy requirements for protein transport across or into thylakoid membranes. Two lumenal proteins are transported in the absence of ATP. J. Biol. Chem. *267*, 2688-2696.

Cline, K., Henry, R., Li, C., and Yuan, J. (1993). Multiple pathways for protein transport into or across the thylakoid membrane. EMBO J. *12*, 4105-4114.

Creissen, G., Reynolds, H., Xue, Y., and Mullineaux, P. (1995). Simultaneous targeting of pea glutathione reductase and of a bacterial fusion protein to chloroplasts and mitochondria in transgenic tobacco. Plant J. *8*, 167-175.

Dahlin, C. (1993). Import of nuclear-encoded proteins into carotenoid-deficient young etioplasts. Physiol. Plant. *87*, 410-416.

de Kruijff, B., Pilon, R., van't Hof, R., and Demel, R. (1996). Lipid-protein interactions in chloroplast protein import. In Molecular Dynamics of Membranes, J. A. F. Op den Kamp, ed. (Berlin: Springer), pp. 99-136.

della-Cioppa, G., and Kishore, G. M. (1988). Import of a precursor protein into chloroplasts is inhibited by the herbicide glyphosate. EMBO J. *7*, 1299-1305.

Filho, M. D. S., Chaumont, F., Leterme, S., and Boutry, M. (1996). Mitochondrial and chloroplast targeting sequences in tandem modify protein import specificity in plant organelles. Plant Mol. Biol. *30*, 769-780.

Fischer, K., Weber, A., Arbinger, B., Brink, S., Eckerskorn, C., and Fluegge U-I (1994). The 24 kDa outer envelope membrane protein from spinach chloroplasts: Molecular cloning, in vivo expression and import pathway of a protein with unusual properties. Plant Mol Biol *25*, 167-177.

Fluegge, U. I., and Hinz, G. (1986). Energy dependence of protein translocation into chloroplasts. Eur. J. Biochem. *160*, 563-570.

Franklin, A. E., and Hoffman, N. E. (1993). Characterization of a chloroplast homologue of the 54-kda subunit of the signal recognition particle. J. Biol. Chem. *268*, 22175-22180.

Friedman, A. L., and Keegstra, K. (1989). Chloroplast protein import: Quantitative analysis of receptor mediated binding. Plant Physiol. *89*, 993-999.

Gray, J. C., and Row, P. E. (1995). Protein translocation across the chloroplast envelope membranes. Trends Cell Biol *5*, 243-247.

Grossman, A., Bartlett, S., and Chua, N.-H. (1980). Energy-dependent uptake of cytoplasmically synthesized polypeptides by chloroplasts. Nature *285*, 625-628.

Guera, A., America, T., van Waas, M., and Weisbeek, P. J. (1993). A strong protein unfolding activity is associated with the binding of precursor chloroplast proteins to chloroplast envelopes. Plant Mol. Biol. *23*, 309-324.

Halpin, C., Elderfield, P. D., James, H. E., Zimmermann, R., Dunbar, B., and Robinson, C. (1989a). The reaction specificities of the thylakoidal processing peptidase and Escherichia coli leader peptidase are identical. EMBO J. *8*, 3917-3921.

Halpin, C., Musgrove, J. E., Lord, J. M., and Robinson, C. (1989b). Import and processing of proteins by castor bean leucoplasts. FEBS Lett. *258*, 32-34.

Hauke, V., and Schatz, G. (1997). Import of proteins into mitochondria and chloroplasts. Tr. Cell. Bio. *x*, x.

Henry, R., Carrigan, M., McCaffery, M., Ma, X., and Cline, K. (1997). Targeting determinats and proposed evolutionary basis for the Sec and Delta pH protein import systems in chloroplast thylakoid membranes. J. Cell Biol. *136*, 823-832.

Hirsch, S., and Soll, J. (1995). Import of a new chloroplast inner envelope protein is greatly stimulated by potassium phosphate. Plant Mol. Biol. *27*, 1173-1181.

Hirsch, S., Muckel, E., Heemeyer, F., Von Heijne, G., and Soll, J. (1994). A receptor component of the chloroplast protein translocation machinery. Science *266*, 1989-1992.

Hoffman, N. E., and Franklin, A. E. (1994). Evidence for a stromal GTP requirement for the integration of a chlorophyll a/b-binding polypeptide into thylakoid membranes. Plant Physiol. *105*, 295-304.

Kessler, F., and Blobel, G. (1996). Interaction of the protein import and folding machineries in the chloroplast. Proc. Natl. Acad. Sci. USA *93*, 7684-7689.

Kessler, F., Blobel, G., Patel, H. A., and Schnell, D. J. (1994). Identification of two GTP-binding proteins in the chloroplast protein import machinery. Science *266*, 1035-1039.

Killian,, de Jong,, Bijvelt, J., Verkleij, A. J., and de Kruijff, B. (1990). Induction of non-bilayer lipid structures by functional signal peptides. EMBO J. *9*, 815-819.

Knight, J. S., and Gray, J. C. (1995). The N-terminal hydrophobic region of the mature phosphate translocator is sufficient for targeting to the chloroplast inner envelope membrane. Plant Cell *7*, 1421-1432.

Knott, T. G., and Robinson, C. (1994). The secA inhibitor, azide, reversibly blocks the translocation of a subset of proteins across the chloroplast thylakoid membrane. J. Biol. Chem. *269*, 7843-7846.

Kouranov, A., and Schnell, D. J. (1996). Protein translocation at the envelope and thylakoid membranes of chloroplasts. J. Biol. Chem. *271*, 31009-31012.

Laidler, V., Chaddock, A. M., Knott, T. G., Walker, D., and Robinson, C. (1995). A secy homolog in arabidopsis thaliana - sequence of a full-length cdna clone and import of the precursor protein into chloroplasts. J. Biol. Chem. *270*, 17664-17667.

Li, H. M., Moore, T., and Keegstra, K. (1991). Targeting of proteins to the outer envelope membrane uses a different pathway than transport into chloroplasts. Plant Cell *3*, 709-717.

Li, H. M., Sullivan, T. D., and Keegstra, K. (1992). Information for targeting to the chloroplastic inner envelope membrane contained in the mature region of the maize Bt1-encoded protein. J. Biol. Chem. *267*, 18999-19004.

Li, X., Henry, R., Yuan, J., Cline, K., and Hoffman, N. E. (1995). A chloroplast homologue of the signal recognition particle subunit SRP54 is involved in the posttranslational integration of a protein into thylakoid membranes. Proc. Natl. Acad. Sci. USA *92*, 3789-3793.

Lorkovic, Z. J., Schroder, W. P., Pakrasi, H. B., Irrgang, K. D., Herrmann, R. G., and Oelmuller, R. (1995). Molecular characterization of psbw, a nuclear-encoded component of the photosystem II reaction center complex in spinach. Proc. Natl. Acad. Sci. USA *92*, 8930-8934.

Madueno, F., Napier, J. A., and Gray, J. C. (1993). Newly imported Reiskeiron-sulfur protein associates with both cpn60 and hsp70 in the chloroplast stroma. Plant Cell *5*, 1865-1876.

Meadows, J. W., and Robinson, C. (1991). The full precursor of the 33 kDa oxygen-evolving complex protein of wheat is exported by Escherichia coli and processed to the mature size. Plant Mol Biol *17*, 1241-1243.

Michl, D., Robinson, C., Shackleton, J. B., Herrmann, R. G., and Klosgen, R. B. (1994). Targeting of proteins to the thylakoids by bipartite presequences: CFoII is imported by a novel, third pathway. EMBO J. *13*, 1310-1317.

Nielsen, E., Akita, M., DavilaAponte, J., and Keegstra, K. (1997). Stable association of chloroplastic precursors with protein translocation complexes that contain proteins from both envelope membranes and a stromal Hsp 100 molecular chaperone. EMBO J. *16*, 935-946.

Nohara, Y., Nakai, M., Goto, A., and Endo, T. (1995). Isolation and characterization of the cDNA for pea chloroplast SecA. Evolutionary conservation of the bacterial-type SecA-dependent protein transport within chloroplasts. FEBS Lett. *364*, 305-308.

Olsen, L. J., and Keegstra, K. (1992). The binding of precursor proteins to chloroplasts requires nucleoside triphosphates in the intermembrane space. J. Biol. Chem. *267*, 433-439.

Pain, D., and Blobel, G. (1987). Protein import into chloroplasts requires a novel chloroplast ATPase. Proc. Natl. Acad. Sci. USA *84*, 3288-3292.

Perry, S. E., and Keegstra, K. (1994). Envelope membrane proteins that interact with chloroplastic precursor proteins. Plant Cell *6*, 93-105.

Pfanner, N., Rassow, J., Van der Klei, I. J., and Neupert, W. (1992). A dynamic model of the mitochondrial protein import machinery. Cell *68*, 999-1002.

Robinson, C., Cai, D., Hulford, A., Brock, I. W., Michl, D., Hazell, L., Schmidt, I., Herrmann, R. G., and Klosgen, R. B. (1994). The presequence of a chimeric construct dictates which of two mechanisms are utilized for translocation across the thylakoid membrane: evidence for the existence of two distinct translocation systems. EMBO J. *13*, 279-285.

Robinson, D., Karnauchov, I., Herrmann, R. G., Klosgen, R. B., and Robinson, C. (1996). Protease-sensitive thylakoidal import machinery for the sec-, delta-ph- and signal recognition particle-dependent protein targeting pathways, but not for cfoii integration. Plant J. *10*, 149-155.

Salomon, M., Fischer, K., Fluegge, U. I., and Soll, J. (1990). Sequence analysis and protein import studies of an outer chloroplast envelope polypeptide. Proc. Natl. Acad. Sci. USA *87*, 5778-5782.

Schatz, G., and Dobberstein, B. (1996). Common principles of protein translocation across membranes. Science *271*, 1519-1525.

Schindler, C., Hracky, R., and Soll, J. (1986). Protein transport in chloroplasts: ATP is prerequisit. Z. Naturforsch. *42*, 103-108.

Schnell, D. J., and Blobel, G. (1993). Identification of intermediates in the pathway of protein import into chloroplasts and their localization to envelope contact sites. J. Cell Biol. *120*, 103-115.

Schnell, D. J., Kessler, F., and Blobel, G. (1994). Isolation of components of the chloroplast protein import machinery. Science *266*, 1007-1012.

Schoenknecht, G., Althoff, G., and Junge, W. (1990). The electric unit size of thylakoid membranes. FEBS Lett. *277*, 65-68.

Scott, S. V., and Theg, S. M. (1996). A new chloroplast protein import intermediate reveals distinct reveals distinct translocation machineries in the two envelope membranes: energetics and mechanistic implications. J. Cell Biol. *132*, 63-75.

Seedorf, M., Waegemann, K., and Soll, J. (1995). Aconstituent of the chloroplast import complex represents a new type of GTP-binding protein. Plant J. *7*, 401-411.

Seidler, A., and Michel, H. (1990). Expression in Escherichia coli of the psbO gene encoding the 33 kd protein of the oxygen-evolving complex from spinach. EMBO J. *9*, 1743-1748.

Soll, J., and Waegemann, K. (1992). A functionally active protein import complex from chloroplasts. Plant J. *2*, 253-256.

Theg, S. M., and Geske, F. J. (1992). Biophysical characterization of a transit peptide directing chloroplast protein import. Biochemistry *31*, 5053-5060.

Theg, S. M., Bauerle, C., Olsen, L. J., Selman, B. R., and Keegstra, K. (1989). Internal ATP is the only energy requirement for the translocation of precursor proteins across chloroplastic membranes. J. Biol. Chem. *264*, 6730-6736.

Tranel, P. J., and Keegstra, K. (1996). A novel, bipartite transit peptide targets oep75 to the outer membrane of the chloroplastic envelope. Plant Cell *8*, 2093-2104.

Tsugeki, R., and Nishimura, M. (1993). Interaction of homologues of Hsp70 and Cpn60 with ferredoxin-NADP+ reductase upon its import into chloroplasts. FEBS Lett. *320*, 198-202.

Van't Hof, R., Demel, R. A., Keegstra, K., De, K., and B (1991). Lipid-peptide interactions between fragments of the transit peptide of ribulose-1,5-bisphosphate carboxylase/oxygenase and chloroplast membrane lipids. FEBS Lett. *291*, 350-354.

Voelker, R., and Barkan, A. (1995). Two nuclear mutations disrupt distinct pathways for targeting proteins to the chloroplast thylakoid. EMBO J. *14*, 3905-3914.

Voelker, R., Mendel-Hartvig, J., and Barkan, A. (1997). Transposon-disruption of a maize nuclear gene, tha1, encoding a chloroplast secA homolog: in vivo role of secA in thylakoid protein targeting. Genetics *145*, 467-478.

von Heijne, G., and Nishikawa, K. (1991). Chloroplast transit peptides the perfect random coil. FEBS Lett. *278*, 1-3.

von Heijne, G., Steppuhn, J., and Hermann, R. G. (1989). Domain structure of mitochondrial and chloroplast targeting peptides. Eur. J. Biochem. *180*, 535-545.

Waegemann, K., and Soll, J. (1996). Phosphorylation of the transit sequence of chloroplast precursor proteins. J. Biol. Chem. *217*, 6545-6554.

Whelan, J., Knorpp, C., and Glaser, E. (1990). Sorting of precursor proteins between isolated spinach leaf mitochondria and chloroplasts. Plant Mol Biol *14*, 977-982.

Wickner, W., Driessen, A. M., and Hartl, F.- U. (1991). The enzymology of protein translocation across the Escherichia coli plasma membrane. Annu. Rev. Biochem. *60*, 101-124.

Wu, C., Seibert, S. S., and Ko, K. (1994). Identification of chloroplast envelope proteins in close physical proximity to a partially translocated chimeric precursor. J. Biol. Chem. *269*, 32264-32271.

Yuan, J., Henry, R., McCaffery, M., and Cline, K. (1994). SecA homolog in protein transport within chloroplasts: evidence for endosymbiont derived sorting. Science *266*, 796-798.

Chaperone Action in Folding Newly-Translated Cytosolic Proteins in Bacteria and Eukaryotes

Arthur L. Horwich

Howard Hughes Medical Institute and
Dept. Genetics, Yale School of Medicine
New Haven, CT 06510

During the past decade, a group of specialized proteins that assist protein folding in the cell, molecular chaperones, has been the subject of intensive study (for review see monograph, Biology of Heat Shock Proteins and Molecular Chaperones, 1994). In general, these components act by binding non-native conformations, preventing them from irreversibly misfolding and aggregating; then, through the action of nucleotide or cofactor binding, substrate protein is discharged from chaperone, either reaching the native state or pursuing additional biogenesis steps. While different non-native conformations appear to be preferred by specific classes of chaperone, it seems that a common feature is probably recognized by all of them, namely, hydrophobic surfaces, specifically exposed in non-native states and ultimately buried and no longer recognizable in the native state. Such hydrophobic interactions do not appear to supply steric information, which is provided, as recognized early by Anfinsen, by the primary amino acid sequence (Anfinsen, 1973). Chaperones thus serve to provide kinetic assistance to the folding process, limiting off-pathway steps that can occur under physiologic conditions. The essential nature of this action has been revealed in vivo, where there is failure of many newly-translated or newly-translocated proteins to reach native form in the setting of mutational alteration of chaperone function.

For the two major families of chaperone, the Hsp60 and Hsp70 families, the context of substrate protein recognition is beginning to be understood. In the case of the Hsp60 family, known as the chaperonins, recognition involves binding of a collapsed globular folding intermediate inside the central channel of a large cylindrical assembly (Langer et al, 1992; Braig et al, 1993). In the case of the bacterial chaperonin, GroEL, the folding intermediate is surrounded by 7 flexible hydrophobic binding sites (Braig et al, 1994). Upon binding GroES, the smaller, seven-membered ring co-chaperonin, to the same ring as polypeptide, in the presence of ATP, substrate protein held by these sites is discharged into the GroES-enclosed GroEL channel where folding commences ($t_{1/2}$< 1sec)(Weissman

NATO ASI Series, Vol. H 106
Lipid and Protein Traffic
Pathways and Molecular Mechanisms
Edited by Jos A. F. Op den Kamp

et al, 1995; Weissman et al, 1996; Mayhew et al, 1996; Rye et al, 1997). Such discharge has recently been shown to be the result of large ATP/GroES-promoted en bloc movements of the peptide binding (apical) domains of GroEL, directing the binding sites away from the central channel and replacing the hydrophobic channel-facing surface with one hydrophilic in character (Xu et al, 1997). This favors both release of peptide and the burial of its exposed hydrophobic surface, promoting the folding process.

In the case of Hsp70 class proteins, an extended segment of polypeptide of 7 amino acids with at least four central hydrophobic residues is bound locally by a single molecule of chaperone (Flynn et al, 1989; Rüdiger et al, 1997), contacted extensively by a hydrophobic arch structure that encloses it (Zhu et al, 1996). Here, the action of ATP binding to chaperone leads to a lower affinity state favoring release of peptide (Schmid et al, 1994; McCarty et al, 1995). Biophysical measurements in vitro have not so far revealed major conformational changes in the substrate protein associated with release from Hsp70 (see e.g.Palleros et al, 1994), but presumably an extended chain released from Hsp70 would be free to pursue folding driven by its primary structure. As such, the action of Hsp70s appears to be more one of stabilizing/holding exposed hydrophobic segments as opposed to directly promoting the native state.

Recent evidence suggests that two other families of chaperone also recognize exposed hydrophobic surfaces, the Hsp90 family of multi-protein complexes, involved with biogenesis of steroid receptors and a number of signal transduction molecules (Sullivan et al, 1997; Rutherford et al, 1994; Bohen et al, 1995), and the Hsp20 family, involved with, among other actions, protection against denaturation by heat and other stresses (Lee et al, 1997; Ehrnsperger et al, 1997). In the case of the Hsp90 family, the conformations recognized may be very near the native state, awaiting binding of a substrate ligand to trigger release and completion of folding. The topology of polypeptide binding by these components remains to be determined. Hsp20 chaperones are found normally as spherical-shaped assemblages of 12-40 or more chaperone monomers, and under stress conditions they appear to "sponge" non-native proteins to their surfaces, at a stoichiometry of nearly one non-native protein per chaperone monomer, producing even larger-sized assemblages. Following relief of stress, substrate proteins are released from Hsp20 and refolded by the assisting action of cooperating Hsp70 or other chaperone systems (Lee et al, 1997; Ehrsperger et al, 1997).

What is the nature of the chaperone "circuitry" in vivo that determines the fate of a newly-translated protein? Where chaperone action is necessary for a protein to reach the native state, does it occur in a linear, ordered, fashion, commencing at the ribosome and proceeding unidirectionally from one chaperone to another, ultimately producing the native state? Such a model has been proposed (Langer et al, 1992; Frydman et al, 1994; Hartl, 1996), and contends not only that there is an ordered set of interactions but also that these interactions occur in a large multi-component complex containing many of the different chaperone family members (Frydman et al, 1994). An alternative model proposes that chaperone interactions occur in a parallel, network-like, fashion, with proteins partitioning among the various chaperone components in a manner dependent on relative affinities of any given conformation for the different chaperones (Weissman et al, 1994; Diamond et al, 1995; Buchberger et al, 1996; Farr et al, 1997). While this latter partitioning model does not exclude cotranslational interactions, the notion of kinetic partitioning of fully-translated proteins necessarily indicates that posttranslational interactions are important. Here, I review the evidence concerning the two models. While early observations concerning chaperone action in translocating systems, such as mitochondria, prompted serious consideration of a linear model for explaining chaperone action on cytosolic proteins (e.g. Langer et al, 1992), more recent evidence examining chaperone action, particularly in the bacterial cytoplasm (Gragerov et al, 1992; Kandror et al, 1994), argues for kinetic partitioning as governing the nature of chaperone interactions with newly-translated proteins (summarized in Fig.1). In the eukaryotic cytosol, our understanding is less well-developed, with uncertainty remaining about which, if any, components play a general role in assisting folding under normal conditions and whether folding in this system might occur cotranslationally without assistance from chaperonin (see Fig.2).

1 Bacterial cytoplasm

Both Hsp60 and Hsp70 systems were originally uncovered in the bacterial cytoplasm by impairment of λ phage biogenesis due to mutations in the resident genes, with GroEL (Hsp60)-deficient bacteria producing aggregated λ phage heads (Georgopoulos et al, 1972; Takano and Kakefuda, 1972) and DnaK (Hsp70)-deficient bacteria unable to support λ DNA replication (Georgopoulos, 1977; Sunshine et al, 1977). It was clear from these earliest studies that the function of these systems extended to host biogenesis, as the original mutants affecting either GroEL or DnaK were observed to be inviable at 42°C and above, even in the absence of phage infection.

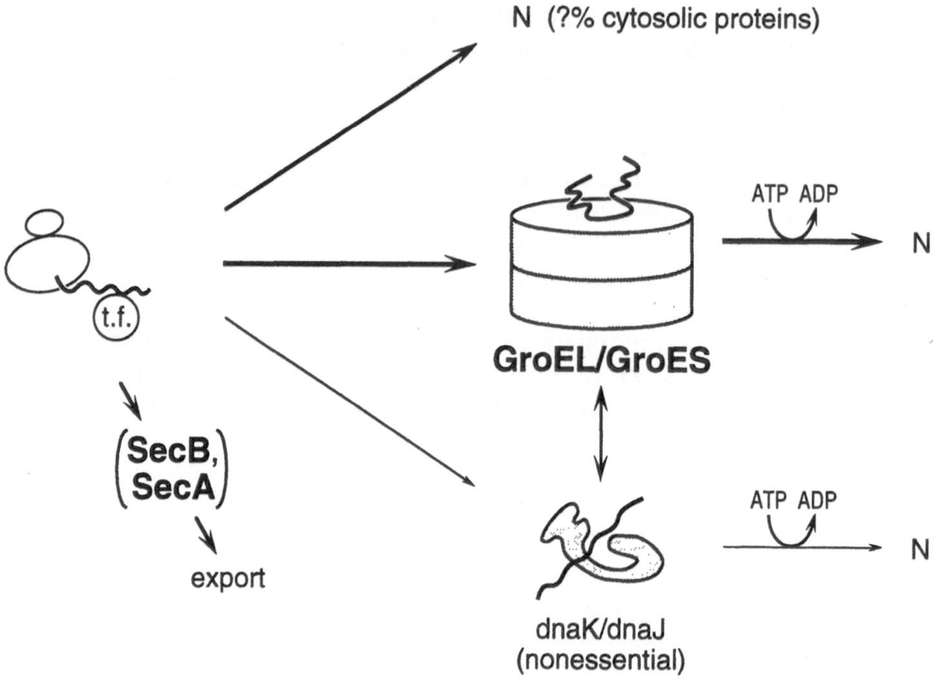

Figure 1. Network of chaperone interactions in the bacterial cytoplasm. Newly-translated proteins in the bacterial cytoplasm reach the native state through several routes. Under normal conditions, the GroEL(Hsp60)/GroES system is essential whereas the DnaK (Hsp70) system is not. The only established cotranslational interaction is with trigger factor (t.f.). See text for discussion.

1.1 GroEL. A further study of the late 80's established that the GroEL system was essential at all temperatures, unable to tolerate gene deletion (Fayet et al, 1989). At about the same time, a primary structural relationship was recognized between GroEL and components inside chloroplasts and mitochondria, the RUBISCO binding protein (Hemmingsen et al, 1988), implicated in assembly of the abundant CO_2-fixing enzyme (Barraclough and Ellis, 1980), and Hsp60 in the mitochondrial matrix (Cheng et al, 1989; Reading et al, 1989), respectively. The function of Hsp60/GroEL proteins in facilitating polypeptide chain folding to the native state was recognized by studies both in vivo studying mitochondrial protein biogenesis (Cheng et al, 1989), and in vitro, studying renaturation of a guanidine HCl-denatured dimeric RUBISCO by purified GroEL-GroES (Goloubinoff et al, 1989). In the mitochondrial studies, a yeast mutant was identified in which mitochondrial precursor proteins were imported and processed but the mature proteins failed to reach native conformation and were found in insoluble aggregates. The mutation was found to affect a mildly heat inducible mitochondrial matrix protein, called Hsp60. Gene disruption studies showed that, despite its name, Hsp60 is essential at all temperatures. In the in vitro studies of Lorimer and coworkers, purified GroEL from *E.coli* was shown to be able to bind the subunit of a homodimeric RUBISCO protein following its dilution from denaturant, preventing it from aggregation; subsequent addition of GroES and ATP to this binary complex then reconstituted the native active enzyme.

Recognizing the action of Hsp60/GroEL in mediating folding to the native state, and considering in physiologic terms here the essential role of GroEL in the bacterial cytoplasm at both normal and heat shock temperatures, one is led to conclude that at least one essential protein of the bacterial cytoplasm requires the assistance of GroEL in reaching its active form. The number of such dependent substrate proteins may be significantly larger, however, as later studies of a severe temperature-sensitive mutant of GroEL suggested that a significant percentage of newly-translated *E.coli* proteins, totaling ~30%, were affected at nonpermissive temperature and either localized to the insoluble fraction or were proteolytically degraded (Horwich et al, 1993). However, since the onset of the growth defect of this conditional mutant was slow (~45 min), it was difficult to distinguish which proteins might be affected in a primary way and which might be secondarily affected.

1.2 DnaK. In contrast with the absolute requirement for GroEL, bacterial strains deleted of DnaK or its cooperating component, DnaJ, are viable if maintained at temperatures of 30°C (Bukau and Walker, 1989; Sell et al, 1990). Notably, however, these mutants tend to pick up suppressor mutations that stabilize their growth, typically mildly upregulating the other heat shock proteins. The implication from these growth studies is that DnaK and DnaJ are required only under heat shock conditions, whereas under non-stress conditions they are not required for the biogenesis of any single protein. Nevertheless, under non-stress conditions the DnaK system seems likely to play a "supporting" role, since the DnaK and DnaJ proteins are already abundant under these conditions (Herendeen et al, 1982), and since deleted cells acquire suppressor mutations that compensate for lack of these components. What are the actions mediated by DnaK-DnaJ?

Studies both in vitro and in vivo have provided insight into the nature of DnaK (Hsp70) action. In the case of λ DNA replication, studies in the late-1980s established that the role of DnaK-DnaJ is to "pry apart" an association between the λO protein and the host protein, DnaB, at the λ replication origin, allowing entry of DNA gyrase and the commencement of the replication process (Liberek et al, 1988). Similarly, the eukaryotic constitutive Hsc70 protein was observed to mediate dissociation of clathrin heavy chain triskelions from each other in vitro in a reaction whose counterpart in the cell would ostensibly remove the clathrin coat from endocytic vesicles (Chappell et al, 1986). Subsequently, studies of ER and mitochondrial protein translocation revealed a role of the Hsp70 system at both *cis* and *trans* sides of these organellar membranes (Chirico et al, 1988; Deshaies et al, 1988; Kang et al, 1990; Vogel et al, 1990). On the cytosolic side, Hsp70s were shown to maintain precursor proteins in extended conformations required for translocation. In the case of mitochondria, inside the organelles, a distinct Hsp70 family member was found to interact with translocating chains, facilitating their entry by binding to extended segments of polypeptide (Kang et al, 1990; Scherer et al, 1990; Schneider et al, 1994; Horst et al, 1997).

Complementing such actions of Hsp70s under normal conditions is its role under heat shock conditions. Here, as originally recognized by Pelham and coworkers, Hsp70s appear to be recruited to sites of "damage," e.g. as originally observed, to heat shocked nucleoli, where they bind non-native proteins (Pelham, 1984; Pelham, 1986). More recent studies indicate that such binding during thermal stress is a protective action, that does not, during the time of exposure, lead to renaturation, but which is required for successful renaturation after downshift of temperature (Schroder et al, 1993). While Hsp70 proteins have thus been identified to mediate a host of crucial cellular actions under both normal and stress conditions, the question has remained as to whether they play a role in de novo cytosolic protein folding.

A recent study of Hesterkamp and Bukau (personal communication), carried out with a DnaK-deleted *E.coli* mutant, addresses the issue, concluding that DnaK does not play a critical role in de novo protein folding in the bacterial cytoplasm. Pulse-labeling experiments were carried out at both 30° C and 42° C, and the cells were fractionated into soluble and insoluble fractions. At both temperatures it was observed that the collective of proteins produced and their solubility was identical to that of an isogenic wild-type strain. In particular, even at heat shock temperature there was no excessive aggregation of newly-translated proteins in the absence of DnaK. Only during a chase period of 8 min at 42° C did some of the radiolabeled proteins become insoluble, but this occurred in parallel with the aggregation of preexistent proteins as revealed by Coomassie staining of total protein. It thus appears that DnaK does not play an essential role in de novo folding even at heat shock temperature but, rather, plays a critical role in repairing already-folded proteins that become thermally damaged.

1.3 Trigger factor. A number of models have proposed that DnaK interacts cotranslationally with nascent polypeptides (Langer et al, 1992; Hartl, 1996), but the foregoing experiment with the DnaK deletion mutant argues against any importance to such an interaction if it occurs. After all, in the deletion strain, even under heat shock conditions, fully translated soluble proteins were produced that must necessarily have left the ribosome and achieved a soluble native or native-like state without the presence of DnaK. DnaK action was found to be required only under heat shock and only at this later point, after chain synthesis/release. Furthermore, studies of the past two years have identified a different component, known as trigger factor, as interacting cotranslationally with nascent chains at the bacterial ribosomes. In these studies, a peptidyl prolyl isomerase activity associated with bacterial ribosomes was discovered to be trigger factor (Stoller et al, 1995), and the component was observed to be associated with nascent polypeptides as judged by release from ribosomes following addition of puromycin (Valent et al, 1995; Hesterkamp et al, 1996). These findings suggest that, in bacterial protein biogenesis, the isomerization of newly-made prolines may be catalyzed at the level of the nascent chain. A further study indicates that the efficiency of the prolyl isomerase activity of trigger factor on nascent chains relies on its high affinity for non-native protein, a chaperone property (Scholz et al, 1997). Interestingly, neither prolyl isomerase nor chaperone activity appear to be essential under normal conditions, because cells depleted of trigger factor exhibit normal growth (Guthrie and Wickner, 1990). At cold temperature, however, the function of trigger factor appears to be essential, as, at these temperatures, depleted cells were inviable (Kandror and Goldberg, 1997). Thus proline isomerization of nascent chains may not proceed efficiently at low temperature in the absence of catalyst. But under these conditions, by contrast, hydrophobic interactions are relatively reduced in strength, reducing the requirement for chaperone action. Thus the requirements for prolyl isomerase activity and chaperone actions appear to be obversely related with respect to temperature.

In sum, under normal conditions of bacterial growth, it appears that the only chaperone function required in folding newly-translated cytoplasmic proteins is the post-translational action of GroEL/GroES (see Fig.1). This state of affairs in the bacterial cytoplasm finds a striking parallel in the thermophilic archaebacterium, *Methanococcus janaschii*, where a complete genomic DNA sequence has revealed that only a chaperonin coding sequence is present and not one for a homologue of DnaK, DnaJ, or GrpE (Bult et al, 1996). Thus a "minimalist" complement of chaperones in either eubacteria or thermophilic archaea appears to incorporate a chaperonin but not an Hsp70 system.

1.4 Kinetic partitioning of non-native forms in the bacterial cytoplasm.

While GroEL/GroES may be the only system that is stringently required in de novo protein folding in bacteria, there are a variety of situations where partitioning of non-native forms to or from the chaperonin system may be critical (see Fig.1). First, it has become clear that such partitioning is an intrinsic feature of the GroEL folding mechanism itself: there is a finite lifetime to the folding-active state of the GroEL-GroES-substrate ternary complex (~15sec), after which GroES and polypeptide are released (Weissman et al, 1996; Burston et al, 1996). For many proteins, only a fraction of released polypeptide chains will have reached native form or a form no longer recognizable by chaperonin, which is committed to reaching the native state in solution. The remaining molecules fail to reach native form and must be bound by the same or another molecule of GroEL, where a further attempt at reaching the native state occurs (Todd et al, 1994; Weissman et al, 1994; Taguchi et al, 1995; Smith and Fisher, 1995; Mayhew et al, 1996; Burston et al, 1996; Ranson et al, 1997). For molecules like rhodanese or RUBISCO, as many as 8-10 such "jumps" may be necessary under standard in vitro conditions for refolding all of the input substrate protein.

Such partitioning can also occur to other components, and seems particularly important in a setting where the substrate protein is damaged or mutationally altered. Proteins incapable of reaching the native state must be released from the chaperonin system or they would tie it up to the exclusion of normal proteins. One fate for such defective proteins is proteolysis, and a study of Kandror, Goldberg, and coworkers (1994) indicates that partitioning from GroEL to proteolytic machinery occurs in vivo. They programmed the inducible synthesis in *E.coli* of a fusion protein that is short-lived, called CRAG, containing a segment of *cro* repressor, a protein A moiety, and a segment of β-galactosidase. They observed that CRAG was efficiently bound by GroEL, then degraded by the ClpP protease, a homotetradecamer resembling the 20S proteasome complexes. (That is, in the setting of ClpP deficiency, CRAG exhibited a prolonged lifetime and was found associated with GroEL.) In the setting of mutations affecting GroES, the CRAG product was stabilized against turnover, remaining at GroEL, apparently due to reduced release and partitioning to the protease.

Partitioning of non-native forms back and forth between GroEL and Hsp70 has also been observed. This may be critical, for example, in the setting of recovery from stress, e.g. after thermal exposure, during which Hsp70 binds misfolded proteins. Some fraction of these bound proteins must in some cases be released and transferred to the GroEL system for refolding. Conversely, under normal conditions other proteins may not be able to be assisted in reaching the native state by GroEL but, instead, must be helped productively by Hsp70 proteins. An example of the latter is the extensively studied test protein, firefly luciferase (Buchberger et al, 1996). This eukaryotic peroxisomal protein is efficiently bound in non-native form by either DnaK or GroEL but is only productively released from DnaK/J. Indeed when luciferase was diluted from denaturant into a mixture containing a molar excess of DnaK/J, it was fully refolded with a $t_{1/2}$ of 6 minutes, but if a GroEL trap mutant, able to bind but not release non-native proteins, was added to the mixture, it immediately quenched refolding, indicating that the non-native protein released from DnaK had transferred to trap. In a reverse order-of-addition, luciferase bound at GroEL could be refolded upon addition of GroES/ATP and the Hsp70 components, reflecting that polypeptide can partition in either direction.

Further evidence for ability of DnaK and GroEL to act on the same substrates, here in an <u>in vivo</u> study, was obtained earlier (Gragerov et al, 1992). A mutant cell deficient of the heat shock transcription factor, σ32, was examined. As the result of diminished Hsp gene transcription, these cells contain low levels of all of the heat shock proteins, including the Hsp60 and Hsp70 systems. In this setting, wholesale protein aggregation occurred, but it was reversed by overexpression of either the GroEL system or the DnaK system. This further reflects the ability of the two systems to recognize the same substrates, albeit they likely do so through recognition of different non-native conformations of the same polypeptides. Indeed, in light of the recent conclusions on the role of DnaK <u>in vivo</u> one can further speculate on the mechanics of prevention of aggregation in this experiment. Where GroEL is overexpressed, it would efficiently assist folding of both the group of newly-translated proteins that it normally acts on, as well as that group of incipiently misfolded proteins that would normally have been most immediately recognized by DnaK. Where DnaK/DnaJ is overexpressed, it seems likely that for obligatory GroEL substrates DnaK/DnaJ acts as a "sink," that dynamically holds GroEL substrates until they are transferred and folded by the limited amount of chaperonin that is still present in these cells. For less stringent GroEL substrates it may also be possible that interaction with the DnaK/DnaJ machinery, probably through multiple cycles of binding and release, can produce the native state. Thus while the GroEL system is absolutely essential under normal conditions, as compared with the DnaK system, it appears that there is kinetic partitioning between the two systems.

More generally, the fate of any given non-native conformation in the non-equilibrium conditions of the bacterial cytoplasm would be determined by the relative concentration of the respective chaperones and the relative rate constants of association. In sum, newly-translated bacterial proteins do not appear to pursue an ordered progression of interaction but are, rather, partitioned among the various chaperone components (see Fig.1). Those proteins with fast-folding kinetics, which may comprise a substantial percentage of cytosolic proteins, will thus be unlikely to interact with or require assistance from any of these components, whereas the class of proteins with slower kinetics and liability to off-pathway steps appears to be subject to such partitioning.

2 Eukaryotic cytosol

A large collective of essential chaperones has been identified in the eukaryotic cytosol but, as yet, the relative roles of the various components in mediating de novo folding of, for example, garden variety metabolic enzymes remains unclear (see Fig.2). These uncertainties are summarized as follows:

1) While there is an essential chaperonin in the eukaryotic cytosol, known as TCP1 complex or CCT (or TRiC), it is unclear whether it has a broad role in de novo folding, like that of GroEL in the bacterial system, or whether its action is more confined. To date, the chaperonin has been established as assisting actin, tubulin, and a few nonstructural proteins including the heterotrimeric G_α protein, transducin, as recently demonstrated (Farr et al, 1997; see also Lewis et al, 1996 and Kubota et al, 1995). But whether it has additional substrates is unclear. Its abundance appears to be considerably less than that of GroEL in some cell types, suggesting that it may not be able to handle a large volume of non-native proteins. Notably also, it is not a heat-inducible component.

2) Hsp90 in the eukaryotic cytosol is generally more abundant than the chaperonin complex, and it is essential. It is found as a component of dynamic multiprotein assemblies that in "mature" form, able to bind substrate protein,

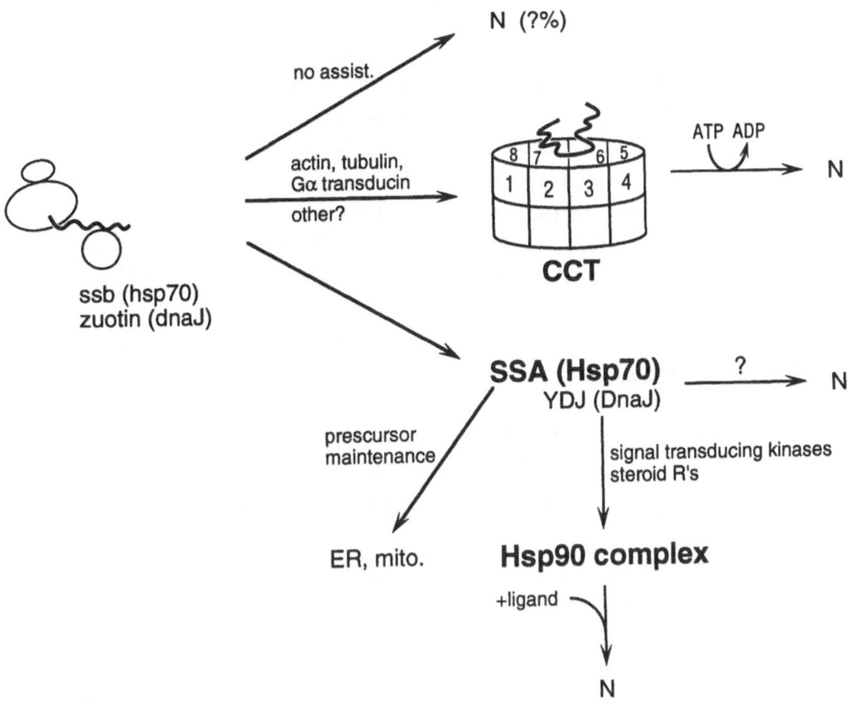

Figure 2. Chaperone network in the eukaryotic cytosol. Newly-translated proteins in the eukaryotic cytosol have been observed to reach the native state through a variety of chaperone interactions. Components playing an essential role, determined by genetic studies in yeast, are indicated in bold.

contain a number of proteins including a cyclophilin (Pratt, 1993; Duina et al, 1996). In vivo studies indicate a role of these assemblies so far confined to the biogenesis of a number of nuclear receptors and signal transduction molecules (Picard et al, 1990; Xu and Lindquist, 1993; Bohen et al, 1995; Rutherford and Zuker, 1994). Recent in vitro experiments have shown that two additional isolated component parts of the Hsp90 assemblies can suppress aggregation of test proteins diluted from denaturant (Bose et al, 1996; Freeman et al, 1996), raising a question of whether there are other substrates.

3) Whereas DnaK and DnaJ in bacteria have been shown to be dispensable to de novo folding, the SSA family of four cytosolic Hsp70 proteins is essential in the yeast cytosol. These proteins appear to function posttranslationally. They have been recognized to play a role in maintaining precursor proteins in an unfolded, extended, state necessary for import into the ER and mitochondria (Chirico et al, 1988; Deshaies et al, 1988), and the strong heat inducibility of several SSA family members suggests a role in protein repair following heat shock. This action is complemented, at least in yeast, by the heat shock protein Hsp104, a hexameric ring assembly whose subunits have a shape resembling GroEL in EM, composed of two globular ATPase-containing domains connected by a hinge. In intact yeast, Hsp104 appears to be able to promote disaggregation of proteins following heat shock (Parsell et al, 1994).

4) It has recently been suggested, based on evolutionary considerations and on studies with a designed fusion protein, that folding of eukaryotic cytosolic proteins, in contrast with posttranslational folding in bacteria, occurs at the ribosome, and may require little or no assistance from chaperones (Netzer and Hartl, 1997). Thus there are three essential components that have been identified to influence the folded state of newly-made proteins in the eukaryotic cytosol, and, more generally, there is a range of experimental evidence and opinion concerning the role of the individual components and even the cotranslational vs. posttranslational nature of folding of newly-translated cytosolic proteins. Here I suggest means by which the picture might be clarified, considering each of the foregoing items in order.

2.1 CCT

This chaperonin assembly, whose rings are each composed of eight distinct essential subunits, has been clearly established from studies both in vivo and in vitro to play an essential role in folding actin and tubulin, two aggregation-prone components that normally reside in polymers, where much of the surface area of a subunit makes contact with neighboring subunits or with specific binding proteins. Recent studies have observed that the mechanism of action of CCT is very similar to that of GroEL, with folding initiating and proceeding (posttranslationally) in the central channel of CCT in the presence of ATP (Tian et al, 1995; Farr et al, 1997). As with GroEL, there is a timed discharge of both native and non-native forms of substrate protein, with non-native forms shown to partition to other CCT molecules both in in vitro experiments in reticulocyte lysate, and in vivo (Farr et al, 1997). Yet the question of whether CCT acts on additional, nonstructural, substrates has remained unsettled. Recently, however, another notoriously aggregation-prone protein, the G_α protein transducin, has been identified as a substrate, both in vitro and in vivo, in cultured retinal cells (Farr et al, 1997). On the other hand, in assays that detect binding of the known substrates, carried out either in buffer with protein diluted from denaturant or in reticulocyte lysate examining association of newly-translated protein, only this limited set of proteins has been shown to interact. A number of garden variety cytosolic proteins including globin, ras, and yeast OTC, have failed to interact (Melki and Cowan, 1994; S.Kim, unpublished). Yet when total message populations are translated, a significant number of additional species interact with CCT. Both the identity of these proteins and whether they can be productively released with ATP is unknown, however. While earlier considerations have suggested that CCT is in some cases nonabundant, this remains open to question because, for example, turnover of the chaperonin complex during isolation from *S.cerevisiae* is considerable unless a pep4, protease-deficient, strain is used, in which case the recovery is substantially greater. Notably, the amounts of CCT in *Xenopus* oocytes (0.5% soluble protein) and bovine liver (0.3%) approaches that of GroEL. Thus it remains unclear whether CCT has a larger role in cytosolic folding than has been demonstrated so far. Further studies carried out in vivo may settle this question, e.g. using available mutant yeast strains or perhaps using *Xenopus* oocytes. Alternatively, in vitro studies providing an understanding of the nature of chaperonin specificity might settle the issue. That is, it would seem informative to know why the cytosolic CCT is unable to bind such GroEL substrates as DHFR and rhodanese, while on the other hand, GroEL is able to bind but not to refold actin or tubulin, which nonproductively recycle off and on GroEL in the presence of ATP/GroES.

2.2 Hsp90

The role of Hsp90 has also remained unclear. Here in vivo experiments have supported a role for Hsp90 complexes in the activation of a specific set of proteins that includes the glucocorticoid, estrogen, and progesterone receptors, and a number of signal transduction molecules (Rutherford and Zuker, 1994; Bohen et al, 1995). It appears that "mature" Hsp90 complexes bind conformations of these components that may be already partially folded and that, at least in some cases, have already interacted with the Hsp70 system, in particular with a DnaJ protein (Kimura et al, 1995). In the well-studied case of the glucocorticoid receptor, binding of the hormonal ligand leads to dissociation of receptor from the Hsp90 complex, followed by ligand-receptor entry into the nuclear compartment and transcriptional activation. Presumably, Hsp90 complexes similarly influence other signal transduction molecules, which may in many cases have already undergone de novo folding to a state that is near-native and which await a final conformational transition driven by ligand binding. Nevertheless, the high abundance of Hsp90 and strong response to heat shock leave one to consider whether there could be a more general role in de novo folding.

Such a role has been proposed based on in vitro studies, taking individual components of the Hsp90 complex including Hsp90 itself, the cyclophilin component (p45), and a third protein of 23 kDa size, and incubating them separately in molar excess with such proteins as β-galactosidase or citrate synthase diluted from denaturant (Bose et al, 1996; Freeman et al, 1996). The presence of Hsp90, cyclophilin, or p23 was found to suppress aggregation in this context. Reactivation of β-galactosidase on the time scale of several hours was then promoted if Hsp70 and DnaJ were supplied to the respective reaction mixtures. The question must be raised as to the physiologic significance of such experiments. It would seem that without an in vivo correlate there is no way to evaluate several concerns about these studies. First, dissecting apart a functional complex into component parts and demonstrating that they can individually exhibit an activity of suppressing aggregation in vitro seems problematic. In such a context it would seem possible that suppression of aggregation is accomplished simply by binding of refolding protein to a surface that is normally interactive with other components of the Hsp90 complex as opposed to a surface that normally contacts non-native protein. Clearly a reconstituted Hsp90 complex should be able to likewise suppress aggregation in vitro if these surfaces are relevant to action in vivo. Second, it would seem necessary to carry out experiments on physiologic substrates whose dependence on Hsp90 complexes has previously been established in vivo. If foreign proteins are going to be employed, it would seem important to test them in an in vivo context (or even in a reticulocyte lysate system) to see whether genetic defects of Hsp90 or cyclophilin, for example, affect the biogenesis of these proteins. Indeed, null mutation in cyclophilin has been shown to affect biogenesis of a known substrate, the gluococorticoid receptor, when its expression was induced in yeast (Duina et al, 1996). Thirdly, why does recovery of activity depend on addition of other chaperones and require so many hours? There is no precedent for such dependence by physiologic substrates in vivo, and, where folding by Hsp70 or Hsp60 systems has been reconstituted in vitro, the action is generally complete within minutes. In sum, then, as with CCT, cellular studies are needed to gain an understanding of the spectrum of action of Hsp90 complexes. As temperature-sensitive lethal mutants of yeast Hsp90 are available, such tests are feasible. But such experiments may also be approachable in a mammalian system as well, where the inhibitor, geldanamycin, which competes for ATP binding and blocks receptor complex maturation, is available. Alternatively, if complexes can be reconstituted efficiently in vitro, tests of binding and folding, examining candidate molecules in parallel with established substrates, could be carried out.

2.3 Hsp70 proteins

The role of eukaryotic cytosolic Hsp70 proteins has remained as elusive as that of the foregoing components. For lack of better understanding, it seems convenient to divide the action of these chaperones into cotranslational and post-translational modes. A number of experiments carried out both in vivo and in cell free systems have suggested that cotranslational interactions occur between nascent cytosolic proteins and Hsp70 proteins. In the earliest study, Beckmann, Welch, and coworkers (1990) carried out immunoprecipitation of extracts of pulse-radiolabeled cells with anti-Hsc70 antisera, and recognized a cohort of proteins. This collective was not recovered if the extract was first treated with puromycin, suggesting early interaction of Hsc70 with translating chains. In a second study, Nelson, Craig, and colleagues (1993) found that two closely-related Hsp70 proteins of the yeast cytosol, the SSB proteins, colocalize with translating ribosomes and were released by puromycin treatment. In a more recent study from Craig, Wiedemann and coworkers (personal communication), nascent chains bearing a photochemical crosslinker were found to specifically crosslink SSB but not SSA proteins in a yeast translation mixture. The nature of SSB function has also been probed in vivo, examining cells disrupted of both SSB genes. Cell growth was unaffected at normal and heat shock temperature, but was impaired at cold temperature. Thus, as with trigger factor in the bacterial system, the role of SSB proteins in assisting nascent chains at the ribosome is dispensable under normal growth conditions.

A second class of Hsp70 proteins in the yeast cytosol, the SSA proteins, appears to participate in posttranslational interactions with a number of non-native proteins. These include newly-made precursor proteins destined for import into mitochondria and precursors that undergo posttranslational import into the ER (non-SRP-mediated import). Here the role of SSA proteins appears to involve maintenance of the precursors in an extended, "loose," state, that permits recognition and passage through the translocation machinery (Chirico et al, 1988; Deshaies et al, 1988). Whether SSA proteins can also interact posttranslationally with proteins destined to reside in the cytosol is unclear, but under conditions of thermal stress, where two members of this family are strongly heat-induced, this seems a likely possibility. Presumably, as in the early studies of Pelham, the stress-induced Hsp70s might also be recruited to the nuclear/nucleolar compartments. It remains, however, that under non-stress conditions the SSA proteins are highly abundant, to a level parallel with Hsp90, and that, for as yet unknown reasons, there are four family members. As with the other components, a broader role can really only be established by studies in vivo.

2.4 Cotranslational vs. posttranslational folding

A recent study has argued from both evolutionary/size considerations and experimental data that, while prokaryotic proteins, only ~10% of which are larger in size than 500 amino acids, are folded post-translationally, eukaryotic proteins, with ~30% larger in size than 500 amino acids, employ cotranslational folding (Netzer and Hartl, 1997). The experimental evidence is drawn from fusing together the coding sequences for two small cytosolic polypeptides, ras and DHFR, and examining folding in both in vitro translation mixtures and in vivo. In these experiments it was observed in the eukaryotic reticulocyte lysate system that the NH_2-terminal domain of the fusion reached a native folded state before the COOH-terminal portion completed its translation, whereas in a bacterial S30 extract, the NH_2-terminal region did not fold until the COOH-terminal region had been synthesized, and the two domains thus folded together temporally. A second example, the bacterial transcription factor OmpR, also appeared to behave in this same differential fashion in the two lysate systems. Kinetic studies carried out in such broken cell systems may not be able to reliably report on the situation in vivo, however, since these systems may be limiting for factors required for translation or for chaperones required for folding. The authors were apparently worried about this problem and attempted an in vivo study with the fusion expressed in COS cells, where they sought to examine the time course of folding of the two domains in relation to translation. At various times after pulse radiolabeling, they added cycloheximide and digitonin with or without protease to assess protection of the respective domains as a measure of the native state. They concluded that the ras domain could fold while the DHFR portion was still being translated. Alternatively, while synthesis of the COOH-terminal portion of the fusion would be prevented by cycloheximide, folding of the translated NH_2-terminal domain would likely proceed to the native state prior to the time the cell was disrupted by digitonin and its contents became susceptible to protease. Even at 4°C, the ras domain might be able to fold very rapidly, prior to its exposure to protease. Whether the ras-DHFR fusion behaves as suggested, additional experiments with endogenous proteins, particularly in vivo, seem warranted to resolve whether this artificially constructed protein is an isolated example, following the model presented, or whether other proteins pursue such differing behavior in bacterial vs. eukaryotic systems. Ideally, one would like to carry out such in vivo studies with real time imaging to locate folding chains with respect to the ribosome and view a signal from them reflecting acquisition of the native state.

Beyond this, the relative roles of chaperones in the prokaryotic and eukaryotic settings needs to be resolved. The recent study suggests that the need for a chaperonin is reduced if a cotranslational mode of folding is employed. Yet aren't eukaryotic proteins composed of domain folds similar to those of prokaryotes, liable to the same problems of misfolding? In sum, while our mechanistic understanding of the major chaperones has substantially advanced during the past few years, as the foregoing discussion would indicate, there is clearly much that needs to be resolved concerning the physiology of protein synthesis and folding in the cell.

References

Anfinsen, C.B. (1973) Principles that govern the folding of protein chains. Science 181: 223-230.

Barraclough, R. and Ellis, R.J. (1980) Protein synthesis in chloroplasts. IX. Assembly of newly-synthesized large subunits into ribulose bisphosphate carboxylase in isolated pea chloroplasts. Biochimica et Biophysica Acta 608: 19-31.

Beckmann, R.P., Mizzen, L.A., and Welch, W.J. (1990) Interaction of Hsp70 with newly-synthesized proteins: Implication for protein folding and assembly. Science 248: 850-854.

Biology of heat shock proteins and molecular chaperones. (1994) Morimoto, R.I., Tissieres, A., & Georgopoulos, C., eds. Cold Spring Harbor Laboratory Press.

Bohen, S.P., Kralli, A., and Yamamoto, K.R. (1996) Hold 'em and fold 'em: Chaperones and signal transduction. Science 268: 1303-1304.

Bose, S., Weikl, T., Bügl, and Buchner, J. (1997) Chaperone function of Hsp90-associated proteins. Science 274: 1715-1717.

Braig, K., Simon, M., Furuya, F., Hainfeld, J.F., and Horwich, A.L. (1993) A polypeptide bound by the chaperonin GroEL is localized within a central cavity. Proc.Natl.Acad.Sci. USA 90: 3978-3982.

Braig, K., Otwinowski, Z., Hegde, R., Boisvert, D.C., Joachimiak, A., Horwich, A.L., and Sigler, P.B. (1994) The crystal structure of the bacterial chaperonin GroEL at 2.8 Å. Nature 371: 578-586.

Buchberger, A., Schröder, H., Hesterkamp, T., Schönfeld, H.-J., and Bukau, B.

(1996) Substrate shuttling between the DnaK and GroEL systems indicates a chaperone network promoting protein folding. J.Mol.Biol. 261: 328-333.

Bukau, B. and Walker, G.C. (1989) Cellular defects caused by deletion of the *Escherichia coli dnaK* gene indicate roles for heat shock protein in normal metabolism. J.Bacteriol. 171: 2337-2346.

Bult, C.J. et al (1996) Complete genome sequence of the methanogenic archaeon, *Methanococcus jannaschii*. Science 273: 1058-1073.

Burston, S.G., Weissman, J.S., Farr, G.W., Fenton, W.A., and Horwich, A.L. (1996) Release of both native and non-native proteins from a *cis*-only GroEL ternary complex. Nature 383: 96-99.

Chappell, T.G., Welch, W.J., Schlossman, D.M., Palter, K.B., Schlesinger, M.J., and Rothman, J.E. (1986) Uncoating ATPase is a member of the 70 kilodalton family of stress proteins. Cell 45: 3-13.

Cheng, M.Y., Hartl, F.U., Martin J., Pollock, R.A., Kalousek, F., Neupert, W., Hallberg, E.M., Hallberg, R.L., and Horwich, A.L. (1989) Mitochondrial heat-shock protein hsp60 is essential for assembly of proteins imported into yeast mitochondria. Nature 337: 620-625.

Chirico, W.J., Waters, M.G., and Blobel, G. (1988) 70 K heat shock related proteins stimulate protein translocation into microsomes. Nature 332: 805-810.

Deshaies, R.J., Koch, B.D., Werner Washburne, M., Craig, E.A., and Schekman, R. (1988) A subfamily of stress proteins facilitates translocation of secretory and mitochondrial precursor polypeptides. Nature 332: 800-805.

Diamond, D.L., Strobel, S., Chun, S.Y., and Randall, L.L. (1995) Interaction of SecB with intermediates along the folding pathway of maltose-binding protein. Prot.Sci. 4: 1118-1123.

Duina, A.A., Chang, H.-C.J., Marsh, J.A., Lindquist, S., and Gaber, R.F. (1996) A cyclophilin function in Hsp90-dependent signal transduction. Science 274: 1713-1715.

Ehrnsperger, M., Gräber, S., Gaestel, M., and Buchner, J. (1997) Binding of non-native protein to Hsp25 during heat shock creates a reservoir of folding intermediates for reactivation. EMBO J. 16: 221-229.

Farr, G.W., Scharl, E.C., Schumacher, R.J., Sondek, S., and Horwich, A.L. (1997) Chaperonin-mediated folding in the eukaryotic cytosol proceeds through rounds of release of native and nonnative forms. Cell 89: 927-937.

Fayet, O., Ziegelhoffer, T., and Georgopoulos, C. (1989) The *groES* and *groEL* heat shock gene products of *Escherichia coli* are essential for bacterial growth at all temperatures. J.Bacterol. 171: 1379-1385.

Flynn, G.C., Chappell, T.G., and Rothman, J.E. (1989) Peptide binding and release by proteins implicated as catalysts of protein assembly. Science 245: 385-390.

Freeman, B.C., Toft, D.O., and Morimoto, R.I. (1996) Molecular chaperone machines: chaperone activities of the cyclophilin Cyp-40 and the steroid aporeceptor-associated protein p23. Science 274: 1718-1720.

Frydman, J., Nimmesgern, E., Ohtsuka, K., and Hartl, F.U. (1994) Folding of nascent polypeptide chains in a high molecular mass assembly with molecular chaperones. Nature 370: 111-117.

Georgopoulos, C.P. (1977) A new bacterial gene (*groPC*) which affects λ DNA replication. Molec.Gen.Genet. 151: 35-39.

Georgopoulos, C.P., Hendrix, R.W., and Kaiser, A.D. (1972) Role of the host cell in bacteriophage morphogenesis: Effects of a bacterial mutation on T4 head assembly. Nature New Biol. 239: 38-41.

Goloubinoff, P., Christeller, J.T., Gatenby, A.A., and Lorimer, G.H. (1989) Reconstitution of active dimeric ribulose bisphosphate carboxylase from an unfolded state depends on two chaperonin proteins and MgATP. Nature 342: 884-889.

Gragerov, A., Nudler, E., Komissarova, N., Gaitanaris, G.A., Gottesman, M.E., and Nikiforov, V. (1992) Cooperation of GroEL/GroES and DnaK/DnaJ heat shock proteins in preventing protein misfolding in *Escherichia coli*. Proc.Natl.Acad.Sci.USA 89: 10341-10344.

Guthrie, B. and Wickner, W. (1990) Trigger factor depletion or overproduction causes defective cell division but does not block protein export. J.Bacteriol. 172: 5555-5562.

Hartl, F.U. (1996) Molecular chaperones in cellular protein folding. Nature 381: 571-580.

Hemmingsen, S.M., Woolford, C., van der Vies, S.M., Tilly, K., Dennis, D.T., Georgopoulos, C.P., Hendrix, R.W., and Ellis, R.J. (1988) Homologous plant and bacterial proteins chaperone oligomeric protein assembly. Nature 333: 330-334.

Herendeen, S.L., VanBogelen, R.A., and Neidhardt, F.C. (1982) Levels of major proteins of *Escherichia coli* during growth at different temperatures. J.Bacteriol. 139: 185-194.

Hesterkamp, T., Hauser, S., Lütcke, H., and Bukau, B. (1996) *Escherichia coli* trigger factor is a prolyl isomerase that associates with nascent polypeptide chains. Proc.Natl.Acad.Sci.USA 93: 4437-4441.

Horst, M., Oppliger, W., Rospert, S., Schönfeld, H.-J., Schatz, G. and Azem, A. (1997) Sequential action of two hsp70 complexes during protein import into mitochondria. EMBO J. 16: 1842-1849.

Horwich, A.L., Low, K.B., Fenton, W.A., Hirshfield, I.N., and Furtak, K. (1993)

Folding in vivo of bacterial cytoplasmic proteins: Role of GroEL. Cell 74: 909-917.

Kandror, O., Busconi, L, Sherman, M., and Goldberg, A.L. (1994) Rapid degradation of an abnormal protein in *Escherichia coli* involves the chaperones GroEL and GroES. J.Biol.Chem. 269: 23575-23582.

Kandror, O. and Goldberg, A.L. (1997) Trigger factor is induced upon cold shock and enhances viability of *Escherichia coli* at low temperatures. Proc.Natl.Acad.Sci.USA 94: 4978-4981.

Kang, P.-J., Ostermann, J., Shilling, J., Neupert, W., Craig, E.A., and Pfanner, N. (1990) Requirement for hsp70 in the mitochondrial matrix for translocation and folding of precursor proteins. Nature 348: 137-142.

Kimura, Y., Yahara, I., and Lindquist, S. (1995) Role of the protein chaperone YDJ1 in establishing Hsp90-mediated signal transduction pathways. Science 268: 1362-1365.

Kubota, H., Hynes, G., and Willison, K. (1995) The chaperonin containing t-complex polypeptide 1 (TCP-1): multisubunit machinery assisting in protein folding and assembly in the eukaryotic cytosol. Eur.J.Biochem. 230: 3-16.

Langer, T., Pfeifer, G., Martin, J., Baumeister, W., and Hartl, F.-U. (1992) Chaperonin-mediated protein folding: GroES binds to one end of the GroEL cylinder, which accomodates the protein substrate within its central cavity. EMBO J. 11: 4757-4765 (1992).

Langer, T., Lu, C., Echols, H., Flanagan, J., Hayer, M.K., and Hartl, F.U. (1992) Successive action of DnaK, DnaJ and GroEL along the pathway of chaperone-mediated protein folding. Nature 356: 683-689.

Lee, G.J., Roseman, A.M., Saibil, H.R., and Vierling, E. (1997) A small heat shock protein stably binds heat- denatured model substrates and can maintain a substrate in a folding-competent state. EMBO J. 16:659-671.

Lewis, S.A., Tian, G., Vainberg, I.E., and Cowan, N.J. (1996) Chaperonin-mediated folding of actin and tubulin. J.Cell Biol. 132: 1-4.

Liberek, K., Georgopoulos, C., and Zylicz, M. (1988) Role of the *Escherichia coli* DnaK and DnaJ heat shock proteins in the initiation of bacteriophage λ DNA replication. Proc.Natl.Acad.Sci.USA 85: 6632-6636.

Mayhew, M., da Silva, A.C.R., Martin, J., Erdjument-Bromage, H., Tempst, P., and Hartl, F.-U. (1996) Protein folding in the central cavity of the GroEL-GroES chaperonin complex. Nature 379: 420-426.

McCarty, J.S., Buchberger, A., Reinstein, J., and Bukau, B. (1995) The role of ATP in the functional cycle of the DnaK chaperone system. J.Mol.Biol.249: 126-137.

Melki, R. and Cowan, N.J. (1994) Facilitated folding of actins and tubulins occurs via a nucleotide-dependent interaction between cytoplasmic chaperonin and distinctive folding intermediates. Mol.Cell.Biol. 14: 2895-2904.

Nelson, R.J., Ziegelhoffer, T., Nicolet, C., Werner-Washburne, M., and Craig, E.A. (1993) The translation machinery and 70 kd heat shock protein cooperate in protein synthesis. Cell 71: 97-105.

Netzer, W.J. and Hartl, F.U. (1997) Recombination of protein domains facilitated by co-translational folding in eukaryotes. Nature 388: 343-349.

Palleros, D.R., Shi, L., Reid, K.L., and Fink, A.L. (1994) hsp70-Protein complexes. J.Biol.Chem. 269: 13107-13114.

Parsell, D.A., Kowal, A.S., Singer, M.A., and Lindquist, S. (1994) Protein disaggregation mediated by heat-shock protein Hsp104. Nature 372: 475-478.

Pelham, H.R.B. (1984) Hsp70 accelerates the recovery of nucleolar morphology after heat shock. EMBO J. 3: 3095-3100.

Pelham, H.R.B. (1986) Speculations on the function of the major heat shock and glucose-regulated proteins. Cell 46: 959-961.

Picard, D., Khursheed, B., Garabedian, M.J., Fortin, M.G., Lindquist, S., and Yamamoto, K.R. (1990) Reduced levels of hsp90 compromise steroid receptor action in vivo. Nature 348: 166-168.

Pratt, W.B. (1993) The role of heat shock proteins in regulating the function, folding, and trafficking of the glucocorticoid receptor. J.Biol.Chem. 268: 21455-21458.

Ranson, N.A., Burston, S.G., and Clarke, A.R. (1997) Binding, encapsulation, and ejection: substrate dynamics during a chaperonin-assisted folding reaction. J.Mol.Biol. 266: 656-664.

Reading, D.S., Hallberg, R.L., and Myers, A.M. (1989) Characterization of the yeast HSP60 gene coding for a mitochondrial assembly factor. Nature 337: 655-659.

Rüdiger, S., Germeroth, L., Schneider-Mergener, J., and Bukau, B. (1997) Substrate specificity of the DnaK chaperone determined by screening of cellulose-bound peptide libraries. EMBO J. 16: 1501-1507.

Rutherford, S.L. and Zuker, C.S. (1994) Protein folding and the regulation of signaling pathways. Cell 79: 1129-1132.

Rye, H.S., Burston, S.G., Fenton, W.A., Beechem, J.M., Xu, Z., Sigler, P.B., and Horwich, A.L. (1997) Distinct actions of *cis* and *trans* ATP within the double ring of the chaperonin GroEL. Nature 388: 792-798.

Scherer, P.E., Krieg, U.C., Hwang, S.T., Vestweber, D., and Schatz, G. (1990) A precursor protein partly translocated into yeast mitochondria is bound to a 70 kDa mitochondrial stress protein. EMBO J. 9: 4315-4322.

Schmid, D., Baici, A., Gehring, H., and Christen, P. (1994) Kinetics of molecular chaperone action. Science 263: 971-973.

Schneider, H.-C., Berthold, J., Bauer, M.F., Dietmeier, K., Buiard, B., Brunner, M., and Neupert, W. (1994) Mitochondrial hsp70/MIM44 complex facilitates protein import. Nature 371: 768-774.

Schroder, H., Langer, T., Hartl, F.-U., and Bukau, B. (1993) DnaK, DnaJ and GrpE form a cellular chaperone machinery capable of repairing heat-induced protein damage. EMBO J. 12: 4137-4144.

Sell, S.M., Eisen, C., Ang, D., Zylicz, M., and Georgopoulos, C. (1990) Isolation and characterization of *dnaJ* null mutants of *Escherichia coli*. J.Bacteriol. 172: 4827-4835.

Scholz, C., Stoller, G., Zarnt, T., Fischer, G., and Schmid, F.X. (1997) Cooperation of enzymatic and chaperone functions of trigger factor in the catalysis of protein folding. EMBO J. 16: 54-58.

Smith, K.E. and Fisher, M.T. (1995) Interactions between the GroE chaperonins and rhodanese. Multiple intermediates and release and rebinding. J.Biol.Chem. 270: 21517-21523.

Stoller, G., Rücknagel, K.P., Nierhaus, K., Schmid, F.X., Fischer, G., and Rahfeld, J.-U. (1995) Identification of the peptidyl-prolyl cis/trans isomerase bound to the *Escherichia coli* ribosome as the trigger factor. EMBO J. 14: 4939-4948.

Sullivan, W., Stensgard, B., Caucutt, G., Bartha, B., McMahon, N., Alnemri, E.S., Litwack, G., and Toft, D. (1997) Nucleotides and two functional states of hsp90. J. Biol. Chem. 272: 8007-8012.

Sunshine, M., Feiss, M., Stuart, J., and Yochem, J. (1977) A new host gene (*groPC*) necessary for lambda DNA replication. Molec.Gen.Genet. 151: 27-34.

Taguchi, H. and Yoshida, M. (1995) Chaperonin releases the substrate protein in a form with tendency to aggregate and ability to rebind to chaperonin. FEBS Lett 359: 195-198.

Takano, T. and Kakefuda, T. (1972) Involvement of a bacterial factor in morphogenesis of bacteriophage capsid. Nature New Biol. 239: 34-37.

Tian, G., Vainberg, I.E., Tap, W.D., Lewis, S.A., and Cowan, N.J. (1995) Quasi-native chaperonin-bound intermediates in facilitated protein folding. J.Biol.Chem. 270: 23910-23913.

Todd, M.J., Viitanen, P.V., and Lorimer, G.H. (1994) Dynamics of the chaperonin ATPase cycle: Implications for facilitated protein folding. Science 265: 659-666.

Valent, Q.A., Kendall, D.A., High, S., Kusters, R., Oudega, B., and Luirink, J. (1995) Early events in preprotein recognition in *E.coli*: interaction of SRP and trigger factor with nascent polypeptides. EMBO J. 14: 5494-5505.

Vogel, J.P., Misra, L.M., and Rose, M.D. (1990) Loss of BiP/GRP78 function blocks translocation of secretory proteins in yeast. J.Cell Biol. 110: 1885-1895.

Weissman, J.S., Kashi, Y., Fenton, W.A., and Horwich, A.L. (1994) GroEL-mediated protein folding proceeds by multiple rounds of binding and release of nonnative forms. Cell 78: 693-702.

Weissman, J.S., Hohl, C.M., Kovalenko, O., Kashi, Y., Chen, S., Braig, K., Saibil, H.R., Fenton, W.A., and Horwich, A.L. (1995) Mechanism of GroEL action: Productive release of polypeptide from a sequestered position under GroES. Cell 83: 577-587.

Weissman, J.S., Rye, H.S., Fenton, W.A., Beechem, J.M., and Horwich, A.L. (1996) Characterization of the active intermediate of a GroEL-GroES-mediated protein folding reaction. Cell 84: 481-490.

Xu, Y. and Lindquist, S. (1993) Heat-shock protein hsp90 governs the activity of $pp60^{v\text{-}src}$ kinase. Proc.Natl.Acad.Sci.USA 90: 7074-7078.

Xu, Z., Horwich, A.L., and Sigler, P.B. (1997) The crystal structure of the asymmetric GroEL-GroES-$(ADP)_7$ chaperonin complex. Nature 388: 741-750.

Zhu, X., Zhao, X., Burkholder, W.F., Gragerov, A., Ogata, C.M., Gottesman, M., and Hendrickson, W.A. (1996) Structural analysis of substrate binding by the molecular chaperone DnaK. Science 272: 1606-1614.

Biogenesis of Peroxisomes

Suresh Subramani
Department of Biology, Rm 3230 Bonner Hall,
University of California, San Diego, La Jolla, CA 92093-0322, USA.

Abstract. Peroxisomes are ubiquitous subcellular organelles found in all eukaryotic cells. This paper summarizes the current status of research in the areas of peroxisomal protein import and biogenesis. Much is known about the peroxisomal targeting signals (PTSs) that transport proteins to the peroxisomal matrix and membrane. The use of multiple strategies has led to the isolation and characterization of yeast and mammalian *pex* mutants deficient in peroxisomal protein import, biogenesis or segregation of the organelle to daughter cells. Biochemical and genetic analyses of wild-type and *pex* mutant cells has led to the characterization of several PTS receptors, and the peroxisomal docking proteins for these receptors. Details of the translocation of proteins across the peroxisomal membrane remain to be elucidated. An investigation of the early stages of peroxisome biogenesis has revealed some unusual features such as the involvement of vesicles, and perhaps the endoplasmic reticulum, in peroxisome biogenesis. The conservation of many *PEX* genes between yeasts and mammals is leading to insights about the molecular basis of a number of human peroxisomal disorders.

Keywords. Peroxisomal protein import, peroxisome biogenesis, peroxisomal disorders

1. Introduction.

The transport of polypeptides and metabolites into or across the peroxisomal membrane serves as an excellent paradigm for the more general problem of trafficking of macromolecules and small molecules across biological membranes. In addition, because functional peroxisomes are essential for human survival and the failure to import proteins faithfully into peroxisomes is responsible for many devastating and fatal human disorders, much attention has been devoted to the mechanism of protein import into this subcellular compartment.

Peroxisomes, the last of the subcellular organelles to be discovered, house multiple metabolic pathways many of which are connected with the metabolism of lipids (Van den Bosch et al., 1992). The organelle is found in virtually all eukaryotic cells and is devoid of DNA. Consequently, all the protein constituents of this organelle are believed to be synthesized in the cytoplasm and then imported post-translationally to the peroxisomes (Lazarow and Fujiki, 1985).

NATO ASI Series, Vol. H 106
Lipid and Protein Traffic
Pathways and Molecular Mechanisms
Edited by Jos A. F. Op den Kamp

The specificity of import is governed by the recognition of several peroxisomal targeting signals (PTSs) by receptors that shuttle proteins from the cytosol to the peroxisome, where the proteins either traverse the peroxisomal membrane to reside in the organelle matrix, or remain in the membrane.

2. PTSs for matrix and membrane proteins.

Two or more sequences targets proteins to the peroxisome matrix. PTS1 is a C-terminal tripeptide (SKL in the one letter amino acid code, or its functional variants) found in the majority of peroxisomal matrix proteins. PTS2 is an N-terminal or internal sequence comprised of a bipartite nonapeptide consensus motif (R/K)(L/V/I) $(X)_5$ (H/Q) (L/A). It is found in about half a dozen proteins. Both PTSs are conserved in evolution from yeast to man (Subramani, 1996).

Other proteins such as acyl-CoA oxidase from *Candida* sp. lack PTS1 and PTS2 and are still targeted to peroxisomes (Small et al., 1988), but it is not clear whether these proteins have classical targeting sequences, or whether they enter peroxisomes in association with other proteins containing a genuine PTS. Such "piggyback" import of protein subunits lacking their own PTS has been documented (Glover et al., 1994; McNew and Goodman, 1994). At least one protein, *S. cerevisiae* carnitine acetylransferase, has an internal sequence required for peroxisomal matrix targeting and yet its import into peroxisomes requires a functional PTS1 receptor, Pex5p, and this internal PTS also interacts with Pex5p in the yeast two-hybrid system (Elgersma et al., 1995b).

Membrane proteins contain distinct mPTSs which lie adjacent to, but do not include hydrophobic transmembrane domains (TMDs), and they reside on the matrix face of the peroxisomal membrane in the few proteins analyzed to date (Dyer et al., 1996; Wiemer et al., 1996). A consensus sequence consisting of a basic amino acid stretch followed by a small hydrophobic region has been found in all three proteins in which mPTSs have been studied (Elgersma et al., manuscript in preparation).

It has been suggested that peroxisomal membrane proteins might be targeted to the peroxisomes via two pathways - one from the cytosol directly to the peroxisomes, and the other from the cytosol to the peroxisomes via the endoplasmic reticulum (ER) (Subramani, 1996). This would require two classes of mPTSs, mPTS1 and mPTS2, which would have to act in the cytosol and ER lumen, respectively, in order to maintain the observed topology of the proteins in the peroxisomal membrane. Little is known regarding the rules that govern the topology of proteins in the peroxisomal membrane.

3. PTS receptors and peroxisomal receptor-docking proteins.

The receptors for the PTS1 and PTS2 sequences were discovered during the analysis of *pex* mutants compromised in peroxisome import, biogenesis or segregation (Distel et al., 1996). These *pex* mutations, now available in multiple yeasts, and in cells from CHO and humans, have served as the cornerstone for the isolation and characterization of almost 20 *PEX* genes (Subramani, 1997). The *PEX* genes from *Pichia pastoris* are shown in Table 1.

The *pex5* mutants of several yeasts are deficient only in the import of PTS1-containing polypeptides, and not in the import of PTS2- or mPTS-containing proteins (McCollum et al., 1993; Van der Leij et al., 1993; Van der Klei et al., 1995; Szilard et al., 1995). Complementation of these mutants led to the cloning of the *PEX5* gene encoding the PTS1 receptor. Pex5p, is a member of the tetratricopeptide repeat (TPR) family of proteins. This protein is conserved in yeasts and in mammals (Dodt et al., 1995; Fransen et al., 1995; Wiemer et al., 1995). It binds the SKL peptide and its variants specifically (Terlecky et al., 1995; Fransen et al., 1995). It is both cytosolic and peroxisome-associated and shuttles during the import cycle from the cytosol to the peroxisome, and then presumably back to the cytosol for another round of import (Dodt and Gould, 1996).

In contrast with the phenotype of yeast cells lacking Pex5p, mammalian cells missing this receptor are also deficient in the import of PTS2-containing proteins such as thiolase (Wiemer et al., 1995). A possible explanation for this observation was suggested by the known interaction between proteins containing TPR motifs with partners possessing WD40 repeats. It was proposed (Rachubinski and Subramani, 1995) that mammalian Pex5p might interact with the putative PTS2 receptor, Pex7p, which in yeast is a member of the WD40 protein family (Marzioch et al., 1994; Zhang and Lazarow, 1995). These predictions have been borne out. Pex7p from mammals is indeed a member of the WD40 family (Braverman et al., 1997; Purdue et al., 1997; Motley et al., 1997). Furthermore, there is a two-hybrid interaction between *S. cerevisiae* Pex5p and Pex7p, but the physiological significance of this interaction is unclear (Rehling et al., 1996). This particular interaction may also be an indirect one mediated by another protein, Pex14p, to be described later (W.-H. Kunau, personal communication).

In some organisms such as *Y. lipolytica*, Pex5p has been reported to be intraperoxisomal (Szilard et al., 1995). It remains to be seen whether this location is a normal consequence of receptor cycling from the cytosol into the peroxisome from where it may be able to exit back into the cytosol (Rachubinski and Subramani, 1995). One cannot rule out, however, that the intraperoxisomal pool of Pex5p might be a dead-end byproduct of the "piggyback" entry of the receptor/PTS1 protein complex into the organelle matrix.

The *pex7* mutants of yeasts and humans are affected only in the import of PTS2-containing proteins, and not PTS1- or mPTS-containing proteins. Complementation of this set of mutants led to the discovery of Pex7p, the PTS2 receptor, which is a member of the WD40 protein family (Mazioch et al., 1994; Zhang and Lazarow, 1995; 1996; Rehling et al., 1996, Braverman et al., 1997; Purdue et al., 1997; Motley et al., 1997). This receptor has been reported to be mostly

Table 1 - PEROXINS AND *PEX* GENES FROM *Pichia pastoris*

PEX GENE	FUNCTION	REFERENCE
PEX1	127 kD AAA ATPase; associated with vesicles; involved in biogenesis	Heyman et al., 1994; Heyman et al., manuscript in preparation
PEX2	52 kD C3HC4 zinc-binding IMP; human homolog complements CG10	Waterham et al., 1996; Shimozawa et al., 1992
PEX3	52 kD IMP involved in biogenesis; mPTS in first 40 aa	Wiemer et al., 1996
PEX4	24 kD peroxisome-associated ubiquitin-conjugating enzyme	Crane et al., 1994
PEX5	68 kD PTS1 receptor; 7 TPR domains; peroxisome-associated; human homolog complements CG2	McCollum et al., 1993; Terlecky et al., 1995; Dodt et al., 1995; Wiemer et al., 1995
PEX6	127 kD AAA ATPase; associated with vesicles; involved in biogenesis; human homolog complements CG4	Spong et al., 1993; Tsukamoto et al., 1995; Yahraus et al., 1996; Heyman et al., manuscript in preparation
PEX7	42 kD peroxisomal PTS2 receptor; 7 WD repeats; human homolog complements RCDP lines	Elgersma et al., manuscript in preparation; Braverman et al., 1997; Motley et al., 1997; Purdue et al., 1997
PEX8	81 kD peroxisome-associated protein; has a PTS1	Liu et al., 1995
PEX10	48 kD C3HC4 zinc-binding IMP	Kalish et al., 1995
PEX12	48 kD C3HC4 zinc-binding IMP; human homolog complments CG3	Kalish et al., 1996; Chang et al., 1997
PEX13	43 kD SH3-domain-containing IMP; binds PTS1 receptor	Gould et al., 1996

cytosolic and partially peroxisomal (Marzioch et al., 1994) or entirely intraperoxisomal in *S. cerevisiae* (Zhang and Lazarow, 1995; 1996). In mammalian cells, the protein is mostly cytosolic (Braverman et al., 1997). All of these localization studies have been performed with epitope-tagged constructs, accompanied in some cases by protein overexpression. In *P. pastoris*, both epitope-tagged and wild-type Pex7p expressed at normal levels are cytosolic and intraperoxisomal (Elgersma et al., unpublished data). Thus, this receptor, like Pex5p, probably shuttles between the cytosol (along with its cargo) to (and perhaps into) the peroxisomes (Marzioch et al., 1994). Whether the intraperoxisomal pool of Pex7p plays an active role in import by pulling PT2-containing proteins into the organelle (Zhang and Lazarow, 1995), or by participating in subsequent rounds of import by recycling to the cytosol, remains to be determined.

Very little is known about the biogenesis of peroxisomal membrane proteins. As mentioned earlier, a few of the mPTSs required for the targeting of proteins to the peroxisomal membrane have been identified, but the receptors for these are unknown. Based on the studies of the import of peroxisomal membrane proteins into peroxisomes *in vitro*, it is predicted that proteinaceous receptors must exist, because mild pretreatment of the peroxisomes with proteases abolishes the membrane targeting (Diestelkotter and Just, 1991; Imanaka et al., 1996).

The shuttling PTS receptors must dock with proteins on the peroxisomal membrane. The *PEX13* gene encodes an SH3-domain-containing protein that has been shown to reside in the peroxisomal membrane and docks with Pex5p via its SH3 domain (Elgersma et al., 1996; Gould et al., 1996; Erdmann and Blobel, 1996). Another peroxisomal membrane protein, Pex14p, docks with both Pex5p and Pex7p, as well as with Pex13p and a different peroxisomal membrane protein, Pas9p (Albertini et al., 1997). These interactions define the early events in peroxisomal protein import, but the details of the translocation step of import are still unknown.

4. Other proteins and cofactors involved in import.

The biochemical analysis of PTS1 protein import using *in vitro* systems shows that hsp70 (Walton et al., 1994), hsp 40 and ATP are required for import of PTS1-containing (Terlecky et al., 1996). ATP is necessary for translocation of proteins across the peroxisomal membrane, but not for binding to peroxisomes (Imanaka et al., 1987), or for the insertion of membrane proteins into the peroxisomal membrane *in vitro* (Diestelkotter and Just, 1993; Imanaka et al., 1996). The analysis of *pex* mutants has revealed several additional genes and proteins likely to be involved in peroxisomal protein import (Table 1) but more work is necessary to gain an understanding of the biochemical roles of these proteins.

5. Folding and assembly of peroxisomal matrix proteins in the cytosol prior to import.

Folded and cross-linked proteins, as well as oligomeric complexes, are imported into peroxisomes, glyoxysomes and glycosomes (Glover et al., 1994; McNew and Goodman, 1994; Walton et al., 1995; Häusler et al., 1996; Lee et al., 1997). An interesting consequence of this is that proteins or subunits of a complex lacking a PTS can hitch a ride into peroxisomes in association with other subunits that possess a PTS.

Many peroxisomal proteins are oligomeric. Yet not all of these assemble into oligomers in the cytosol. Several proteins such as tetrameric catalase, trimeric chloramphenicol acetyltransferase and dimeric thiolase, carnitine acetylransferase, malate dehydrogenase or alanine glyoxylate aminotransferases appear to be transported as oligomers (Wanders et al., 1984; McNew and Goodman, 1994; Glover et al., 1994; Elgersma et al., 1995b; 1996b; Leiper et al., 1996). However, alcohol oxidase in methylotrophic yeasts is transported as monomers and then assembles with cofactors (and the aid of chaperones) in the organelle matrix before octamerization (Bellion and Goodman, 1987; Evers et al., 1994; 1996).

6. Two pathways for peroxisomal membrane proteins?

The dogma that all peroxisomal membrane and matrix proteins are targeted directly from the cytosol to the peroxisomes has been questioned recently in the light of new data and a reexamination of old experimental results that were difficult to explain (Subramani, 1996). Certain peroxisomal membrane proteins, such as *S. cerevisiae* Pas21p (Elgersma, 1995a) and rat liver PMP50 (Bodnar and Rachubinski, 1991) are targeted both to the ER and the peroxisome. The localization of several cellular proteins to dual locations is well documented and not surprising (Danpure, 1995). What is new is the possibility that some peroxisomal membrane proteins may go to the peroxisomes via the ER (Elgersma et al., manuscript in preparation; Richard Rachubinski, personal communication). This suggestion was actually made by Novikoff and Shin (1964) three decades ago based on the morphological juxtaposition of ER and peroxisomal membranes, but was discarded in the mid 1980s because most matrix, and the few membrane proteins, studied at that time appeared to be transported directly from the cytosol to the peroxisomes (Lazarow and Fujiki, 1985). The emerging data suggest that a revised model of peroxisome biogenesis may be necessary (Elgersma, 1995a; Subramani, 1996). In such a model, all matrix proteins would be imported directly from the cytosol to the peroxisomes. Consistent with this is the fact that deletions or mutations inactivating the PTS1 and PTS2 sequences result in the cytosolic localization of these proteins. Membrane proteins are predicted to use two import pathways - one from the cytosol to the peroxisome, and the other from the cytosol to the peroxisome via the ER (Subramani, 1996). Such a model would require one type of mPTS (called mPTS1) to direct proteins along the first pathway. Proteins with the mPTS1 sequence would remain in the cytosol when their mPTS is mutated or deleted, as has been shown for *C. boidinii* PMP47 (Dyer et al., 1996). A second sequence (called mPTS2) would be required to achieve targeting of proteins from the ER membrane to the peroxisomal membrane. Such proteins would be targeted to the ER if their mPTS is rendered non-functional by mutation or deletion. This is indeed the case for Pas21p (Elgersma et al., manuscript in preparation). This model for peroxisome biogenesis also predicts that there must be ER-derived vesicles carrying certain ER-inserted membrane proteins destined for the peroxisomes, and that the lumen of the ER is topologically equivalent to the peroxisome matrix.

Evidence for the role of vesicles in peroxisome biogenesis comes from the discovery that two *P. pastoris* proteins Pexlp and Pex6p, are found on vesicles that behave differently from peroxisomes during differential centrifugation and fractionation on gradients. These proteins are AAA-family proteases that interact with each other in an ATP-dependent manner, and the two proteins appear to be on different vesicles. Based on the similarity of these proteins to NSF/Sec18p, it is proposed that they play an indirect or direct role in heterotypic vesicle fusion events during peroxisome biogenesis (Heyman et al., manuscript in preparation).

This alternative view of peroxisome biogenesis is just emerging and a number of important experiments remain to be done to confirm various aspects of the model. It is also possible that the model will have to be revised as we learn more about the biogenesis process. Despite some of these shortcomings, the model does explain how one can regenerate peroxisomes (apparently *de novo*) by complementation of *pex* mutants that have no signs of peroxisomal remnants (Waterham et al., 1993). It also explains how peroxisomes might derive lipids and membranes for their growth and/or proliferation in response to nutritional cues.

In summary, the study of peroxisomal protein import and biogenesis has uncovered many aspects that are distinct from those found in other organelles. Furthermore, many of the *PEX* genes identified in lower eukaryotes are conserved in mammals (Subramani, 1997). This is of special significance because of the existence of a dozen or so devastating human peroxisomal disorders in which the underlying molecular defect is either in peroxisomal protein import and/or biogenesis. Of the 16 *PEX* genes published to date, 8 have mammalian homologs and 6 are already implicated in human disease (Subramani, 1997). This fortunate convergence of basic and medical biology is likely to keep the peroxisomal field interesting and thriving for some time.

Acknowledgements.

Work in my lab is supported by a grant from NIHDK41737. I thank all the past and present members of my lab who contributed to the work described here.

References.

Albertini M, Rehling P, Erdmann R, Girzalsky W, Kiel JA, Veenhuis M and Kunau WH (1997) Pex14p, a peroxisomal membrane protein binding both receptors of the two PTS-dependent import pathways. Cell 89: 83-92

Bellion E and Goodman JM (1987) Proton ionophores prevent assembly of a peroxisomal protein. Cell 48: 165-173

Bodnar AG and Rachubinski RA (1991) Characterization of the integral membrane polypeptides of rat liver peroxisomes isolated from untreated and clofibrate-treated rats. Biochem. Cell Biol. 69: 499-508

Braverman N, Steel G, Obie C, Moser A, Moser H, Gould SJ and Valle D (1997) Human PEX7 encodes the peroxisomal PTS2 receptor and is responsible for rhizomelic chondrodysplasia punctata. Nat. Genet. 15: 369-376

Chang CC, Lee WH, Moser H, Valle D and Gould SJ (1997) Isolation of the human PEX12 gene, mutated in group 3 of the peroxisome biogenesis disorders. Nat Genet 15: 385-388

Crane DI, Kalish JE and Gould SJ (1994) The Pichia pastoris PAS4 gene encodes a ubiquitin-conjugating enzyme required for peroxisome assembly. J. Biol. Chem. 269: 21835-21844

Danpure JC (1995) How can the products of a single gene be localized to more than one intracellular compartment? Trends Cell Biol. 5: 230-238

Diestelkotter P and Just WW (1993) In vitro insertion of the 22-kD peroxisomal membrane protein into isolated rat liver peroxisomes. J. Cell Biol. 1717-1725

Distel B, Erdmann R, Gould SJ, Blobel G, Crane DI, Cregg JM, Dodt G, Fujiki Y, Goodman JM, Just WW, Kiel JA, Kunau WH, Lazarow PB, Mannaerts GP, Moser HW, Osumi T, Rachubinski RA, Roscher A, Subramani S, Tabak HF, Tsukamoto T, Valle D, van der Klei I, van Veldhoven PP and Veenhuis M (1996) A unified nomenclature for peroxisome biogenesis factors. J. Cell Biol. 135: 1-3

Dodt G, Braverman N, Wong C, Moser A, Moser HW, Watkins P, Valle D and Gould SJ (1995) Mutations in the PTS1 receptor gene, PXR1, define complementation group 2 of the peroxisome biogenesis disorders. Nat. Genet. 9: 115-125

Dodt G and Gould SJ (1996) Multiple PEX genes are required for proper subcellular distribution and stability of Pex5p, the PTS1 receptor: evidence that PTS1 protein import is mediated by a cycling receptor. J. Cell Biol. 1763-1774

Dyer JM, McNew JA and Goodman JM (1996) The sorting sequence of the peroxisomal integral membrane protein PMP47 is contained within a short hydrophilic loop. J. Cell Biol. 133: 269-280

Elgersma Y (1995a) Transport of proteins and metabolites across the peroxisomal membrane in *Saccharomyxes cerevisiae*. Ph. D. Thesis, University of Amsterdam

Elgersma Y, Kwast L, Klein A, Voorn-Brouwer T, van den Berg M, Metzig B, America T, Tabak H and Distel B (1996a) The SH3 domain of the peroxisomal membrane protein

Pex13p functions as a docking site for Pex5p, a mobile receptor for peroxisomal proteins. J. Cell Biol. 135: 97-109

Elgersma Y, van Roermund CW, Wanders RJ and Tabak HF (1995b) Peroxisomal and mitochondrial carnitine acetyltransferases of Saccharomyces cerevisiae are encoded by a single gene. EMBO J. 14: 3472-3479

Elgersma Y, Vos A, van den Berg M, van Roermund CW, van der Sluijs P, Distel B and Tabak HF (1996b) Analysis of the carboxyl-terminal peroxisomal targeting signal 1 in a homologous context in Saccharomyces cerevisiae. J. Biol. Chem. 271: 26375-82

Erdmann R and Blobel G (1996) Identification of Pex13p a peroxisomal membrane receptor for the PTS1 recognition factor. J. Cell Biol. 135: 111-121

Evers ME, Titorenko V, Harder W, van der Klei IJ and Veenhuis M (1996) Flavin adenine dinucleotide binding is the crucial step in alcohol oxidase assembly in the yeast Hansenula polymorpha. Yeast 12: 917-923

Evers ME, Titorenko VI, van der Klei IJ, Harder W and Veenhuis M (1994) Assembly of alcohol oxidase in peroxisomes of the yeast Hansenula polymorpha requires the cofactor flavin adenine dinucleotide. Mol. Biol. Cell 5: 829-837

Fransen M, Brees C, Baumgart E, Vanhooren JC, Baes M, Mannaerts GP and Van Veldhoven PP (1995) Identification and characterization of the putative human peroxisomal C-terminal targeting signal import receptor. J. Biol. Chem. 270: 7731-7736

Glover JR, Andrews DW and Rachubinski RA (1994) Saccharomyces cerevisiae peroxisomal thiolase is imported as a dimer. Proc. Natl. Acad. Sci. U. S. A. 91: 10541-10545

Gould SJ, Kalish JE, Morrell JC, Bjorkman J, Urquhart AJ and Crane DI (1996) Pex13p is an SH3 protein of the peroxisome membrane and a docking factor for the predominantly cytoplasmic PTS1 receptor. J. Cell Biol. 135: 85-95

Häusler T, Stierhof Y, Wirtz E and Clayton C (1996) Import of DHFR hybrid protein into glycosomes in vivo is not inhibited by the folate-analogue aminopterin. J. Cell Biol. 132: 311-324

Heyman JA, Monosov E and Subramani S (1994) Role of the PAS1 gene of Pichia pastoris in peroxisome biogenesis. J. Cell Biol. 127: 1259-1273

Imanaka T, Shiina Y, Takano T, Hashimoto T and Osumi T (1996) Insertion of the 70-kDa peroxisomal membrane protein into peroxisomal membranes in vivo and in vitro. J. Biol. Chem. 271: 3706-3713

Imanaka T, Small GM and Lazarow PB (1987) Translocation of acyl-CoA oxidase into peroxisomes requires ATP hydrolysis but not a membrane potential. J. Cell Biol. 105: 2915-2922

Kalish JE, Keller GA, Morrell JC, Mihalik SJ, Smith B, Cregg JM and Gould SJ (1996) Characterization of a novel component of the peroxisomal protein import apparatus using fluorescent peroxisomal proteins. EMBO J. 15: 3275-3285

Kalish JE, Theda C, Morrell JC, Berg JM and Gould SJ (1995) Formation of the peroxisome lumen is abolished by loss of Pichia pastoris Pas7p, a zinc-binding integral membrane protein of the peroxisome. Mol. Cell. Biol. 15: 6406-6419

Lazarow PB and Fujiki Y (1985) Biogenesis of peroxisomes. Annu Rev. Cell Biol. 1: 489-530

Lee MS, Mullen RT and Trelease RN (1997) Oilseed isocitrate lyases lacking their essential type I peroxisomal targeting signal are piggybacked to glyoxysomes. The Plant Cell 9: 185-197

Leiper JM, Oatey PB and Danpure CJ (1996) Inhibition of alanine:glyoxylate aminotransferase 1 dimerization is a prerequisite for its peroxisome-to-mitochondrion mistargeting in primary hyperoxaluria type 1. J. Cell Biol. 135: 939-951

Liu H, Tan X, Russell KA, Veenhuis M and Cregg JM (1995) PER3, a gene required for peroxisome biogenesis in Pichia pastoris, encodes a peroxisomal membrane protein involved in protein import. J. Biol. Chem. 270: 10940-10951

Marzioch M, Erdmann R, Veenhuis M and Kunau WH (1994) PAS7 encodes a novel yeast member of the WD-40 protein family essential for import of 3-oxoacyl-CoA thiolase, a PTS2-containing protein, into peroxisomes. EMBO J. 13: 4908-4918

McCollum D, Monosov E and Subramani S (1993) The pas8 mutant of Pichia pastoris exhibits the peroxisomal protein import deficiencies of Zellweger syndrome cells--the PAS8 protein binds to the COOH-terminal tripeptide peroxisomal targeting signal, and is a member of the TPR protein family. J. Cell Biol. 121: 761-774

McNew JA and Goodman JM (1994) An oligomeric protein is imported into peroxisomes in vivo. J. Cell Biol. 127: 1245-1257

Motley AM, Hettema EH, Hogenhout EM, Brites P, ten Asbroek AL, Wijburg FA, Baas F, Heijmans HS, Tabak HF, Wanders RJ and Distel B (1997) Rhizomelic chondrodysplasia punctata is a peroxisomal protein targeting disease caused by a non-functional PTS2 receptor. Nat. Genet. 15: 377-380

Novikoff AB and Shin W (1964) The endoplasmic reticulum in the Golgi zone and its relations to microbodies, Golgi apparatus and autophagic vacuoles in rat liver cells. J. Microscopie 3: 187-206

Purdue PE, Zhang JW, Skoneczny M and Lazarow PB (1997) Rhizomelic chondrodysplasia punctata is caused by deficiency of human PEX7, a homologue of the yeast PTS2 receptor. Nat. Genet. 15: 381-384

Rachubinski RA and Subramani S (1995) How proteins penetrate peroxisomes. Cell 83: 525-528

Rehling P, Marzioch M, Niesen F, Wittke E, Veenhuis M and Kunau WH (1996) The import receptor for the peroxisomal targeting signal 2 (PTS2) in Saccharomyces cerevisiae is encoded by the PAS7 gene. EMBO J. 15: 2901-2913

Shimozawa N, Tsukamoto T, Suzuki Y, Orii T, Shirayoshi Y, Mori T and Fujiki Y (1992) A human gene responsible for Zellweger syndrome that affects peroxisome assembly. Science 255: 1132-4

Small GM, Szabo LJ and Lazarow PB (1988) Acyl-CoA oxidase contains two targeting sequences each of which can mediate protein import into peroxisomes. EMBO J. 7: 1167-1173

Spong AP and Subramani S (1993) Cloning and characterization of PAS5: A gene required for peroxisome biogenesis in the methylotrophic yeast Pichia pastoris. J. Cell Biol. 123: 535-548

Subramani S (1996) Protein translocation into peroxisomes. J. Biol. Chem. 271: 32483-32486

Subramani S (1997) PEX genes on the rise [news]. Nat. Genet. 15: 331-333

Szilard RK, Titorenko VI, Veenhuis M and Rachubinski RA (1995) Pay32p of the yeast Yarrowia lipolytica is an intraperoxisomal component of the matrix protein translocation machinery. J. Cell Biol. 131: 1453-1469

Terlecky SR, Nuttley WM, McCollum D, Sock E and Subramani S (1995) The Pichia pastoris peroxisomal protein PAS8p is the receptor for the C-terminal tripeptide peroxisomal targeting signal. EMBO J. 14: 3627-3634

Terlecky SR, Nuttley WM and Subramani S (1996) The cytosolic and membrane components required for peroxisomal protein import. Experientia 52: 1050-1054

Tsukamoto T, Miura S, Nakai T, Yokota S, Shimozawa N, Suzuki Y, Orii T, Fujiki Y, Sakai F, Bogaki A, Yasumo H and Osumi T (1995) Peroxisome assembly factor-2, a putative ATPase cloned by functional complementation on a peroxisome-deficient mammalian cell mutant. Nat. Genet. 11: 395-401

Van den Bosch H, Schutgens RBH, Wanders RJA and Tager JM (1992) Biochemistry of peroxisomes. Annu. Rev. Biochem. 61: 157-197

Van der Klei IJ, Hilbrands RE, Swaving GJ, Waterham HR, Vrieling EG, Titorenko VI, Cregg JM, Harder W and Veenhuis M (1995) The Hansenula polymorpha PER3 gene is essential for the import of PTS1 proteins into the peroxisomal matrix. J. Biol. Chem. 270: 17229-17236

Van der Leij I, Franse MM, Elgersma Y, Distel B and Tabak HF (1993) PAS10 is a tetratricopeptide-repeat protein that is essential for the import of most matrix proteins into peroxisomes of Saccharomyces cerevisiae. Proc. Natl. Acad. Sci. U. S .A. 90: 11782-11786

Walton PA, Hill PE and Subramani S (1995) Import of stably folded proteins into peroxisomes. Mol. Biol. Cell 6: 675-683

Walton PA, Wendland M, Subramani S, Rachubinski RA and Welch WJ (1994) Involvement of 70-kD heat-shock proteins in peroxisomal import. J. Cell Biol. 125: 1037-1046

Wanders RJ, Kos M, Roest B, Meijer AJ, Schrakamp G, Heymans HS, Tegelaers WH, van den Bosch H, Schutgens RB and Tager JM (1984) Activity of peroxisomal enzymes and intracellular distribution of catalase in Zellweger syndrome. Biochem. Biophys. Res. Commun. 123: 1054-1061

Waterham HR, Titorenko VI, Swaving GJ, Harder W and Veenhuis M (1993) Peroxisomes in the methylotrophic yeast Hansenula polymorpha do not necessarily derive from pre-existing organelles. EMBO J. 12: 4785-4794

Waterham HR, de Vries Y, Russel KA, Xie W, Veenhuis M and Cregg JM (1996) The Pichia pastoris PER6 gene product is a peroxisomal integral membrane protein essential for peroxisome biogenesis and has sequence similarity to the Zellweger syndrome protein PAF-1. Mol. Cell. Biol. 16: 2527-2536

Wiemer EA, Nuttley WM, Bertolaet BL, Li X, Francke U, Wheelock MJ, Anne UK, Johnson KR and Subramani S (1995) Human peroxisomal targeting signal-1 receptor restores peroxisomal protein import in cells from patients with fatal peroxisomal disorders. J. Cell Biol. 130: 51-65

Wiemer EAC, Luers G, Faber KN, Wenzel T, Veenhuis M and Subramani S (1996) Isolation and characterization of Pas2p, a peroxisomal membrane protein essential for peroxisome biogenesis in the methylotrophic yeast Pichia pastoris. J. Biol. Chem. 271: 18973-18980

Yahraus T, Braverman N, Dodt G, Kalish JE, Morrell JC, Moser HW, Valle D and Gould SJ (1996) The peroxisome biogenesis disorder group 4 gene, PXAAA1, encodes a cytoplasmic ATPase required for stability of the PTS1 receptor. EMBO J. 15: 2914-2923

Zhang JW and Lazarow PB (1995) PEB1 (PAS7) in Saccharomyces cerevisiae encodes a hydrophilic, intra-peroxisomal protein that is a member of the WD repeat family and is essential for the import of thiolase into peroxisomes. J. Cell Biol. 129: 65-80

Zhang JW and Lazarow PB (1996) Peb1p (Pas7p) is an intraperoxisomal receptor for the NH_2-terminal, type 2, peroxisomal targeting signal of thiolase: Peb1p itself is targeted to peroxisomes by an NH_2-terminal peptide. J. Cell Biol. 132: 325-334

Molecular Dissection of Clathrin Coat Formation and Receptor Sorting

Tom Kirchhausen [1], Chris Brunner[2] and Anja Renold [2]
[1] Harvard Medical School, Dept. of Cell Biology and Center for Blood Research, 200 Longwood Av., Boston, MA 02115 e-mail kirchhausen@crystal.harvard.edu
[2] Biozentrum, Dept. of Biochemistry, Klingelbergstrasse 70, CH-4056 Basel

Mechanism of clathrin coat formation

Introduction

Traffic between intracellular compartments requires a special mechanism to transport proteins, lipids and cargo molecules from a donor to an acceptor organelle. This process is mediated by vesicles that bud off the donor membrane and fuse with an acceptor membrane. This solves the problem of selection, because the cargo molecules are concentrated into the vesicles before the vesicle separates from the donor membrane. The first step in cargo concentration is the binding of cargo molecules to the lumenal domain of a transmembrane receptor. This is followed by the interaction of the cytoplasmic domain of the receptor with a protein complex that is believed to trigger the formation of a protein coat, in turn leading to the formation of a pit in the membrane that will eventually become a vesicle. This coupling of cargo concentration and coat formation ensures efficient cargo loading into the assembling vesicles. The neck of the vesicle then narrows, and the vesicle finally pinches off the membrane. At this point the protein coat is removed, and the vesicle is targeted to the acceptor compartment, eventually fusing with the target membrane. Among the best understood processes for moving receptors and ligands are clathrin coated pits and vesicles. In what follows, we review the current state of knowledge of the clathrin-dependent pathway. Specific primary references to the work cited here can be found in the published reviews listed at the end of this lecture notes.

NATO ASI Series, Vol. H 106
Lipid and Protein Traffic
Pathways and Molecular Mechanisms
Edited by Jos A. F. Op den Kamp

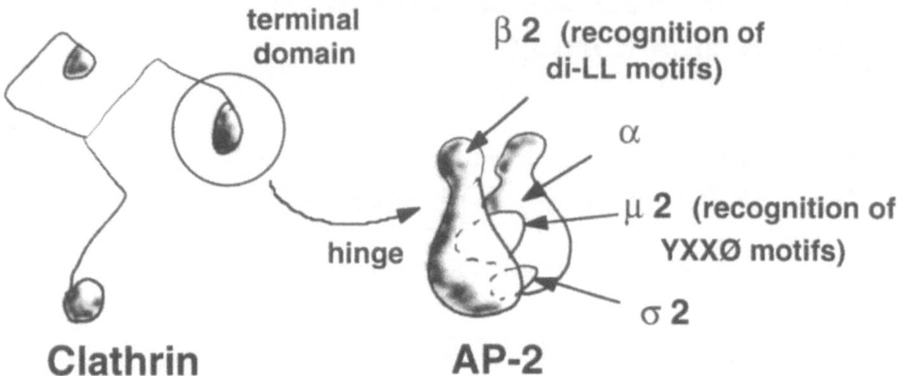

Figure 1. Schematic representation of clathrin and its adaptor complex AP-2.

Coat components

The coat of the vesicles in the clathrin-dependent pathway is composed of many units, which assemble in a cooperative manner. The major structural component of these vesicles is clathrin, a protein with a three-legged "triskelion" shape, which organizes itself into cage-like lattices. Each of the legs of the clathrin trimer is made up of a heavy and a light chain, separately encoded in the genome. The total molecular weight of the trimer is 640 kDa.

The three legs are extended and relatively stiff, with a well-defined bend near the center of each leg. They are held together by interactions towards the carboxyl termini of the heavy chains, which form the hub of the clathrin molecule. Each leg is divided into proximal and distal segments by the bend in the center, the distal segment ending with a globular structure called the 'terminal domain'. Each vertex of the clathrin cage contains a hub of a single clathrin triskelion and the bends of three legs from other clathrin molecules. Each lattice edge is therefore formed by the superposition of the proximal domains of the clathrin molecules at two adjacent vertices, plus the distal domains of two clathrin molecules centered at vertices one step further along the lattice. This arrangement naturally forms the characteristic open hexagonal and pentagonal facets of the clathrin coat.

Clathrin is an organizing framework for receptor sorting, membrane budding and the other steps in the cycle of vesicle assembly, uncoating and fusion, both at the plasma membrane and in the trans-Golgi network (TGN).

Coat formation is driven by two distinct but related heterotetramers of adaptor proteins in the two different membrane compartments, AP-1 for TGN budding and AP-2 for plasma membrane vesicles. Both adaptor complexes have a molecular weight of 270 kDa and are composed of four subunits, γ, $\beta1$, $\sigma1$, $\mu1$ for AP-1 and $\alpha2$, $\beta2$, $\sigma2$, $\mu2$ for AP-2 complex. The functions of α and γ are so far unknown. The μ subunits are crucial for sorting, as we will discuss later, and the β subunits have recently been shown to be responsible for directing clathrin polymerization. The β subunit has a characteristic shape containing a N-terminal core and C-terminal ear, linked by a hinge where clathrin binding occurs. Truncation of the C-terminal ear/hinge abolishes clathrin interaction but not oligomerization with the other AP subunits, and phosphorylation of the β hinges prevents interaction with clathrin. The portion of the clathrin molecule involved in binding appears to be the terminal domain.

The interaction between the β-hinge and clathrin is very weak, though detectable. In the *in vivo* situation, many AP complexes will be collected in one place by the membrane-bound receptors, and a large number of β–ear/hinge domains will be arrayed on the cytoplasmic face of the membrane. The affinity of the β–ear/hinge for the clathrin trimer will therefore appear to be much higher than the *in vitro* measurements suggest, due to avidity effects.

Complex though this picture is, the clathrin-dependent pathway may be more complex still. Sequences related to the known adaptor complexes have recently been identified (AP-3, AP-4), although these appear to lack the β-ear region that is required in AP-1 and AP-2 for clathrin binding. They are known not to be present in clathrin coats, but may be involved in other aspects of receptor selection and coat initiation. In contrast, an apparently quite unrelated protein, NP180, which is monomeric and highly expressed in brain, has been shown to interact with clathrin and be present in clathrin coats. The function of this protein and its relationship to the "classical" clathrin-dependent pathway is also unknown.

Coat assembly

Purified clathrin triskelions can be induced to form lattices spontaneously in low salt concentrations (10-20 mM). Stop-flow measurements coupled to electron microscopy have shown that this process is very rapid (occurring within 1 second), and the fact that the lattices form spontaneously shows that coat formation is, or can be, energy-independent. Coat formation in physiological buffers requires AP complexes, however, and there is increasing evidence to suggest that coat formation in membranes is a much more complex process, requiring ATP, GTP, small GTP-binding proteins and possibly additional protein components.

Steps *in vivo* for the clathrin pathway

The entire process of membrane protein transport, from assembly of the clathrin coat to fusion of the uncoated vesicle with the target membrane takes only 1-2 minutes. The steps in the process can be summarized as follows:

1) Recruitment of adaptor complexes to the membrane. The AP-1 and AP-2 complexes segregate to the TGN and plasma membrane, respectively, despite the fact that the sorted proteins in both membranes are the same. Targeting of the AP complexes from cytosol to the membrane therefore cannot be explained by the interaction with the cytoplasmic tails of the sorted proteins alone. Several lines of evidence suggest that a putative membrane-bound docking complex may be involved in this step.

2) Recognition of sorting signals. Once bound to the membrane, the adaptor proteins are able to interact with sorting signals in the cytosolic domains of membrane receptors. It is currently unclear whether this interaction occurs before or after clathrin recruitment.

3) Initiation of coat assembly. The adaptor proteins, either bound to the putative docking complex or to the sorted membrane proteins, recruit cytosolic clathrin to the membrane. At this point, additional adaptor complexes may also be recruited to the membrane, and additional proteins involved in coat formation may also integrate into the coat.

4) Propagation of clathrin lattice. The nucleus of the clathrin coat grows by incorporating additional membrane receptors, AP complexes and clathrin into the nascent vesicle.

5) Membrane constriction and budding leading to vesicle disengagement. Recent studies have shown that this step involves additional proteins including dynamin, amphiphysin and synaptogamin.

6) Release of coat components. The disassembly is still poorly understood. Clathrin lattices can be depolymerized *in vitro* by the addition of Hsc70, auxilin and ATP in a reaction that involves the terminal domain, but under these

conditions the AP complexes are not released. The release of AP's requires the addition of cytosol and ATP. It is unclear whether the *in vivo* uncoating reaction is a two-step process, as the *in vitro* observations suggest, or whether the AP complexes and the clathrin coat are simultaneously removed in a reaction that requires a so-far unidentified cytoplasmic protein.

Receptor sorting by clathrin coated pits

The endocytic and secretory pathways of eukaryotic cells consist of an array of membrane bound compartments, each of which contains a characteristic cohort of transmembrane proteins. How are these transmembrane proteins targeted to and maintained within the appropriate membrane compartment?

A common event in the sorting of many transmembrane proteins is the interaction of a sorting signal in the cytosolic domains and a component of an organellar protein coat, such as the clathrin adaptors AP-1 and AP-2 of clathrin coated pits or the COP-I/COP-II components of coatomers. There are several different classes of sorting signals; some examples are listed in Table 1.

Table 1

Motif	Sorted proteins containing motif
tyrosine-based (YXXØ-type)	CI-M6PR, TfR, EGFR, TGN38/41, gp41
tyrosine based (NPXY-type)	LDLR
dileucine-based	CD4, Glut4, CD3γ, CI-M6PR, Ii
Acidic clusters	Furin
Dilysine (KKFF-type)	VIP36
Ubiquitin addition	Yeast α-factor, Eps15

These sorting signals are diverse and highly degenerate, and they function only in the appropriate context. For example, when the TfR is in the endosomal compartment it is not recognized by AP-2, even though AP-2 recognizes this receptor extremely efficiently when it is at the plasma membrane. Moreover, each sorting signal acts as an addressing code: for example, TGN38 is targeted to the TGN, whereas the TfR, which contains a very similar motif, recycles to the plasma membrane. Switching the sorting motif alters the fate of the protein.

To date, more and more different internalization signals are being discovered. tyrosine-based signals, however, continue to receive the major attention because they are most commonly found among rapidly internalized proteins. Tyrosine-based signals are characterized by the presence of a critical tyrosine residue within an otherwise degenerate sequence context. The most common tyrosine motifs are NPXY and YXXØ, where X represents any amino acid and Ø represents a bulky hydrophobic amino acid. Using combinatorial selection methods and surface plasmon resonance as well as the two-hybrid approach to narrow the requirements for the YXXØ interaction with AP-2 and μ2, it has been shown that the optimal sequence for interaction has tyrosine as an anchor, prefers arginine at position +2 and leucine at position +3. In contrast, no preferred sequence surrounding the YXXØ is detected indicating that recognition of the YXXØ endocytosis signal does not require a pre-folded structure.

Different biochemical and genetic approaches demonstrated interaction of the μ chain of the AP complexes with many different tyrosine signals. The fine specificity of interaction of tyrosine-based signals with μ2 correlates with the interaction of tyrosine-based signals with the complete AP-2 complex. Phosphorylation of the critical tyrosine residue abrogates interaction with μ2/AP-2 (e.g. CTLA4). The avidity for membrane-immobilized AP's might be dramatically increased by the oligomeric state of membrane proteins linking one or more signals. Endocytic receptors might be recruited to pre-formed coated pits since a conformational switch of AP's induced by the interaction with clathrin modulates the affinity.

It has recently been found that phosphoinositide-3'-phosphate (PI3P) enhances the interaction of AP-2 with YXXØ-based endocytic signals. Experiments using the PI3-kinase inhibitor Wortmannin, which induces a loss of about 70% of PI3P showed no effect on endocytosis but abolished endosomal traffic between TGN, endosomes and lysosomes. PI3P regulates the affinity of binding of μ2 and YXXØ-based sorting signals, whereas inositol has no effect. This observation indicates that there might be a link between signal transduction and membrane traffic, and stresses the importance of phosphoinositide-kinases in vesicular traffic.

Decoding process

Although we cannot yet explain all of the details of the decoding process, some basic principles have recently emerged. At least in the case of the tyrosine YXXØ motif and the dileucine motif, distinct subunits of the AP-1 complex recognize the different motifs. Furthermore, different motifs are bound by the AP-1 and AP-2 complexes with different affinities. By varying the site to which the motif binds, and presumably therefore the consequences of the binding event, one can imagine that the sorting machinery would be able to target different motifs to different acceptor membranes.

Multiple coats, different locations and differential but redundant recognition are the constraints of tyrosine-based signals. The characteristics are differences in avidity of specificity of interaction of YXXØ-based signals with AP-1 and AP-2, where AP-2 shows higher/broader specificity. The order of encounter in the biosynthetic pathway is AP-1 at the TGN, AP-3 (AP-4) at the endosome and AP-2 at the plasma membrane.

Acknowledgment

The lecture notes correspond to two lectures given by one of us (T.K.) which were then edited to its present form with the help of two participating students (C.B. and A.R.) in the course.

Recent reviews [

1. Kirchhausen, T.: **Coated pits and coated vesicles - sorting it all out.** *Curr Opin Struc Biol,* 1993, **3:** 182-188.

2. Schmid, S. L.: **Coated-vesicle formation in vitro: Conflicting results using different assays.** *Trends Cell Biol,* 1993, **3:** 145-148.

3. Robinson, M. S.: **The role of clathrin, adaptors and dynamin in endocytosis.** *Curr Opin Cell Biol,* 1994, **6:** 538-544.

4. Ohno, H., Kirchhausen, T., and Bonifacino, J. S.: **Recognition of tyrosine-based sorting signals by the medium chains of clathrin- associated protein complexes.** *J NIH Res,* 1995, **7:** 50-51.

5. Schmid, S. L., and Damke, H.: **Coated vesicles: A diversity of form and function.** *Faseb J,* 1995, **9:** 1445-1453.

6. Bonifacino, J. S., Marks, M. S., Ohno, H., and Kirchhausen, T.: **Mechanisms of signal-mediated protein sorting in the endocytic and secretory pathways.** *Proc Assoc Am Phys,* 1996, **108:** 285-295.

7. De Camilli, P., Emr, S. D., McPherson, P. S., and Novick, P.: **Phosphoinositides as regulators in membrane traffic.** *Science,* 1996, **271:** 1533-1539.

8. Robinson, M. S., Watts, C., and Zerial, M.: **Membrane dynamics in endocytosis.** *Cell,* 1996, **84:** 13-21.

9. Seaman, M. N., Burd, C. G., and Emr, S. D.: **Receptor signalling and the regulation of endocytic membrane transport.** *Curr Opin Cell Biol,* 1996, **8:** 549-556.

10. Kirchhausen, T., Bonifacino, J., and Riezmann, H.: **Linking cargo to vesicle formation: Receptor tail interactions with coat proteins.** *Curr Opin Cell Biol,* 1997, .

11. Marks, M. S., Ohno, H., Kirchhausen, T., and Bonifacino, J. S.: **Protein sorting by tyrosine-based signals: Adapting to the Ys and wherefores.** *Trends Cell Biol,* 1997, **7:** 124-128.

12. Robinson, M. S.: **Coats and vesicle budding.** *Trends Cell Biol,* 1997, **7:** 99-102.

A Function for EGF-Induced Eps15 Ubiquitination in Endocytosis

Sanne van Delft, Arie J. Verkleij and Paul M.P. van Bergen en Henegouwen

Department of Molecular Cell Biology

Institute of Biomembranes

Utrecht University

The Netherlands

1. Introduction

Binding of growth factors, such as the epidermal growth factor (EGF), to their specific receptors on the cell surface causes the initiation of a signal transduction cascade which leads to changes in gene expression and finally to cell division. Inactivation of the EGF-receptor can occur via several mechanisms, such as receptor mediated transmodulation (Northwood and Davis, 1990), receptor dephosphorylation (Faure *et al*, 1992) and receptor down-regulation (for review see Sorkin and Waters, 1993). The importance of down-regulation as a negative regulatory mechanism of receptor tyrosine kinase signaling is stressed by the observation that defects in this regulation can facilitate cellular transformation (Wells *et al.*, 1990) and tumor formation (Masui *et al.*, 1991). Receptor down-regulation results in the loss of EGF binding sites from the plasma membrane by internalization of the receptors. EGF-receptors enter the cell via receptor mediated endocytosis, a process involving clathrin coated pits and vesicles. The coat is composed of a number of proteins, such as the adaptor proteins (Aps), the heavy and light chain of clathrin, forming the clathrin lattice (for review see Schmid, 1992) and, as recently has been demonstrated, Eps15 (Tebar *et al.*, 1996; van Delft *et al.*, 1997).

Eps15 has been identified as a 142 kDa substrate of the EGF-receptor (Faziolo *et al.* 1993). In quiescent cells Eps15 is associated to the EGF-receptor, and upon EGF stimulation this association has been found to increase dramatically (van Delft *et al.* 1997). Eps15 has been shown to bind to both adaptor protein-2 (AP-2) and clathrin (Benmerah *et al.* 1995; van

NATO ASI Series, Vol. H 106
Lipid and Protein Traffic
Pathways and Molecular Mechanisms
Edited by Jos A. F. Op den Kamp

Delft *et al.* 1997). Subcellular fractionation and immunolocalization studies have shown that Eps15 is present in clathrin-coated pits and vesicles but not in early endosomes (Tebar *et al.*, 1996; van Delft *et al.*, 1997). Eps15 has been shown to share homology with the yeast proteins End3p and Pan1p. End3p and Pan1p both contain multiple EH (Eps15 homology) domains, a motif proposed to mediate protein-protein interaction, and have been implicated in the endocytosis of the α-factor and lipids respectively (Benedetti *et al.*, 1994; Wendland *et al.*, 1996).

Tyrosine phosphorylation of Eps15 is transient and occurs within 2 minutes of EGF stimulation (Fazioli *et al.*, 1993). In addition, EGF stimulation results in the appearance on SDS polyacrylamide gels of a slowly migrating band of Eps15 of approximately 150 kD. Tyrosine kinase activity of the EGF-receptor was found to be required for this apparent increase in molecular weight of Eps15 (van Delft *et al.*, 1997b). Expression of Eps15 cDNA in bacteria shows the presence of only the 142 kD form suggesting that Eps15 is undergoing an EGF-induced posttranslational modification (Faziolo *et al.* 1993).

We investigated the nature of this posttranslational modification of Eps15 and found that the appearance of the high molecular weight form of Eps15 is not due to EGF-induced hyperphosphorylation. Instead, we found that the 8 kD increase in molecular weight was caused by ubiquitination of Eps15. This indicates that EGF induces two different modes of posttranslational modification of Eps15: tyrosine phosphorylation and ubiquitination.

2. Eps15 Mobility Shift is not due to Tyrosine-Phosphorylation.

To determine the nature of the EGF-induced increase in the molecular mass of Eps15, we investigated whether this increase is caused by tyrosine-phosphorylation. HER14 fibroblasts, expressing the human EGF-receptor, were stimulated with EGF and Eps15 was immunoprecipitated from the cell lysates. One Eps15 immunoprecipitate was treated with alkaline phosphatase while two controls were either left untreated or treated with heat-inactivated phosphatase. The proteins were separated on 8% SDS-polyacrylamide gels and blotted onto PVDF-membrane. After detection of tyrosine-phosphorylated Eps15 by anti-phosphotyrosine antibodies, two Eps15 bands of 142 and 150 kDa were visible in untreated cells. This indicates that both forms of Eps15 are phosphorylated (Fig. 1, CON). The two Eps15 bands of 142 and 150 kDa are each resolved into a tightly spaced doubled, and both forms of the doubled are tyrosyl phosphorylated. The reason for the slight difference in molecular weight is not known but could be due to differential splicing. Treatment of Eps15 with alkaline phosphatase resulted in the complete dephosphorylation of Eps15 (Fig. 1, AP), while treatment with heat-inactivated alkaline phosphatase did not change the phosphorylation state of Eps15 (Fig. 1, HI-AP). Reprobing the same blot with anti-Eps15 antibodies showed that irrespective of the phosphorylation state of Eps15, the 142 and 150 kDa form were present (Fig. 1). These results demonstrate that the appearance of high molecular weight form of Eps15 is not the result of tyrosine-phosphorylation.

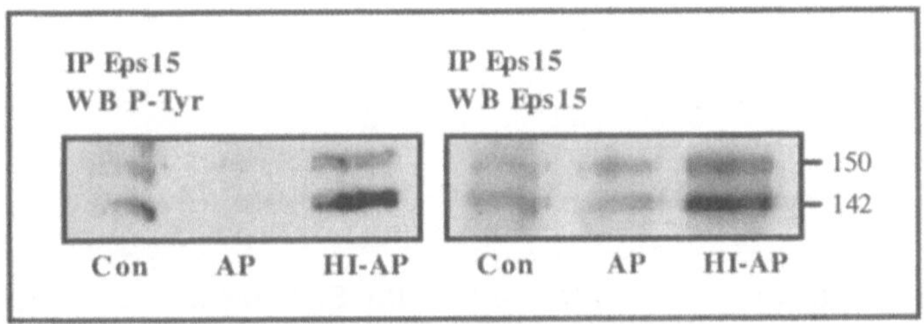

Figure 1. **Alkaline phosphatase treatment of Eps15 immunoprecipitates.**
HER14 cells were stimulated for 10 min with 50 ng/ml EGF. Eps15 was immunoprecipitated from the cell lysates and the immunoprecipitates were either left untreated (CON), treated with alkaline phosphatase (AP) or heat-inactivated alkaline phosphatase (HI-AP). Proteins were separated on an 8% SDS polyacrylamide gel and the Western blot was incubated with anti-phospho-tyrosine antibodies (WB p-tyr). Subsequently, the blot was stripped and incubated with anti-Eps15 antibodies (WB Eps15). Immunoprecipitations were performed as previously described (van Delft et al., 1995).

3. Eps15 Mobility Shift is due to Mono-ubiquitination.

The approximate increase of 8 kDa in the modified form of Eps15 stimulated us to investigate the possible mono-ubiquitination of Eps15. Ubiquitin is a highly conserved protein of about 8 kDa, which is abundant in eukaryotes. Ubiquitin is found free or covalently linked via its carboxy-terminus to $-NH_2$ groups of one or more lysine residues of a variety of cytoplasmic, nuclear and integral membrane proteins (Finley and Chau, 1991).

To investigate the possible mono-ubiquitination of Eps15, Eps15 was immunoprecipitated from HER14 cells that were either left unstimulated or stimulated with 50 ng/ml EGF. The protein samples were separated on 8% SDS-polyacrylamide gels and the Western blot was probed with anti-Eps15 antibodies. A clear mobility shift was seen after

EGF stimulation but not in unstimulated cells (Fig. 2). Subsequently, the Western blot was stripped and reprobed with anti-ubiquitin antibodies. In this case only the 150 kDa from of Eps15 was detected, demonstrating that Eps15 becomes mono-ubiquitinated upon EGF stimulation (Fig. 2). In addition to the appearance of the 150 kDa band, a slight staining of

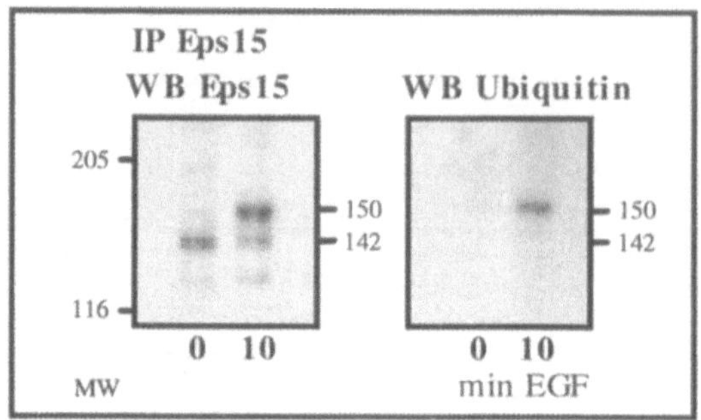

Figure 2. **Stimulation of HER14 cells with EGF induces mono-ubiquitination of Eps15.** HER14 cells were left unstimulated or stimulated for 10 min with 50 ng/ml EGF. Eps15 was immunoprecipitated from the cell lysates and the proteins were separated on an 8% SDS polyacrylamide gel. The Western blot was incubated with anti-Eps15 antibodies (WB Eps15). Subsequently, the blot was stripped and incubated with anti-ubiquitin antibodies (WB Ubiquitin). Immunoprecipitations were performed as previously described (van Delft et al., 1995).

higher molecular weight Eps15 was detected upon EGF addition. This phenomenon was better visible upon longer exposures (data not shown). Eps15 of higher molecular weight was previously also found on Western blots containing immunoprecipitated Eps15 that were stained for phospho-tyrosine residues (van Delft *et al.*, 1997). These observations suggest that Eps15 is not only mono-ubiquitinated but that a minority of Eps15 may also be multi-ubiquitinated.

Furthermore we previously determined the N-terminal amino acid sequence by Edman-degradation (van Delft *et al.*, 1997b). Sequencing of the 142 kDa form of Eps15 did not result in any signal due to N-terminal blocking. Sequencing of the 150 kDa form of Eps15 resulted in a single protein sequence. Comparison of these 10 amino acids with the published sequence of bovine ubiquitin revealed that the obtained amino acids were identical to the first amino acids of ubiquitin (van Delft *et al.*, 1997b). Therefore we conclude that Eps15 becomes ubiquitinated after stimulation of the cell with EGF. Because the increase in molecular weight of Eps15 is similar to the molecular weight of ubiquitin (8 kDa), we conclude that Eps15 becomes predominantly mono-ubiquitinated. Since the approximate ratio of the two Eps15 forms in EGF-stimulated cells was previously determined as 1:1 we estimate that about 50% of Eps15 becomes mono-ubiquitinated after stimulation of cells with EGF (van Delft *et al.*, 1997b). Both forms of Eps15 become phosphorylated on tyrosine residues (Fig. 1) which indicates that ubiquitination of Eps15 is not required for its phosphorylation.

Mono-ubiquitination of proteins has not frequently been reported. Examples of mono-ubiquitination are the T cell antigen receptor (Hou *et al.*, 1994), histone H2A (Davie and Murphy, 1990) and cytochrome c (Sokolik and Cohen, 1991). The yeast α-factor receptor has recently been shown to become either mono- or di-ubiquitinated (Roth and Davis, 1996). Multi-ubiquitination of proteins usually starts on one lysine residue (Finley and Chau, 1991). Subsequently, this ubiquitin becomes ubiquitinated resulting in the formation of multiubiquitin chains. Examples of multi-ubiquitination include cytoplasmic and nuclear proteins but also integral membrane proteins such as receptors for EGF (Galcheva-Gargova *et al.*, 1995), growth hormone (Leung *et al.*, 1987; Strous *et al.*, 1996), PDGF (Mori *et al.*, 1992) and the tumor necrosis factor (Loetscher *et al.*, 1990). Protein ubiquitination has been implicated in many cellular processes (Finley and Chau, 1991). The most widely studied function of ubiquitination lies in the targeting of (multi-ubiquitinated) proteins for degradation to the 26S proteasome. However, not all ubiquitinated proteins are degraded, suggesting additional functions for ubiquitination besides proteolysis. Treatment of cells with transcription inhibitors resulted in a reduced level of ubiquitinated histone H2B, suggesting a role for ubiquitination in chromatin organization (Davie and Murphy, 1990). Recently, it has been suggested that ubiquitination plays a role in the activation of IκBα, a regulator of the transcription factor NFκB (Roff *et al.*, 1996). Interestingly, a new function for ubiquitination has been recently described for the ubiquitination of plasma membrane receptors. Ubiquitination of both the growth hormone receptor and the α-factor receptor in *S. cerevisiae* have been implicated in the endocytosis of these receptors (Roth and Davis, 1996; Strous *et al.*, 1996; Hicke and Riezman, 1996).

An important question is, where in the EGF-induced signaling cascade is Eps15 ubiquitinated? An array of inhibitors of important steps in signal transduction have been used to determine where this takes place. Stimulation of HER14 cells with EGF but without any inhibitors clearly shows ubiquitination of Eps15 (Fig. 3A). However, stimulation with EGF in the presence of phenylarsenoxide (PAO), a potent inhibitor of ubiquitination by ATP scavenging (Berleth *et al.*, 1992), results as expected in an inhibition of Eps15 ubiquitination (Fig. 3A). In the presence of dihydroxycytochalasine B (DHC B), an inhibitor of actin polymerization or nocadazole, which induces depolimerization of the tubulin fibers, no change in ubiquitination has occurred (Fig. 3A). Suggesting that the cytoskeleton is not essential for EGF induced Eps15 ubiquitination. Furthermore two inhibitors of PI-3-kinase, wortmannin and LY 294002, also did not inhibit Eps15 ubiquitination. The activation of PI-3-kinase is not involved in ubiquitination of Eps15, suggesting that Eps15 ubiquitination is either upstream of PI-3-kinase activation or on a different pathway. The binding of Eps15 to both AP-2 and clathrin, and the presence of Eps15 in clathrin-coated pits and vesicles of mammalian cells suggests a role for Eps15 in the endocytosis of the EGF-receptor. Inhibition of the fusion between early and late endosomes by concanamycin A (Drose *et al.*, 1993) did not change the Eps15 ubiquitination (Fig. 3A), suggesting that if Eps15 ubiquitination is

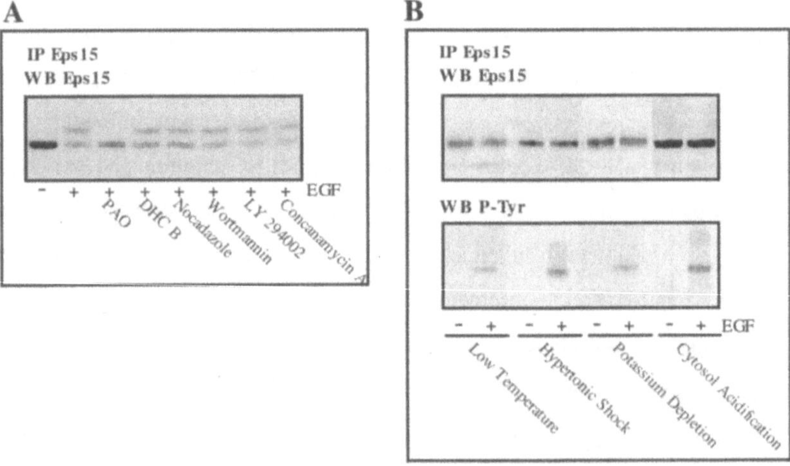

Figure 3. Inhibition of endocytosis prevents Eps15 mono-ubiquitination.
HER14 cells were subjected to different inhibitors of signal transduction steps: phenylarsenoxide (PAO), dihydroxycytochalasine B (DHC B), Nocadazole, Wortmannin, LY294002 and concanamycin M. Eps15 was immunoprecipitated from the lysates and the proteins were separated on a 8% SDS polyacrylamide gel. The Western blot was probed with anti-Eps15 antibodies (A). HER14 cells were subjected to four different methods to inhibit endocytosis: 1) low temperature, 2) hypertonic shock, 3) potassium depletion or 4) cytosol acidification. Subsequently the cells were left unstimulated or stimulated with 50 ng/ml EGF for 10 min. Eps15 was immunoprecipitated and proteins were separated on 8% SDS polyacrylamide gels and the Western blots were incubated with anti-Eps15 antibodies (B, top panel) or with anti-phospho-tyrosine antibodies (B, bottom panel). Immunoprecipitations were performed as previously described (van Delft et al., 1995).

involved in endocytosis, this process takes place early in the endocytotic pathway. To investigate the possible relationship between Eps15 ubiquitination and EGF-receptor internalization further, we examined the effect of blocking EGF-receptor internalization on Eps15 ubiquitination. Internalization of EGF-receptors was inhibited in four different ways: by incubation at low temperature, by depleting potassium from the cytosol, by a hypertonic shock of the cells or by acidification of the cytosol. These methods inhibit different steps in endocytosis: hypertonic shock and incubation at low temperature prevent clustering of receptors (Daukas and Zigmond, 1985), potassium depletion inhibits the assembly of coated pits (Larkin *et al.*, 1983), while acidification of the cytosol is suggested to inhibit pinching off of clathrin-coated pits from the plasma membrane (Sandvig *et al.*, 1987). The effect of these conditions on EGF endocytosis was measured using ^{125}I-labeled EGF. In control cells EGF was rapidly internalized, while under all four endocytosis inhibiting conditions EGF internalization was inhibited for more than 80% (data not shown). When cells were incubated and stimulated at 4°C, ubiquitination of Eps15 was completely abolished (Fig. 3B, top panel). The same results were obtained when endocytosis was inhibited by alternative methods such as potassium depletion, acidification of the cytosol and hypertonic shock (Fig. 3B, top panel). Analysis of Eps15 phosphorylation revealed that in all cases Eps15 became phosphorylated on tyrosine residues after EGF addition (Fig. 3B, bottom panel). These results demonstrate that EGF-receptor activity has not been affected by either of these treatments. Interestingly, it was recently reported by Vieira and coworkers (Vieira *et al.*, 1996) that inhibition of endocytosis using a dynamin mutant resulted in differences in EGF-receptor substrate phosphorylation. This indicates that some substrates are specifically phosphorylated at the plasma membrane while others become phosphorylated at the endosomal membrane. No differences in Eps15 phosphorylation were observed between control cells and cells, which were inhibited in EGF-receptor endocytosis. This indicates that maximum Eps15 phosphorylation is already achieved before the EGF-receptor internalization process is initiated. Together these data show that when endocytosis is inhibited Eps15 mono-ubiquitination is abolished, but tyrosine phosphorylation remains undisturbed.

4. Conclusions

An interesting question is the possible function of the mono-ubiquitination of Eps15. The absence of Eps15 ubiquitination under conditions that inhibit EGF-receptor internalization indicates that either Eps15 ubiquitination is required for Eps15 endocytosis or that EGF-receptor endocytosis is a prerequisite for Eps15 ubiquitination. The first possibility is in analogy to what has been reported for the growth hormone receptor in mammalian cells (Strous *et al.*, 1996) and the α-factor receptor in yeast (Roth and Davis, 1996). In this case Eps15 ubiquitination could be involved in the early steps of endocytosis of the EGF-receptor. However, inhibition of the endocytosis of the α-factor receptor in yeast resulted in an

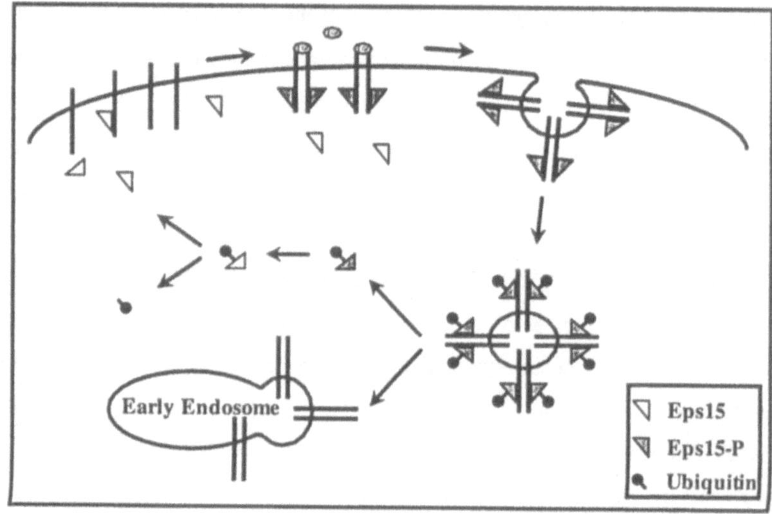

Figure 4. A model for the possible function of EGF-induced ubiquitination of Eps15

increased ubiquitination of the receptor, which is in contrast to the results presented here (Roth and Davis, 1996). Alternatively, our results may indicate that endocytosis is required for Eps15 ubiquitination. This would imply that Eps15 ubiquitination occurs exclusively at a post-surface endocytic transport step. We have shown previously that Eps15 localization is restricted to coated pits and coated vesicles and absent from early endosomes (van Delft *et al.*, 1997). Taken together, these data suggest that ubiquitination of Eps15 occurs after the internalization of EGF-receptors. Based on these data we propose the following model: After binding to the EGF-receptor, Eps15 becomes phosphorylated at the plasma membrane. When the coated pit has pinched of, Eps15 ubiquitination takes place. Both the de-phosphorylation and de-ubiquitination of Eps15 take place before fusion of the vesicle with the early endosome. And thus could mono-ubiquitination of Eps15 be involved in the targeting of coated pits to the early endosome and/or in the uncoating of the coated vesicle (Fig. 4). Preliminary results suggest that in contrast to the EGF-receptor, Eps15 is not rapidly degraded after EGF treatment of the cells. This may indicate that the de-phosphorylated and the de-ubiquitinated form of Eps15 can de reused for another cycle of EGF-receptor endocytosis.

Acknowledgements: We wish to thank Bram Dijker for practical assistance, Fridolin van der Lecq and Ton Aarsman (Sequence Center, Institute of Biomembranes, University Utrecht, The Netherlands) for performing and interpretation of the Edman Degradation experiments, and Theo van der Krift for photographic reproductions.
This work was supported by the Life Sciences Foundations (SLW, grant 17.182), which is subsidized by the Netherlands Organization for Scientific Research (S.v.D.).

References

Berleth, E.S., Kasperek, E.M., Grill, S.P., Braunscheidel, J.A., Graziani, L.A. and Pickart, C.M. (1992) *J. Biol. Chem.* **267**: 16403-16411

Benedetti, H., Raths, S., Crausaz, F., and Riezman, H. (1994) *Mol. Biol. Cell* **5**: 1023-1037

Benmerah, A., Gagnon, J., Bègue, B., Mégarbané, B., Dautry-Varsat, A., and Cerf-Bensussan, N. (1996) *J. Cell Biol.* **131**: 1831-1838

Daukas, G., and Zigmond, S.H. (1985) *J. Cell Biol.* **101**: 1673-1679

Davie, J.R., and Murphy, L.C. (1990) *Biochemistry* **29**: 4752-4757

van Delft, S., Verkleij, A.J., Boonstra, J. and van Bergen en Henegouwen, P.M.P. (1995) *FEBS lett* **357**: 251-254

van Delft, S., Schumacher, C., Hage, W., Verkleij, A.J., and van Bergen en Henegouwen, P.M.P. (1997) *J. Cell Biol.* **136**: 811-823

van Delft, S., Govers, R., Strous, G.J., Verkleij, A.J. and van Bergen en Henegouwen, P.M.P. (1997b) *J. Biol. Chem.* **272**: 14013-14017

Drose, S., Bindseil K.U., Siebers, A., Zeeck, A. and Altendorf, K. (1993) *Biochemistry* **32**: 3902-3906

Faure, R., Baquiran, G., Bergeron, J.J.M. and Posner, B.I. (1992) *J. Biol. Chem.* **267**: 11215-11221

Fazioli, F., Minichiello, L., Matoskova, B., Wong, W.T., and Di Fiore, P.P. (1993) *Mol. Cell. Biol.* **13**: 5814-5828

Finley, D., and Chau, V. (1991) *Annu. Rev. Cell Biol.* **7**: 25-69

Galcheva-Gargova, Z., Theroux, S.J., and Davis, R.J. (1995) *Oncogene* **11**: 2649-2655

Hicke, L., and Riezman, H. (1996) *Cell* **84**: 277-287

Hou, D., Cenciarelli, C., Jensen, J.P., Nguygen, H.B., and Weissman, A.M. (1994) *J. Biol. Chem.* **269**: 14244-14247

Larkin, J.M., Brown, M.S., Goldstein, J.L., and Anderson, R.G.W. (1983) *Cell* **33**: 273-285

Leung, D.W., Spencer, S.A., Cachianes, G., Hammonds, R.G., Collins, C., Henzel,W.J., Barnard, R., Waters, M.J., and Wood, W.I. (1987) *Nature* **330**: 537-543

Loetscher, H., Schlaeger, E.J., Lahm, H.W., Pan, Y.C., Lesslauer, W., and Brockhaus, M (1990) *J. Biol. Chem.* **265**: 20131-20138

Masui, H., Wells, A., Lazar, C.S., Rosenfeld, M.G. and Gill, G.N. (1991) *Cancer Res.* **51**: 6170-6175

Mori, S., Heldin, C.H., and Claesson-Welsh, L. (1992) *J. Biol. Chem.* **267**: 6429-6434

Northwood, I.C. and Davis, R.J. (1990) *Proc. Natl. Acad. Sci. USA* **87**: 6107-6111.

Roff, M., Thompson, J., Rodriquez, M.S., Jacgue, J.M., Baleux, F., Arenzan-Seisdedos, F., and Hay, R.T. (1996) *J. Biol. Chem.* **271**: 7844-7852

Roth, A.F., and Davis, N.G. (1996) *J. Cell Biol.* **134**: 661-674

Sorkin, A. and Waters, C.M. (1993) *BioEssays* **15**: 375-382

Sandvig, K., Olsnes, S., Petersen, O.W., and van Deurs, B. (1987) *J. Cell Biol.* **105**: 679-689

Schmid, S.L. (1992) *BioEssays* **14**: 589-596

Sokolik, C.W., and Cohen, R.E. (1991) *J. Biol. Chem.* **266**: 9100-9107

Strous, G.J., van Kerkhof, P., Govers, R., Ciechanover, A., and Schwartz, A.L. (1996) *EMBO J.* **15**: 3806-3812

Tebar, F., Sorkina, T., Sorkin, A., Ericsson, M., and Kirchhausen, T. (1996) *J. Biol. Chem.* **271**: 28727-28730

Vieira, A.V., Lamaze, C., and Schmid, S. (1996) *Science* **274**: 2086-2089

Wells, A., Welsh, J.B., Lazar, C.S., Wiley, H.S., Gill, G.N. and Rosenfeld, M.G. (1990) *Science* **247**: 962-964

Wendland, B., McCaffery, J.M., Xiao, Q., and Emr, S.D. (1996) *J. Cell Biol.* **135**: 1485-1500

The Role of α_2Macroglobulin Receptor Associated Protein as a Chaperone for Multifunctional Receptors

Lars Ellgaard, Pernille Stage, Michael Etzerodt and Hans Christian Thøgersen

Laboratory of Gene Expression, Department of Molecular and Structural Biology
University of Aarhus, Gustav Wieds Vej 10, DK-8000 Aarhus C, Denmark

Abstract. The $\alpha_2$macroglobulin receptor associated protein (RAP) is a 39 kDa large glycoprotein of 323 amino acid residues, which binds to members of the low density lipoprotein receptor (LDLR) family. These receptors are large multifunctional endocytic receptors capable of binding and internalizing many different ligands, such as lipoproteins and complexes of proteinases and their inhibitors. While RAP is able to inhibit binding of all currently known extracellular ligands of these receptors in *in vitro* assays, the protein is an intracellular ER- and Golgi-resident protein which serves a function as a molecular chaperone for LDLR family members. The intracellular binding of RAP to these receptors assists their folding and prevents premature binding of other ligands in the endocytic pathway. Here we describe the role of RAP as a molecular chaperone for LDLR family members, as well as the interaction between RAP and two recently discovered non-family member receptors, gp95/sortilin and sorLA-1. Based on ligand competition analysis and the recently solved three-dimensional structure of the N-terminal domain of RAP, a model of RAP domain 1/receptor interaction is proposed.

Keywords. $\alpha_2$macroglobulin receptor associated protein (RAP), molecular chaperone, $\alpha_2$macroglobulin receptor/low density lipoprotein receptor-related protein (α_2MR/LRP)

1 RAP Binding Receptors

The ability of RAP to inhibit binding of all other ligands of the family of LDL receptors has made this protein a very useful experimental tool in the investigation of receptor function. However, the physiological role of RAP in relation to receptor interaction has only recently been emerging. The following provides an overview of the structure and function of the LDLR family members, as well as of two non-family member receptors, gp95/sortilin and sorLA-1, that also bind RAP.

The LDLR family comprises large cell surface receptors mediating endocytosis, as reviewed in (Krieger and Herz, 1994). At present it includes five different receptors in mammalian species. Beside LDLR (Yamamoto et al., 1984), it is the $\alpha_2$macroglobulin receptor/low density lipoprotein receptor-related protein, (α_2MR/LRP) (Herz et al., 1988), the very low density lipoprotein receptor (VLDLR) (Oka et al., 1994), the recently cloned gp330/megalin (Hjalm et

NATO ASI Series, Vol. H 106
Lipid and Protein Traffic
Pathways and Molecular Mechanisms
Edited by Jos A. F. Op den Kamp

al., 1996) and the apolipoprotein E receptor 2 (apoER2) (Kim et al., 1996). The structural basis for classifying the receptors as belonging to the same family relies on the fact that they contain several common sequence motifs. The occurrence of these cysteine-rich structural elements in each of the five LDLR family members and in two other recently discovered RAP binding receptors, sorLA-1 and gp95/sortilin (see below), is illustrated schematically in Fig. 1.1.

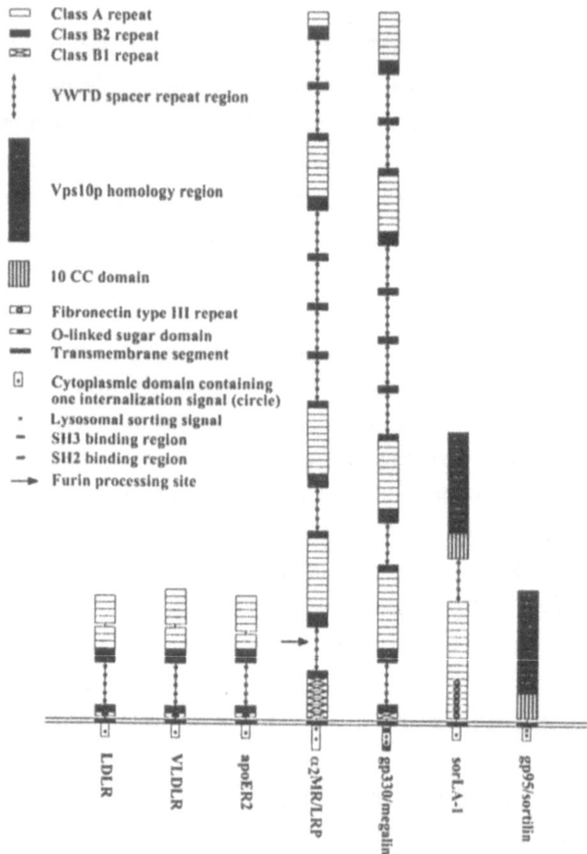

Fig. 1.1. Schematic representation of the overall domain structure of the human members of the LDLR family as well as the two other RAP binding receptors, sorLA-1 and gp95/sortilin.

Of these receptors gp330/megalin and α_2MR/LRP are both giant molecules of approximately 600 kDa, whereas apoER2, LDLR and VLDLR are of more moderate size (approximately 100 kDa). All human members of the LDLR family, except apoER2, have been shown to bind RAP.

The receptors are multifunctional in the sense that they are able to bind and internalize several different and apparently unrelated ligands, such as lipoproteins and proteinase:proteinase-inhibitor complexes. Whereas structural differences and sequence diversities are likely to account for different ligand binding specificities and functional properties of the various receptors, a certain overlap in ligand binding specificity between receptors occurs. However, while the receptors are capable of binding many of the same ligands, their tissue distribution vary. This pattern of differential tissue-distribution has led to the proposal that the receptors could perform similar functions, but in different cell types (Gliemann et al., 1994; Krieger and Herz, 1994; Zheng et al., 1994; Czekay et al., 1995).

Recently, two new RAP binding receptors, sorLA-1 (sorting protein-related receptor containing LDLR class A repeats) (Jacobsen et al., 1996) and gp95/sortilin (Petersen et al., 1997), have been identified (see Fig 1.1). These receptors both contain a yeast Vps10p (Marcusson et al., 1994) homology region including a segment containing 10 conserved cysteine residues (named "ten cysteine consensus" - 10CC). The yeast Vps10p functions as a receptor for carboxypeptidase Y and the binding of the protein to the receptor leads to lysosomal targeting of the protein (Marcusson et al., 1994). In addition gp95/sortilin contains a lysosomal sorting signal and the protein has been suggested to be involved in sorting from the trans-Golgi network of newly synthesized lysosomal enzymes (Petersen et al., 1997).

2 The Function of RAP as a Molecular Chaperone

Originally, RAP was isolated as a protein co-purifying with α_2MR/LRP and the ability of the protein to inhibit binding of all other ligands of this receptor led to suggestions that RAP acts as an extracellular physiological modulator of ligand binding. However, subsequent immunocytochemistry has shown RAP to be localized in the ER and Golgi, with only low levels detectable at the cell surface (Lundstrom et al., 1993; Bu et al., 1994; Zheng et al., 1994; Bu et al., 1995). Likewise, attempts to detect RAP in cell supernatants or in the blood have been unsuccessful (Strickland et al., 1991; Williams et al., 1994). In addition, the carbohydrate composition of the protein has been shown to correlate with that expected from a protein resident of the ER and medial-Golgi, whereas the more complex carbohydrates characteristic of proteins transported to the trans-Golgi are absent (Jensen et al., 1992; Bu et al., 1995). Taken together, these findings suggest an intracellular function of the protein. Moreover, the association of RAP in the ER with gp330/megalin into a complex which remains stable through its transport to the medial-Golgi provided a first clue to a possible function of the protein as a specialized molecular chaperone for members of the LDLR family (Biemesderfer et al., 1993).

Several recent reports have contributed information to support this idea and the role of RAP as an ER-resident protein has now become firmly established. In (Bu et al., 1995) the C-terminal

HNEL tetrapeptide sequence of RAP was shown to be necessary and sufficient for ER retention. In addition, the HNEL sequence was shown to be as effective as the mammalian consensus ER-retention KDEL sequence in retaining a normally secreted protein in the ER (Bu et al., 1995). This result has very recently been corroborated and extended by demonstrating that the ERD2 proteins (ERD2.1 and ERD 2.2), known to function as KDEL receptors, are able to retain proteins containing the HNEL signal at their C-terminal (Bu et al., 1997). Whether ERD2 proteins function as physiological ER-retention receptors for RAP remains to be established.

These results have prompted further investigation of the intracellular function of RAP in relation to receptor interaction. The interaction between RAP and α_2MR/LRP within the cell has been shown to occur transiently early in the secretory pathway, thereby preventing premature binding of other α_2MR/LRP ligands (Bu et al., 1995). Furthermore, the RAP/receptor interaction decreases dramatically at pH-values below 6.6. These results suggest that RAP functions as a regulator of ligand binding to the receptor along the secretory pathway, and that the RAP/receptor complex could dissociate in the medial-Golgi due to the low pH (~6.4) of this compartment.

Gene knock-out studies also provide evidence of the importance of RAP for α_2MR/LRP trafficking (Willnow et al., 1995). Homozygous RAP-deficient mice were shown to be viable without any apparent defects. However, the α_2MR/LRP level in liver and brain was reduced by 75% in these animals compared to the wild-type - possibly as a result of protein aggregation and subsequent degradation, since the level of mRNA for the receptor was similar in both mutant and wildtype mice. These results were extended by the same group in (Willnow et al., 1996), where it was verified that α_2MR/LRP accumulates in the ER as large insoluble aggregates in the liver and brain of RAP $^{-/-}$ mice. Similarly, it was shown that export from the ER of LDLR family members gp330/megalin and VLDLR was impaired in different cell types in the RAP-deficient mice. Interestingly, α_2MR/LRP expression in the liver could be restored to normal levels by the transfer of RAP cDNA encoded by recombinant adenovirus.

To test whether premature binding of other ligands might prevent folding of the receptor thereby leading to its aggregation, the effect of co-expressing the α_2MR/LRP ligand apoE with the receptor was investigated (Willnow et al., 1996). Indeed, co-expression resulted in the aggregation of the receptor and impaired its transport to the cell surface. These ill-effects were shown to be relieved by gene transfer of RAP.

While the above results demonstrate the importance of RAP for the maturation of α_2MR/LRP through prevention of premature interaction of ligands, a more direct role for the protein in the folding of the receptor has also been demonstrated. For this purpose, a system of co-expression of RAP and recombinant soluble mini-receptors containing clusters of α_2MR/LRP class A repeats (see Fig.1.1), shown to be important for ligand binding (Moestrup et al., 1993; Willnow et al., 1994), was developed (Bu & Rennke, 1996). While over-expression of recombinant mini-receptors resulted in their misfolding due to the formation of intermolecular disulfide bridges, co-expression of RAP was shown to prevent this aggregation (Bu & Rennke, 1996).

These reports all support a role for RAP as a specialized molecular chaperone for proteins belonging to the LDLR family, preventing both the premature binding of other receptor ligands and the aggregation of the receptors themselves through intermolecular disulfide bridges. After dissociation of the complex between RAP and receptor in the medial-Golgi due to low pH, RAP is bound to an HNEL/KDEL receptor and recycled to the ER, while the receptor is transported to the cell surface. A model of the proposed action for RAP as a molecular chaperone is shown in Fig. 2.1.

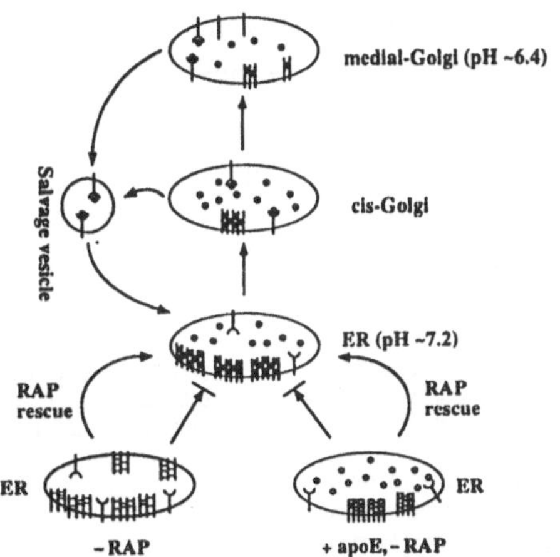

Fig. 2.1. Model of the proposed action for RAP as a molecular chaperone. RAP is found in the ER in a soluble form and in complex with α_2MR/LRP. These large aggregates keep the receptor in an inactive state until it dissociates from RAP in the medial-Golgi due to the low pH of this compartment. The receptor travels to the cell surface, while RAP is retrieved by the recycling receptor for the C-terminal HNEL sequence. In the absence of RAP, α_2MR/LRP aggregates in the ER and further transport is prevented (lower left hand corner). Co-expression of RAP is able to overcome this aggregation. Over-expression of the α_2MR/LRP ligand apoE similarly results in the aggregation of the receptor, while co-expression of RAP is able to abolish the effect of apoE (lower right hand corner). The model is built on results presented in (Willnow et al., 1995), (Bu et al., 1995) and (Bu & Rennke, 1996) and adapted from (Bu et al., 1995).

With respect to the interaction between RAP and the gp95/sortilin receptor, recent investigation has demonstrated the interaction in ligand blotting as well as cross linking experiments between RAP and the lumenal region of this receptor (C. M. Petersen, personal communication). Presently, similar investigations to characterize the interaction between RAP and sorLA-1 are under way along with ligand specificity investigations of both receptors. However, a possible functional role of RAP as a molecular chaperone for the sorLA-1 and gp95/sortilin receptors remains to be established. Likewise, a possible functional role of these receptors in lysosomal sorting, as suggested by their homology to Vps10p and the lysosomal sorting signal of sorLA-1, has not yet been verified.

3 The Interaction between RAP and α_2MR/LRP

With the elucidation of several aspects of the functional role of RAP as a specialized molecular chaperone during recent years, a more detailed understanding of the interaction between RAP and LDLR family members has become an important issue of investigation. While elucidation of the three-dimensional structure of the complex between RAP and α_2MR/LRP ligand binding regions remains a long term goal, their interaction has been analysed in considerable detail by different methods.

The realization that the RAP sequence contains an internal triplication (Bu et al., 1995; Warshawsky et al., 1995; Ellgaard et al., 1997), has directed attention at investigating single domain function. With the domain boundaries defined in (Ellgaard et al., 1997), RAP domains 1, 2 and 3 can be assigned to residues 18-112, 113-218 and 219-323, respectively. The alignment of RAP domains is shown below in Fig. 3.1.

```
1-17     YSREKNQPKPSPKRESG

18-79    EEFRMEKLNQLWEKAQRLHLPPVRLAE.LHA.DLK....IQERDELAWKK.LK.LDGLDEDGEK.E..ARLIR.........
113-186  DDPRLEK...LWHKAKTSGKFSGEELDKLWREFL.HH...KEKV.HEYNVLLETLSRTEEIHENVISPSDLSDIKGSVLHSR
219-287  EEPRV..I.DLWDLAQ.SANLTDKELEA.FREELKHFEAKIEKH.NHYQKQLE.I..AHEKLRHAESVGDGERV......SR

80-112   ......NLNVILAKYGLD.GKKDARQ.VTSNSLSG.TQEDGL
187-218  HTELKEKL..RSINQGLDRLRRVSHQGYSTEAEF
288-323  SREKHALLEGRTKELGY.TVKK.HLQDL.SGRISRARHNE.L
```

Fig. 3.1. Three-domain partition of RAP as proposed in (Ellgaard et al., 1997), dividing RAP into domains comprising residues 18-112, 113-218 and 219-323, respectively.

This alignment suggests the 17 N-terminal residues of the protein, a sequence segment which does not seem to be closely related to any other segment in RAP, to constitute an N-terminal extension of domain 1. Furthermore, C-terminal segments of domains 1 and 3 are proposed to be related, whereas domain 2 appears to differ in this region. It is noteworthy, that according to this sequence alignment C-terminal regions of domains 1 and 3 belong to the proposed α_2MR/LRP recognition motif shared among several ligands (Nielsen et al., 1996). Ligand competition assays performed with RAP domain constructs as well as various domain derivatives have confirmed the importance of the 15 C-terminal residues of domains 1 and 3 in receptor interaction. Thus, the deletion of this region from domain 1 abolished its effect in receptor interaction (Warshawsky et al., 1993) as well as its ability to inhibit binding of the α_2MR/LRP ligand α_2M (Ellgaard et al., 1997; Obermoeller et al., 1997). Likewise, experiments performed with deletion derivatives of domain 3 have verified the importance of the C-terminal region of this domain in receptor interaction (Warshawsky et al., 1993; Ellgaard et al., 1997).

Recently, the ability of individual domain constructs to prevent misfolding of soluble mini-receptors in co-expression experiments has been investigated (Obermoeller et al., 1997). Interestingly, it was demonstrated that a construct comprising residues 191-323, similar to domain 3 as defined in Fig. 3.1, was able to prevent misfolding of a selection of mini-receptors with equal or higher efficiency than full length RAP. In comparison, assistance in folding by constructs similar to domains 1 and 2 was insignificant. While able to inhibit binding of several α_2MR/LRP ligands to the receptor, the construct similar to domain 3 was unable to inhibit α_2M binding to the receptor, thus confirming previous results (Ellgaard et al., 1997). However, inhibition of α_2M was achieved by the domain 1 construct (Ellgaard et al., 1997; Obermoeller et al., 1997).

In combination, these results point to domain 3 as being essential for the ability of RAP to assist α_2MR/LRP folding and for inhibiting the binding of certain receptor ligands. As hypothesized in (Obermoeller et al., 1997), this suggests the domain 3 gene to be the ancestral domain from which the other two domains arose by gene duplication. Sequence homology comparisons between domains seem to confirm this role of the domain 3 gene as the ancestral RAP gene (Obermoeller et al., 1997). With the subsequent evolution of other receptor ligands, such as α_2M, the function of other domains as inhibitors of premature receptor binding became essential. Verification (or rejection) of this theory should follow from evolutionary studies of RAP and α_2MR/LRP ligands in various species.

To further characterize the importance of specific residues of domain 3 for the interaction with α_2MR/LRP we have recently constructed a whole series of alanine mutants of each of the residues of the C-terminal region of this domain. Competition experiments of these domain 3 mutants versus iodinated wildtype domain show, that while the large majority of mutants exhibit little or no difference from the wild type domain in apparent K_i-value (~20 nM), the Leu311Ala mutant shows significantly weaker binding to α_2MR/LRP with a K_i-value of ~500 nM (P. Stage, L. Ellgaard and H. C. Thøgersen, unpublished result).

With respect to the interaction between domain 1 and α_2MR/LRP, important information can be extracted from the recently solved solution structure of this N-terminal domain of RAP (P. R. Nielsen et al., 1997). The structure of the 20 C-terminal residues of the domain was shown to be unordered, and subsequent structure determination was performed on a truncated domain 1 derivative (residues 18-97). This structure contains three helices comprising residues 23-34, 39-65 and 73-88, respectively, arranged in an up-down-up pattern and is presented in Fig. 3.2. The structure reveals a groove on one side of the surface formed by the three helices as an important feature. It could be speculated, that this groove is a binding site for α_2MR/LRP or alternatively for the C-terminal residues important for receptor interaction, but shown not to be organized in a stable structure in domain 1. Furthermore, the deletion of residues 18-23 has been shown to be detrimental to receptor binding (Warshawsky et al., 1993). Inspection of the structure in this region reveals that deletion of residues around the buried Phe20 is likely to destabilize the overall structure of the domain. In combination, the above results indicate that at least two integral parts are important for the interaction between domain 1 and α_2MR/LRP - the C-terminal 15 amino acid residues and another component defined by the tertiary structure of the domain.

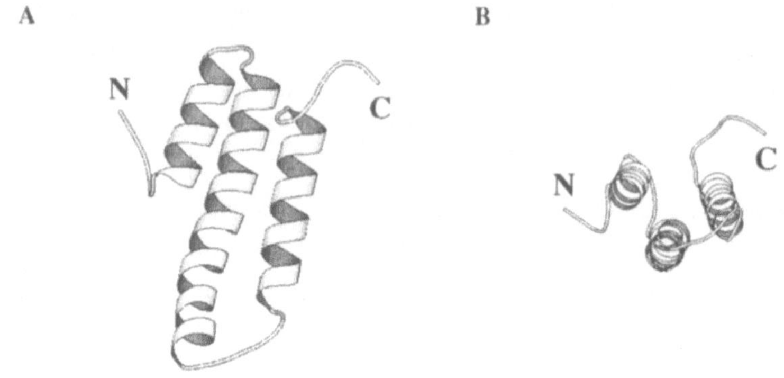

Fig. 3.2. The structure of the truncated domain 1 derivative (residues 18-97) of RAP as solved by NMR-spectroscopy. *Panel A* Side-view *Panel B* End-view, revealing the groove formed by the three helices.

4 Discussion

At present, the function of RAP as a molecular chaperone most closely resembles that of the invariant chain which plays an important role in the export of MHC class II from the ER (Viville et al., 1993). However, the exact mechanism of RAP in the folding process of members of the LDLR family has not yet been established. Experiments to test if the protein like many other molecular chaperones binds and hydrolyses nucleoside triphosphates have shown that this does not seem to be the case for RAP (M. Etzerodt and L. Ellgaard, unpublished observation). Furthermore, as discussed in (Obermoeller et al., 1997), it is presently not clear if the functions of RAP in the folding of α_2MR/LRP and in preventing premature ligand binding are related. With respect to RAP's function in relation to sorLA-1 and gp95/sortilin, further investigation is necessary to establish the physiological role of this interaction as well as of receptor function.

The ongoing characterization of the interaction between RAP and α_2MR/LRP is important to reach a deeper understanding of the mechanism of RAP as a molecular chaperone. Binding and competition assays of various domain constructs of RAP and the structure determination of the N-terminal domain of the protein have provided insight into structure- and sequence-determinants of RAP important for this interaction. The long term objective of solving the three-dimensional structure of a complex between RAP and α_2MR/LRP ligand binding domains will require further characterization of the receptor with respect to specific domains/residues involved in ligand interaction, along with the development of a system for heterologous gene expression of such domains.

Acknowledgements:

This work was supported by the Danish Biotechnology Research Programme. Figure 2.1 was adapted from Bu G, Geuze HJ, Strous GJ and Schwartz AL (1995) EMBO J. 14: 2269-2280 by permission of Oxford University Press.

References:

Biemesderfer D, Dekan G, Aronson PS, and Farquhar MG (1993) Biosynthesis of the gp330/44-kDa Heymann nephritis antigenic complex: assembly takes place in the ER. Am. J. Physiol. 264: F1011-1020

Bu G, Maksymovitch EA, Geuze H and Schwartz AL (1994) Subcellular localization and endocytic function of low density lipoprotein receptor-related protein in human glioblastoma cells. J. Biol. Chem. 269: 29874-29882

Bu G, Geuze HJ, Strous GJ and Schwartz AL (1995) 39 kDa receptor-associated protein is an ER resident protein and molecular chaperone for LDL receptor-related protein. EMBO J. 14: 2269-2280

Bu G and Rennke S (1996) Receptor-associated protein is a folding chaperone for low density lipoprotein receptor-related protein. J. Biol. Chem. 271: 22218-22224

Bu G, Rennke S and Geuze HJ (1997) ERD2 Proteins Mediate ER Retention Of the HNEL Signal Of LRPs Receptor-Associated Protein (RAP). J. Cell Sci. 110: 65-73

Czekay RP, Orlando RA, Woodward L, Adamson ED and Farquhar MG (1995) The expression of megalin (gp330) and LRP diverges during F9 cell differentiation. J. Cell Sci. 108: 1433-1441.

Ellgaard L, Holtet TL, Nielsen PR, Etzerodt M, Gliemann J and Thøgersen HC (1997) Dissection of the domain architecture of the $\alpha_2$macroglobulin-receptor-associated protein. Eur. J. Biochem. 244: 544-551

Gliemann J, Nykjaer A, Petersen CM, Jørgensen KE, Nielsen M, Andreasen PA, Christensen EI, Lookene A, Olivecrona G and Moestrup SK (1994) The multiligand $\alpha_2$macroglobulin receptor/low density lipoprotein receptor-related protein (α_2MR/LRP). Binding and endocytosis of fluid phase and membrane-associated ligands. Ann. N. Y. Acad. Sci. 737: 20-38

Herz J, Hamann U, Rogne S, Myklebost O, Gausepohl H and Stanley KK (1988) Surface location and high affinity for calcium of a 500-kd liver membrane protein closely related to the LDL-receptor suggest a physiological role as lipoprotein receptor. EMBO J. 7: 4119-4127

Hjalm G, Murray E, Crumley G, Harazim W, Lundgren S, Onyango I, Ek B, Larsson M, Juhlin C, Hellman P, Davis H, Akerstrom G, Rask L and Morse B (1996) Cloning and sequencing of human gp330, a Ca(2+)-binding receptor with potential intracellular signaling properties. Eur. J. Biochem. 239: 132-137

Jacobsen L, Madsen P, Moestrup SK, Lund AH, Tommerup N, Nykjaer A, Sottrup-Jensen L, Gliemann J and Petersen CM (1996) Molecular Characterization Of a Novel Human Hybrid-Type Receptor That Binds the α_2Macroglobulin Receptor-Associated Protein. J. Biol. Chem. 271: 31379-31383

Jensen PH, Gliemann J and Ørntoft T (1992) Characterization of carbohydrates in the $\alpha_2$macroglobulin receptor. FEBS Lett 305: 129-132

Kim DH, Iijima H, Goto K, Sakai J, Ishii H, Kim HJ, Suzuki H, Kondo H, Saeki S and Yamamoto T (1996) Human apolipoprotein E receptor 2. A novel lipoprotein receptor of the low density lipoprotein receptor family predominantly expressed in brain. J. Biol. Chem. 271: 8373-8380

Krieger M and Herz J (1994). Structures and functions of multiligand lipoprotein receptors: macrophage scavenger receptors and LDL receptor-related protein (LRP). Annu. Rev. Biochem. 63: 601-37

Lundstrom M, Orlando RA, Saedi MS, Woodward L, Kurihara H and Farquhar MG (1993) Immunocytochemical and biochemical characterization of the Heymann nephritis antigenic complex in rat L2 yolk sac cells. Am. J. Pathol. 143: 1423-1435

Marcusson EG, Horazdovsky BF, Cereghino JL, Gharakhanian E and Emr SD (1994) The sorting receptor for yeast vacuolar carboxypeptidase Y is encoded by the VPS10 gene. Cell 77: 579-86

Moestrup SK, Holtet TL, Etzerodt M, Thøgersen HC, Nykjaer A, Andreasen PA, Rasmussen HH, Sottrup-Jensen L and Gliemann J (1993) $\alpha_2$macroglobulin-proteinase complexes, plasminogen activator inhibitor type-1-plasminogen activator complexes, and receptor-associated protein bind to a region of the $\alpha_2$macroglobulin receptor containing a cluster of eight complement-type repeats. J. Biol. Chem. 268: 13691-13696

Nielsen KL, Holtet TL, Etzerodt M, Moestrup SK, Gliemann J, Sottrup-Jensen L and Thøgersen HC (1996) Identification of residues in α–macroglobulins important for binding to the α_2-macroglobulin receptor/low density lipoprotein receptor-related protein. J. Biol. Chem. 271: 12909-12912

Nielsen PR, Ellgaard L, Etzerodt M, Thøgersen HC and Poulsen FM (1997) The solution structure of the N-terminal domain of $\alpha_2$macroglobulin receptor-associated protein. Proc. Natl. Acad. Sci. USA 94: 7521-7525

Obermoeller LM, Warshawsky I, Wardell MR and Bu G (1997) Differential Functions of Triplicated Repeats Suggest Two Independent Roles for the Receptor-associated Protein as a Molecular Chaperone. J. Biol. Chem. 272: 10761-10768

Oka K, Tzung KW, Sullivan M, Lindsay E, Baldini A and Chan L (1994). Human very-low-density lipoprotein receptor complementary DNA and deduced amino acid sequence and localization of its gene (VLDLR) to chromosome band 9p24 by fluorescence in situ hybridization. Genomics 20: 298-300

Petersen CM, Nielsen MS, Nykjaer A, Jacobsen L, Tommerup N, Rasmussen HH, Røigaard H, Gliemann J, Madsen P and Moestrup SK (1997) Molecular Identification Of a Novel Candidate Sorting Receptor Purified From Human Brain By Receptor-Associated Protein Affinity Chromatography. J. Biol. Chem. 272: 3599-3605

Strickland DK, Ashcom JD, Williams S, Battey F, Behre E, McTigue K, Battey JF and Argraves WS (1991) Primary structure of $\alpha_2$macroglobulin receptor-associated protein. Human homologue of a Heymann nephritis antigen. J. Biol. Chem. 266: 13364-13369

Viville S, Neefjes J, Lotteau V, Dierich A, Lemeur M, Ploegh H, Benoist C and Mathis D (1993) Mice lacking the MHC class II-associated invariant chain. Cell 72: 635-648

Warshawsky I, Bu G and Schwartz A (1993) Identification of domains on the 39-kDa protein that inhibit the binding of ligands to the low density lipoprotein receptor-related protein. J. Biol. Chem. 268: 22046-22054

Warshawsky I, Bu G and Schwartz AL (1995) Sites within the 39-kDa protein important for regulating ligand binding to the low-density lipoprotein receptor-related protein. Biochemistry 34: 3404-3415

Williams SE, Kounnas MZ, Argraves KM, Argraves WS and Strickland DK (1994) The $\alpha_2$macroglobulin receptor/low density lipoprotein receptor-related protein and the receptor-associated protein. An overview. Ann. N. Y. Acad. Sci. 737: 1-13

Willnow TE, Orth K and Herz J (1994) Molecular dissection of ligand binding sites on the low density lipoprotein receptor-related protein. J. Biol. Chem. 269: 15827-15832

Willnow TE, Armstrong SA, Hammer RE and Herz J (1995) Functional expression of low density lipoprotein receptor-related protein is controlled by receptor-associated protein in vivo. Proc. Natl. Acad. Sci. U S A 92: 4537-4541

Willnow TE, Rohlmann A, Horton J, Otani H, Braun JR, Hammer RE and Herz J (1996) RAP, a specialized chaperone, prevents ligand-induced ER retention and degradation of LDL receptor-related endocytic receptors. EMBO J. 15: 2632-2639

Yamamoto T, Davis CG, Brown MS, Schneider WJ, Casey ML, Goldstein JL and Russell DW (1984) The human LDL receptor: a cysteine-rich protein with multiple Alu sequences in its mRNA. Cell 39: 27-38

Zheng G, Bachinsky DR, Stamenkovic I, Strickland DK, Brown D, Andres G and McCluskey RT (1994) Organ distribution in rats of two members of the low-density lipoprotein receptor gene family, gp330 and LRP/α_2MR, and the receptor-associated protein (RAP). J. Histochem. Cytochem. 42: 531-542

Characterization of the Gene VII and Gene IX Minor Coat Proteins from Bacteriophage M13

Cor J.A.M. Wolfs*, M. Chantal Houbiers, Ruud B. Spruijt and Marcus A. Hemminga.

Wageningen Agricultural University
Department of Molecular Physics
Dreijenlaan 3, 6703 HA Wageningen, The Netherlands.

*Corresponding author: Phone +31-317-482039
Fax +31-317-482725
E-mail: Cor.Wolfs@virus.mf.wau.nl

A most interesting question in biology is how viruses and bacteriophages assemble and disassemble thereby utilizing the properties of the plasma membrane of biological cells. To pass a plasma membrane, a virus makes use of receptor molecules within the membrane, and after binding, triggers a sequence of processes that enables it to disassemble and start the production of new copies of the virus. Bacteriophage M13, one of the most studied bacteriophages, and the closely related phages f1 and fd are *Escherichia coli* specific filamentous phages belonging to the genus Inovirus. The virion consists of a circular single-stranded DNA molecule of 6408 nucleotides (coding for 10 identified proteins) encapsulated in a long cylindrical protein coat. The protein coat is composed of about 2700-3000 copies of the major coat protein, the product of gene VIII. Apart from this abundant coat protein, the virions contain at one end 3-5 copies of the adsorption protein (the product of gene III) together with some copies of the bacteriophage particle stabilizing gene VI protein. The opposite end of the virion is composed of about 5 copies each of the protein products of gene VII and gene IX. Detailed descriptions of the viral particle can be found in the reviews of Rasched and Oberer (1986) and Model and Russel (1988).

The reproductive life cycle of bacteriophage M13 includes the adsorption of the virions at the F-pili of the host *Escherichia coli* cells specifically mediated by the viral adsorption protein (gene III protein). The viral particle is brought to the cell surface, by a mechanism that is yet not well understood. The gene III protein is thought to play a key role in facilitating the entry of the viral DNA into the cell by the formation of a pore, enabling passing through both the outer cell envelop and the cytoplasmic or inner membrane (Glaser-Wuttke et al., 1989). During the infectious entry of the virus the hydrophobic gene VI protein, which functions as a sort of sealing agent to stabilize the phage particle, is lost from the virion resulting in a destabilized nucleo protein particle. The major coat protein of this destabilized nucleo protein particle is stripped off from the DNA and dissolved in the inner membrane. The DNA is released in the cytoplasm and new DNA and proteins

NATO ASI Series, Vol. H 106
Lipid and Protein Traffic
Pathways and Molecular Mechanisms
Edited by Jos A. F. Op den Kamp

are produced using the host cell machinery. Newly synthesized coat proteins are inserted into the inner membrane prior to be used in the assembly process.

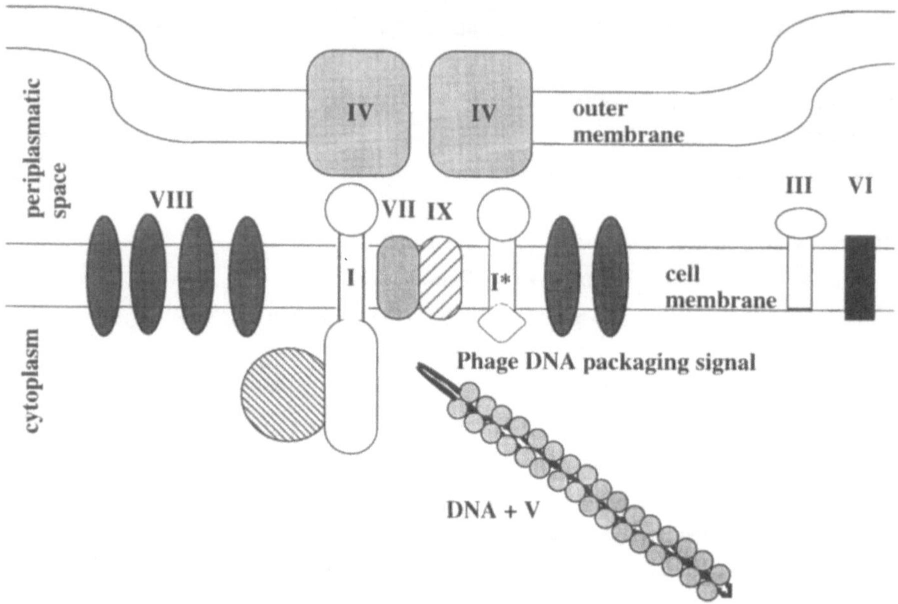

Figure 1: Schematic illustration of the assembly site bacteriophage of M13. In the assembly site the gene IV proteins are thought to constitute an extrusion channel through the outer membrane. The gene I and I* proteins are assumed to form an extrusion channel in the inner membrane. The viral protein coded by genes VII and IX and at least one host protein (thioredoxin) are necessary for the initiation of the viral assembly process, in which the viral DNA and the abundant gene VIII (major) coat proteins are assembled and extruded without lysis of the host cell. The gene V protein is released as dimers in the cell. Assembly of the virus is terminated with the proteins coded by genes III and VI. (Modified after Brissette & Russel, 1990, Guy-Caffey et al., 1992 and Russel, 1991)

Whereas much is known about the structure of the gene VIII major coat protein in the phage particle (Glucksman et al., 1992; Marvin et al., 1994; Opella et al., 1980) and in the membrane-bound state (Henry & Sykes, 1992; Spruijt et al., 1996; Van de Ven et al., 1993), the information about the gene VII and gene IX protein is limited. It has been suggested that these proteins are located at one end of the phage particle (Grant et al., 1981). This end appears to be the end that leaves the cell first (Lopez & Webster, 1985).

Further these two proteins were proposed to be associated with the membrane, prior to assembly, based on the finding that they retain their amino-terminal formyl group (Simons et al., 1981). Simons et al. (1981) suggested that the gene VII and gene IX proteins are incorporated in phage as the primary translational products, so no proteolytic processing occurs. These findings suggest that these proteins might readily insert into the membrane after synthesis. In this respect they behave differently from the gene VIII protein, which is synthesized as a pre-protein and whose signal sequence is cleaved off after integration into the inner membrane (Model & Russel, 1988). Recently it was shown that the gene VII and gene IX proteins are localized in the *E. coli* inner membrane (Endemann & Model, 1995).

The gene VII and gene IX proteins also play an important role in initiating the assembly process by interacting with the packaging signal of the phage genome (Russel & Model, 1989). This is supported by the fact that the gene VII and gene IX proteins are located at the end of the phage particle that emerges first (Lopez & Webster, 1985). It was also found that deletions in the packaging signal could be compensated for by mutations in the gene VII, gene IX and gene I (Russel & Model, 1989), suggesting a role for these proteins in the initiation of assembly (see Figure 1). Furthermore, absence of gene VII or gene IX protein almost completely abolishes production of phage (Webster & Lopez, 1985).

A complete description of assembly requires detailed knowledge of both cellular localization and ultimate location of the coat proteins in phage. Until now assembly models were based only on assumptions regarding the minor coat proteins (Russel, 1991; Webster & Lopez, 1985).

Our aim is to get a more detailed knowledge of gene VII protein and gene IX protein in the membrane bound state, prior to assembly. In our research, firstly a general characterization of the chemically synthesized proteins has been performed, and secondly we have started to overexpress the proteins in *E. coli* to be able to generate mutant proteins for site specific labeling and characterization.

Prediction trans membrane domain

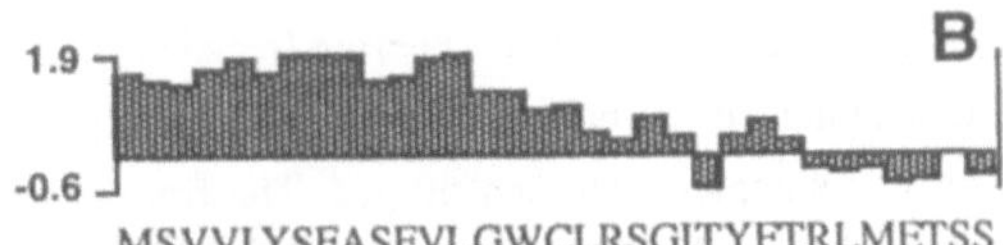

Figure 2: The primary sequence and hydrophobicity plot according to Kyte & Doolittle (Kyte & Doolittle, 1982) of gene VII protein (A) and gene IX protein (B). A window of 9 was used for the hydrophobicity plot.

As can be seen from Figure 2, each protein has a hydrophobic region long enough to form a transmembrane α-helix. Within gene IX protein this region is located near the N-terminus. In contrast to other membrane spanning domains, gene IX protein lacks charged residues in its N-terminal side that can be used to direct and to stabilize a transmembrane orientation of the protein. However, gene IX protein of the closely related IKe and $I_{2\text{-}2}$ phages does have an aspartyl residue at its N-terminal side, which supports the probability of a transmembrane orientation. Looking at the primary sequence of the gene IX protein, a striking feature is the high number and regular distribution of the aromatic amino acids within this protein. Gene VII protein has its hydrophobic region in the C-terminal part of the protein. As this protein has charges on both sides of the hydrophobic stretch and also some aromatic amino acids at the N-terminal side of the expected helix that could function as a membrane anchor, this protein is very likely to be transmembrane. By looking at both sequences and making a comparison with other filamentous phages, such as IKe, the gene IX protein might function as a reversed signal sequence. The N-terminus is translocated across the membrane instead of the C-terminus. Based on its primary sequence gene VII protein is a likely candidate to insert in the membrane in the same way as Pf3 major coat protein (Kuhn, 1995).

Synthesis and purification of the proteins

The gene VII and gene IX proteins were prepared by solid-phase synthesis by means of Fmoc on a Fmoc-His(Trt)-O-[Wang-resin]. During synthesis the cysteine is protected by a S-tert-butyl. Each synthesis was completed by subjecting the peptide adduct to acidolysis in TFA for two hours to remove the sidechain protecting groups. The resulting suspension was filtered with suction and the filtrate was added dropwise to a ten-fold amount of dry ether, which dissolves triphenylmethanol and precipitates the product. Finally the peptides were lyophilized, HPLC purified and analysed by mass spectroscopy.

Reconstitution of protein into SUV's

To incorporate the protein in lipid bilayers a similar protocol as described by Killian (Killian et al., 1994) was used. The desired amount of protein in TFE was added to an equal volume of buffer (10 mM phosphate, pH 7.5), containing the desired amount of lipids. The mixture was briefly vortexed. Next, buffer was added to yield a 16:1 ratio of buffer to TFE by volume. The samples were mixed by vortexing for 2 s and lyophilized overnight after rapid freezing in liquid nitrogen The samples were rehydrated in H_2O to give the same volume as before freeze drying. To convert the liposomes into SUV's the sample was sonicated. To pellet down titanium particles and any residual multilammelar structures the sonicated samples were centrifuged. The final protein concentration was 20 μM and the lipid concentration 600 μM, to give an L/P ratio of 30. Sucrose gradient centrifugation analysis of lipid protein samples prepared in this way showed a clear band lower then the reference band for empty vesicles indicating a tight lipid-protein association for both proteins.

Circular Dichroism

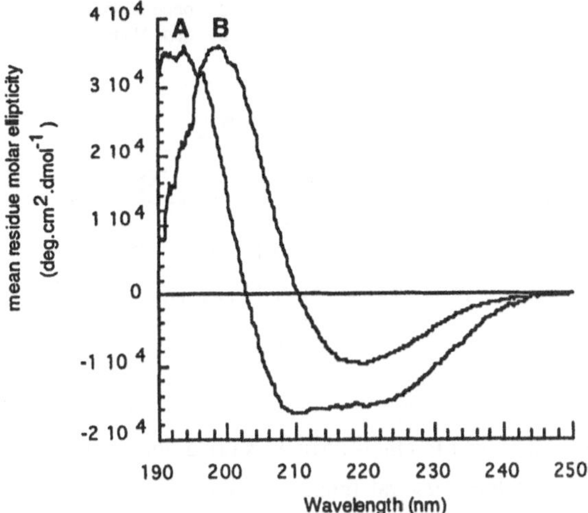

Figure 3: CD spectra of gene IX (A) and gene VII (B) protein reconstituted in phospholipid bilayers composed of DOPC and DOPG 2:1 (mole/mole). The spectra are corrected for the background signal of the phospholipid bilayers. The lipid to protein ratio was 30:1 (mole/mole). CD spectra were recorded on a Jobin-Ivon Dichrograph Mark V at room temperature. For each spectrum four scans were averaged.

Circular dichroism indicates that the gene VII and the gene IX protein both contain a high amount of α-helix when dissolved in TFE or solubilized by SDS (data not shown). Gene IX still contains a high amount of α-helix conformation after reconstitution in lipid bilayers as seen in Figure 3. Gene VII protein shows a strong tendency to adopt a β-sheet conformation, which is a polymeric form of the protein, as seen after analysis on a Tricine SDS Poly Acrylamide gel.

Monolayer insertion

Monolayer pressure was measured by the (paper) Wilhelmy plate method (Demel, 1994) in a thermostatically controlled box, using a Cahn D202 microbalance. The lipid monolayers were spread from a chloroform solution to give an initial surface pressure between 15 and 40 mN/m on a subphase solution of 10 mM Tris hydrochloride, 0.2 mM EDTA, pH 8.0. During the measurements the temperature was kept at 22.5 ± 1 °C and the subphase was continuously stirred with a magnetic bar. A saturating amount of protein was injected into the subphase. As seen in Figure 4 both proteins are surface active and can insert into monolayers composed of phospholipids with neutral or negatively charged headgroups. Gene VII protein has a preference for the charged DOPG lipid system. Insertion occurs up to initial pressure of 37 mN/m in DOPG and only up to 29 mN/m in DOPC. From these results it follows that the interaction of gene IX has a more hydrophobic origin, less dependent on charged lipids, as the maximal initial pressure is 35 mN/m for both lipid systems.

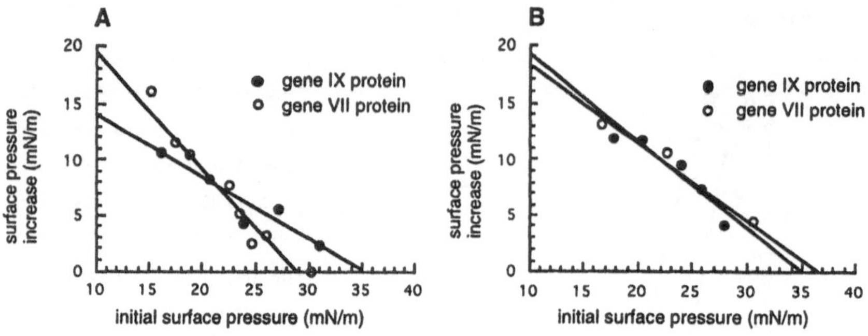

Figure 4: Insertion of gene VII and gene IX protein in DOPC (A) and DOPG (B) monolayers. The final protein concentration was 2.5 μM.

Carboxyfluorescein leakage

An amount of 12.5 μmol of DOPC, dissolved in chloroform, was dried under a stream of nitrogen and subsequent high vacuum for at least several hours. A solution containing 5,6-carboxyfluorescein (121 mM) in 2 ml of 10 mM Tris hydrochloride, 0.2 mM EDTA, pH 8.0 was used to rehydrate the lipids. Small unilamellar vesicles were then prepared by sonication. External carboxyfluorescein was removed by gel filtration. Vesicles were diluted to give a maximal absorption at 493 nm of 0.1. To monitor the effect of addition of protein the fluorescence of carboxyfluorescein was followed at an excitation wavelength of 493 nm and an emission wavelength of 520 nm.

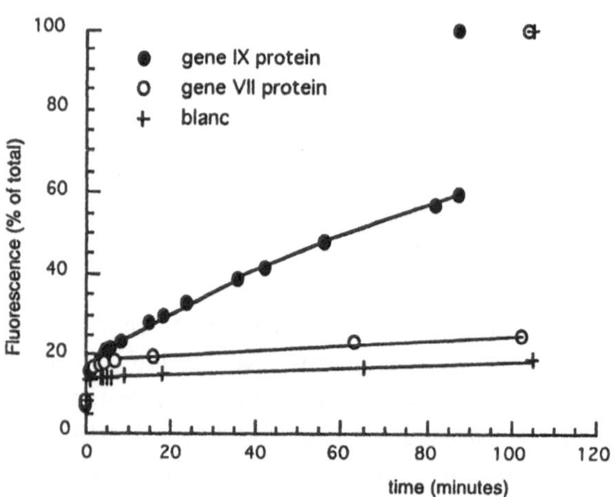

Figure 5: Carboxyfluorescein leakage from DOPC vesicles performed at a lipid to protein ratio of 260 (mole/mole). Maximal (100%) fluorescence was obtained by addition of Triton X-100. The protein concentration during the experiment was 0.35 μM.

As seen in Figure 5 gene IX protein induces a release of carboxyfluorescein, entrapped within the vesicle interior, indicating a destabilization of the vesicle membrane or defects in the phospholipid packing. Gene VII protein hardly shows any leakage. This could be a result of a superficial, non penetrating or destabilization interaction with the bilayer. The gene VII protein might only insert in a loop structure, thereby increasing the surface pressure in insertion experiments. To obtain complete translocation an additional membrane potential might be necessary.

Overexpression in *E. coli*

Because of the low amounts of gene VII and gene IX protein directly available from the phage, overexpression in *E. coli* was started. By using PCR a NdeI and a BamHI sites were introduced before and after the gene coding for gene VII and gene IX protein respectively. M13mp18 rf was used as a template for PCR. After insertion in the plasmid pT7-7 expression was tested after induction with IPTG in *E. coli* BL21(DE3).

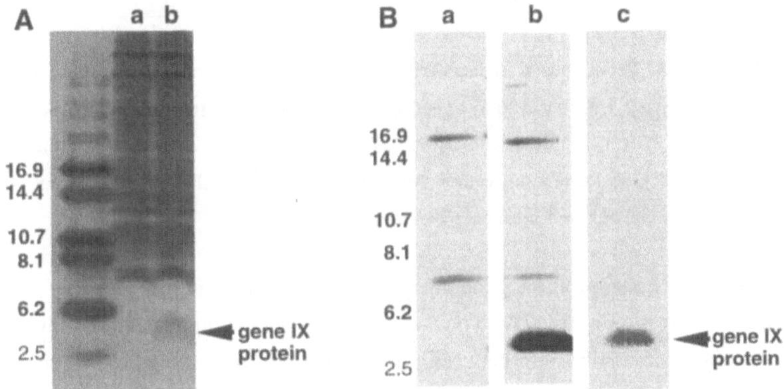

Figure 6: A) Serva Blue G stained Tricine Poly Acrylamide gel with the total *E. coli* protein extract before (a) and 1 hour after induction (b) of the plasmid expressing the gene IX protein. B) Antibody detection of gene IX protein after western blotting before (a) and 1 hour after induction (b). The purified protein, after membrane filtration is visible in lane c.

As seen in Figure 6A a clear extra band is visible after induction corresponding to position found for the synthetically prepared gene IX protein. The level of expression remained the same from 30 minutes up to several hours after induction, apparently due to cessation of the cell growth as no increase or decrease in OD600 was observed. Detection with antibodies, raised against the synthetic protein, proved this band to be gene IX protein. For isolation, cells are harvested 1 hour after induction. After lysis by means of sonication the membrane fraction is isolated, extracted with TFE, and further purified by Source-15 (Pharmacia-LKB) reverse phase chromatography using a water/iso-propanol gradient. The fractions containing the protein were further purified and concentrated by ultra filtration over a membrane with a 10 KDa MW cutoff and a membrane with a 3 KDa MW cutoff respectively. As seen on gel (Figure 7) and western blot (Figure 6B) the protein is pure for more then 95%. The overexpression for gene VII protein is carried out as described for gene IX protein, but still needs more attention.

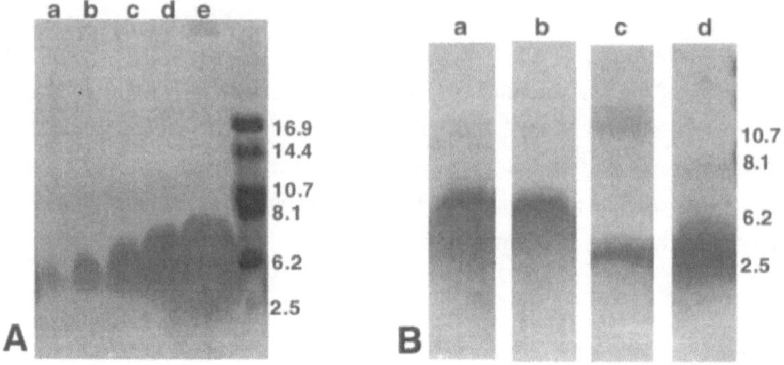

Figure 7: A) Serva Blue G stained Tricine Poly Acrylamide gel with from lane a to e respectively 1, 2.5, 5, 10 and 20 μg of purified gene IX protein. B) 10 μg of gene IX protein in 300 mM SDS (a) incubated with 10 mM DTT (b) and 5% (w/v) H_2O_2 (c). Chemically synthesized gene IX protein, with a S-tert-butyl blocked cysteine, is show in lane (d). Molecular weight markers are indicated by their sizes in KDa.

Increasing the concentration gene IX protein shows that it has a strong tendency to aggregate, as seen from Figure 7A. The smear band that is appearing at higher protein concentration suggest an equilibrium between a monomer and dimer situation. At higher protein concentrations even larger aggregate bands appear (data not shown). Although gene IX protein has a strong tendency to aggregate, up to now it appears to be a reversible aggregation without a concomitant conformational change to β-sheet as observed for gene VII or gene VIII protein (Spruijt et al., 1989). Dimerization is not a result of formation of disulphide bridges, as the smear band is also present in a reducing environment (lane b in Figure 7B). Further, substituting the cysteine for serine by site directed mutagenesis (result not shown) neither prevented aggregation to occur, also indicating that the cysteine plays a minor role in protein aggregation. From Figure 7B it can also be concluded that the aggregation is influenced by a modification of the nature of the side chain size of the cysteine. Introduction of a negatively charged sulfinic acid moiety at the cysteine side chain upon oxidation of the thiol group by H_2O_2 leads to a dramatic increase of monomers present (lane c), which can be explained by reduced protein-protein interaction due to an electrostatic repulsion. An increase in the size of the side chain by blocking the cysteine with a bulky S-tert-butyl group (lane d), results in a much sharper monomeric gene IX protein band with almost no aggregation. This is in agreement with the idea that helix-helix aggregation is regulated by side chain size distribution as suggested for M13 gene VIII protein (Williams et al., 1995) and Glycophorin A (Mingarro et al., 1996). So this knobs into holes model might also fit to the aggregational properties of gene IX protein.

Now we are able to overexpress and purify the gene IX protein form *E. coli*, we will start a biochemical and biophysical characterization of this membrane bound protein. We have already expression of several site specifically cysteine mutants. These mutants will be used for fluorescent and ESR studies, similar as carried out for M13 gene VIII protein (Spruijt et al., 1996; Stopar et al., 1996). This will provide detailed information about the state of the protein.

References

Brissette, J. L. & Russel, M. (1990) Secretion and membrane integration of a filamentous phage-encoded morphogenetic protein, J. Mol. Biol 211, 565-80.

Demel, R. A. (1994) Monomolecular layers in the study of biomembranes in Subcellular Biochemistry.Vol. 23, edited by H.J Hilderson & G.B. Ralston, Plenum Press NY, 83-120.

Endemann, H. & Model, P. (1995) Location of filamentous phage minor coat proteins in phage and in infected cells. J. Mol. Biol. 250, 496-506.

Glaser-Wuttke, G., Keppner, J. & Rasched, I. (1989) Pore-forming properties of the adsorption protein of filamentous phage fd. Biochim. Biophys. Acta 985, 239-47.

Glucksman, M. J., Bhattacharjee, S. & Makowski, L. (1992) 3-Dimensional Structure of a Cloning Vector - X-Ray Diffraction Studies of Filamentous Bacteriophage-M13 at 7-Angstrom Resolution. Journal of Molecular Biology 226, 455-470.

Grant, R. A., Lin, T.-C., Webster, R. E. & Konigsberg, W. (1981) Structure of filamentous bacteriophage: isolation, characterization, and localization of the minor coat proteins and orientation of the DNA. Prog. Clin. Biol. Res. 64, 413-28.

Guy-Caffey, J. K., Rapoza, M. P., Jolley, K. A. & Webster, R. E. (1992) Membrane Localization and Topology of a Viral Assembly Protein. Journal of Bacteriology 174, 2460-2465.

Henry, G. D. & Sykes, B. D. (1992) Assignment of Amide H-1 and N-15 NMR Resonances in Detergent-Solubilized M13 Coat Protein - A Model for the Coat Protein Dimer. Biochemistry 31, 5284-5297.

Killian, J. A., Trouard, T. P., Greathouse, D. V., Chupin, V. & Lindblom, G. (1994) A general method for the preparation of mixed micelles of hydrophobic peptides and sodium dodecyl sulphate. FEBS Letters 348, 161-165.

Kuhn, A. (1995) Major coat proteins of bacteriophage Pf3 and M13 as model systems for Sec-independent protein transport. FEMS Microbiology Reviews 17, 185-190.

Kyte, J. & Doolittle, R. F. (1982) A simple method for displaying the hydropathic character of a protein. J. Mol. Biol 157, 105-32.

Lopez, J. & Webster, R. E. (1985) Assembly Site of Bacteriophage f1 corresponds to Adhesion Zones between the Inner and Outer membrane of the Host Cell. Journal of Bacteriology 163, 1270-74.

Marvin, D. A., Hale, R. D., Nave, C. & Citterich, M. H. (1994) Molecular Models and Structural Comparisons of Native and Mutant Class-I Filamentous Bacteriophages Ff (fd, f1, M13), If1 and IKe. Journal of Molecular Biology 235, 260-286.

Mingarro, I., Whitley, P., Lemmon, M. A. & Von Heijne, G. (1996) Ala-insertion scanning mutagenesis of the glycophorin A transmembrane helix: A rapid way to map helix-helix interactions in integral membrane proteins. Protein Sci. 5, 1339-1341.

Model, P. & Russel, M. (1988) Filamentous Bacteriophage in Series The Viruses; Fraenkel-Conrat, H.//Wagner, R.R. (Eds.) The Bacteriophages Vol. 2; Calender, R. (Ed), New York: Plenum Press.

Opella, S. J., Cross, T. A., DiVerdi, J. A. & Sturm, C. F. (1980) Nuclear magnetic resonance of the filamentous bacteriophage fd. Biophys. J. 32, 531-48.

Rasched, I. & Oberer, E. (1986) Ff coliphages: structural and functional relationships. Microbiol. Rev. 50, 401-27.

Russel, M. (1991) Filamentous phage assembly. Mol. Microbiol 5, 1607-13.

Russel, M. & Model, P. (1989) Genetic analysis of the filamentous bacteriophage packaging signal and of the proteins that interact with it. J. Virol. 63, 3284-95.

Simons, G. F. M., Konings, R. N. H. & Schoenmakers, J. G. G. (1981) Genes VI, VII, and IX of phage M13 code for minor capsid proteins of the virion. Proc. Natl. Acad. Sci. U. S. A. 78, 4194-8.

Spruijt, R. B., Wolfs, C. J. A. M. & Hemminga, M. A. (1989) Aggregation-related conformational change of the membrane-associated coat protein of bacteriophage M13. Biochemistry 28, 9158-65.

Spruijt, R. B., Wolfs, C. J. A. M., Verver, J. W. G. & Hemminga, M. A. (1996) Accessibility and environment probing using cysteine residues introduced along the putative transmembrane domain of the major coat protein of bacteriophage M13. Biochemistry 35, 10383-10391.

Stopar, D., Spruijt, R. B., Wolfs, C. J. A. M. & Hemminga, M. A. (1996) Local dynamics of the M13 major coat protein in different membrane-mimicking systems. Biochemistry 35, 15467-15473.

Van de Ven, F. J. M., Van Os, J. W. M., Aelen, J. M. A., Wymenga, S. S., Remerowski, M. L., Konings, R. N. H. & Hilbers, C. W. (1993) Assignment of H-1, N-15, and Backbone C-13 Resonances in Detergent-Solubilized M13 Coat Protein via Multinuclear Multidimensional NMR - A Model for the Coat Protein Monomer. Biochemistry 32, 8322-8328.

Webster, R. E. & Lopez, J. (1985) Structure and assembly of the class I filamentous bacteriophage in Virus structure and assembly. Casjens (Ed), pp235-267, Boston MA: Jones and Bartelett Publishers Inc.

Williams, K.A., Glibowicka, M., Li, Z.M., Li, H., Khan, A.R., Chen, Y.M.Y., Wang, J., Marvin, D.A., Deber, C.M. (1995) Packing of coat protein amphipathic and transmembrane helices in filamentous bacteriophage M13: Role of small residues in protein oligomerization. J. Mol. Biol. 252, 6-14.

Lipoprotein Sorting in *Escherichia Coli*

Anke Seydel and Anthony P. Pugsley
Unité de Génétique Moléculaire (CNRS UMR321),
Institut Pasteur,
25, rue du Dr. Roux,
75724 Paris Cedex 15,
France
Telephone: 33-145688494; Fax: 33-145688960; E-mail: aseydel@pasteur.fr

What are bacterial lipoproteins?

A special group of bacterial exported proteins, the lipoproteins, have a unique, fatty acylated amino-terminal structure. Like all other envelope proteins, they are synthesised with a signal peptide and are exported across the cytoplasmic membrane via the Sec-machinery, the first step of the general secretory pathway (GSP). Although the signal peptides of lipoproteins are similar to those of the precursors of other envelope proteins, they are generally somewhat shorter and more hydrophobic (von Heijne, 1989), and the cleavage site (↓) consensus (Leu-Leu-Ala-Gly↓Cys) is different from that recognised by the classical signal (leader) peptidase (Ala-Xxx-Ala↓). In Gram-negative bacteria, most lipoproteins are anchored by their fatty acids to the periplasmic face of the outer membrane but some remain anchored to the periplasmic face of the cytoplasmic membrane, while at least one other lipoprotein, pullulanase (PulA) is sorted from this site to the outer face of the outer membrane (Pugsley *et al.*, 1990a; Pugsley *et al.*, 1991). Over 130 bacterial lipoproteins have been identified (Braun & Wu, 1994), including at least 10 that have been characterised in *E. coli* K-12 (Braun & Wu, 1994; Bishop *et al.*, 1995; Lange & Hengge-Aronis, 1994; Ma *et al.*, 1993; Snyder *et al.*, 1995). The function of most of these proteins remains unknown. In Gram-positive bacteria, lipoproteins are exclusively embedded in the outer leaflet of the cytoplasmic membrane, where many of them perform the same functions as periplasmic proteins in Gram-negative bacteria. (e.g., substrate binding components of cytoplasmic membrane permeases (Perego *et al.*, 1991).

In *E. coli* K-12, only one of the characterised lipoproteins, NlpA, is located in the cytoplasmic membrane. The only other lipoprotein of Gram-negative bacteria that has been demonstrated to reside in this membrane is pullulanase of *Klebsiella oxytoca*. Pullulanase can be found in either of two locations in *E. coli*, depending on whether or not expression of its structural gene is accompanied by expression of fourteen other genes specifically required to transport the enzyme from the cytoplasmic membrane to the cell surface (Pugsley *et al.*, 1990b).

NATO ASI Series, Vol. H 106
Lipid and Protein Traffic
Pathways and Molecular Mechanisms
Edited by Jos A. F. Op den Kamp

Processing and modification

Lipoproteins are the most abundant post-translationally modified bacterial secretory proteins. Their characteristic feature is the modified cysteine residue at their amino-terminus. All bacterial lipoproteins are probably modified in exactly the same way (Fig. 1), although only one lipoprotein (the peptidoglycan-associated major lipoprotein Lpp) has been studied in detail (Hantke & Braun, 1973). The first step in the processing pathway is the transfer of a diacylglyceryl moiety to the sulfhydryl group of the cysteine residue in the unmodified prolipoprotein by prolipoprotein diacylglyceryl transferase (Sankaran & Wu, 1994). Diacylglyceryl modification is a prerequisite for cleavage of the signal peptide by the lipoprotein-specific signal peptidase (LspA). The antibiotic globomycin inhibits LspA, thereby causing the accumulation of the precursor forms of all the lipoproteins in the cytoplasmic membrane. Globomycin is lethal to *E. coli* (Inukai *et al.*, 1978) because the Lpp precursor that remains anchored in the cytoplasmic membrane still binds to the peptidoglycan, thereby anchoring this structure to the cytoplasmic- rather than to the outer membrane (Inukai *et al.*, 1979).

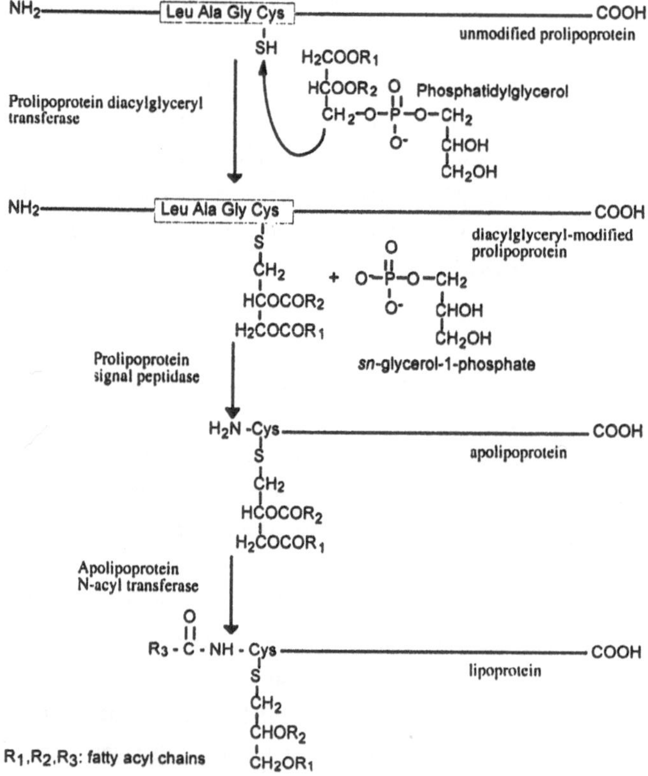

Fig. 1: Post-translational steps in the assembly of bacterial lipoproteins, modified from Sankaran & Wu (1994)

The new amino-terminus of the cleaved product (apolipoprotein), is further N-acylated by apolipoprotein transacylase (Sankaran & Wu, 1994). All of these reactions are carried out by integral cytoplasmic membrane proteins (not all of which have been characterised), and the mature lipoprotein is at least temporarily associated with this membrane due to the high hydrophobicity of the lipid moiety.

How do lipoproteins reach their final destination?

Current interest in lipoproteins centres around how they reach their final destination. Lipoprotein transport from the cytoplasmic membrane to the outer membrane poses unique mechanical problems: how do the fatty acids avoid contact with the aqueous periplasm and how do lipoproteins pass through the peptidoglycan layer? It has been proposed that lipoproteins could transit from the cytoplasmic- to the outer membrane via contact sites or in micelles (Pugsley, 1993). The possible existence of membrane contact sites permitting intermembrane fatty acid transfer is still poorly documented and the pores in the peptidoglycan layer are probably not large enough to permit direct membrane contact or the free diffusion of micelles (Demchick & Koch, 1996; Dijkstra & Keck, 1996). Indeed, the presence of peptidoglycan pores large enough to accommodate large multiprotein transfer machines is often taken for granted but is not proven. Another possibility is that the fatty acids are shielded from the aqueous periplasm by lipoprotein carrier proteins. Whatever the mechanism, it is obvious that a specialised machinery and specific recognition of subclasses of lipoproteins destined for different compartments must be involved, implying the existence of one or more sorting signals in the lipoproteins themselves.

What is the nature of the lipoprotein sorting signal(s)?

If one accepts the unproven idea that all lipoproteins are modified in exactly the same way, it is obvious that the polypeptide sequence of the lipoproteins must contain the sorting signal(s). To determine whether the signal peptide is important, Inouye and his colleagues exchanged the signal peptides of the cytoplasmic membrane protein NlpA and the outer membrane protein Lpp. They found that the localisation of these two proteins was unaffected by the signal peptide exchange. They then constructed a number of other hybrid lipoproteins by fusing different lengths of NlpA or Lpp to ß-lactamase. They found that the first few amino acids of the lipoprotein sequence determined whether ß-lactamase was anchored to the cytoplasmic- or outer membrane (Yamaguchi *et al.*, 1988).

The N-terminal regions of mature lipoproteins are quite dissimilar, but the two lipoproteins known to associate with the cytoplasmic membrane, NlpA and PulA, have an aspartate residue at position +2, whereas all other lipoproteins have a different amino acid at this position (often, though not exclusively a serine). This observation led Inouye and his colleagues to the idea that the aspartate could represent (part of) the sorting signal. This was proven to be the case by constructing and analysing Lpp with an aspartate at position +2 [Lpp(D+2)], and NlpA with a serine +2. Lpp(D+2) was found in the cytoplasmic while NlpA(S+2) was found in the outer membrane. Thus, a single amino acid in a membrane protein determines its final location in the *E. coli* envelope. On the basis of these results, they suggested that D+2 constituted a signal for retention in the cytoplasmic membrane, all other lipoproteins being released from this membrane for insertion into the outer membrane (Yamaguchi *et al.*, 1988).

The role of D+2 in PulA was also examined. Replacement of D+2 by other amino acids, including the structurally-similar N+2 and the negatively-charged E+2, caused PulA to associate with the inner face of the outer membrane. These substitutions reduced PulA secretion in a strain carrying the cognate secretion factors, presumably because the incorrectly localised secretion intermediates could not be efficiently recognised by its secretion machinery (Pugsley & Kornacker, 1991). The sequence immediately downstream from position +2 did not affect either of these phenomena. However, the exact localisation of wild-type (D+2) PulA was unclear: some data suggested that it could be located in membrane fractions distinct from the cytoplasmic- and outer membranes (Poquet *et al.*, 1993).

Interestingly, cytoplasmic membrane anchored lipoproteins of Gram positive bacteria do not have an aspartate at the second position of the mature protein, and at least one of them, the BlaZ ß-lactamase, is sorted to the outer membrane when produced in *E. coli* (Bouvier *et al.*, 1991; Hayashi *et al.*, 1984). This implies that if a specific sorting machinery is necessary for transport of S+2[1] lipoproteins in Gram-negative bacteria, then it should be completely absent from Gram-positive bacteria since it would cause all lipoproteins to be released by these cells.

A periplasmic lipoprotein carrier protein

These data imply the existence of at least one envelope component that recognises the sorting signal. The next major step forward was made by Tokuda and colleagues who, in 1995, reported the discovery of a periplasmic carrier protein that is required for the release of Lpp from the cytoplasmic membrane (Matsuyama *et al.*, 1995). This protein, called P20, forms a soluble complex with Lpp, and is necessary for the assembly of Lpp into the outer membrane (Matsuyama *et al.*, 1995). This complex was also formed between P20 and the Lpp(S+2)-ß-lactamase hybrid but not with Lpp(D+2)-ß-lactamase hybrid. The structural gene coding for P20 (*lplA*) has been identified but has not been subjected to mutation analysis. However, in view of

---

[1] For simplicity, all lipoproteins that are sorted to the outer membrane are collectively referred to as S+2, even if the amino acid in position +2 is not a serine.

the lethality associated with enforced retention of Lpp in the cytoplasmic membrane (see above), loss of P20 function should also be lethal.

On the basis of these results, Tokuda (Matsuyama *et al.*, 1995) proposed a new model in which P20 binds to all lipoproteins anchored in the periplasmic face of the cytoplasmic membrane except those with a D+2, and then transports them through the periplasm to the outer membrane. In this model, D+2 prevents the association between P20 and those lipoproteins destined to remain in the cytoplasmic membrane. It is not clear, however, whether p20 recognises the sorting signal directly or indirectly; it is also conceivable that another protein recognises the cytoplasmic membrane sorting signal (D+2) directly and prevents the formation of a complex with P20.

Lipoprotein sorting: a working model (Fig. 2)

Lpp and other outer membrane lipoproteins are found exclusively in the outer membrane. Furthermore, even in the *in vitro* assay developed by Tokuda, the incorporation of Lpp into the outer membrane was almost quantitative, indicating that Lpp localisation is unidirectional (Matsuyama *et al.*, 1995). Since the lipid composition of the periplasmic faces of the cytoplasmic- and outer membranes are very similar (Lugtenberg & Peters, 1976), the unidirectional transport of outer membrane lipoproteins must be determined by factors other than p20. Thus, in order to establish insertion of lipoproteins into only the outer membrane, P20 and/or the transported lipoprotein must be recognised by a specific receptor protein in this membrane. Lpp was incorporated into outer membranes devoid of Lpp, indicating that the pre-existing Lpp is not the receptor (Matsuyama *et al.*, 1995). The identification and characterisation of this receptor is one of our major objectives. Although it is assumed that P20 interacts with all lipoproteins that are sorted to the outer membrane, this has not been established. If this is not the case, then there might be other periplasmic carrier proteins together with their receptors.

Protein P20 apparently recognises cytoplasmic membrane-associated S+2 lipoproteins, but is this sufficient to extract them from the membrane? Lipoproteins are firmly anchored in membranes and it is therefore quite conceivable that energy would be required to detach them. If this is the case, then additional, cytoplasmic membrane proteins would be required to couple energy (presumably the proton motive force or nucleotide triphosphates) to this process. The specific localisation of these energy-coupling proteins in the cytoplasmic membrane would be another factor that determines undirectionality of lipoprotein sorting.

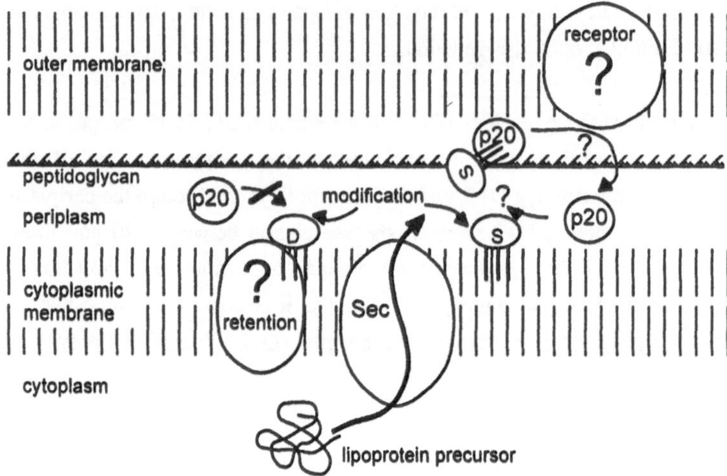

Fig. 2: Model for recognition and sorting of lipoproteins based on the observations of Matsuyama *et al.* (1995) with the outer membrane lipoprotein Lpp.

Another factor that has to be considered is whether or not retention of D+2 lipoproteins in the cytoplasmic membrane is a default pathway (i.e., they do not interact with P20 or with the hypothetical energy-coupling factors). The alternative would be for one or more specific cytoplasmic membrane proteins to remain bound to D+2 lipoproteins to retain them in the cytoplasmic membrane and to prevent their interaction with P20. In the absence of these hypothetical binding proteins, D+2 lipoproteins might interact with P20 and be sorted to the outer membrane.

New experimental systems for studying lipoprotein sorting

A more detailed characterisation of the signal(s) and machinery involved in the sorting of the lipoproteins would represent a considerable step towards understanding how this system works. We plan to use both biochemical and genetic methods for this purpose. First, it should be possible to select mutations that cause lipoproteins to be routed to the wrong membrane, where they may be inactive or acquire new functions. However, the functions of very few lipoproteins are known and none lend themselves to the design of genetic strategies for the selection of such mutations. We tried to overcome this technical difficulty by constructing a new lipoprotein in which the normally periplasmic maltose binding protein, MBP or MalE, was fused to the signal peptide and the first four amino acids of pullulanase. MalE is the periplasmic maltose-binding component of the *E. coli* maltose permease. Its function is to capture maltose that diffuses across the outer membrane from the milieu and to present it to the permease located in the cytoplasmic membrane. In the absence of MalE, the bacteria are unable to use maltose as a carbon source. However, MalE does not have to be free in the periplasm to function: Fikes *et al.* showed that MalE which could not be processed by signal (leader) peptidase (due to two mutations near the cleavage site) remained anchored to the cytoplasmic membrane yet still functioned normally in maltose transport (Fikes & Bassford, 1987). Fractionation of membranes from cells producing the fatty acylated MalE indicated that it too was located in the cytoplasmic membrane. This was indeed what we predicted from the fact that the second amino acid of this hybrid protein is D+2 derived from PulA. Furthermore, the hybrid protein could substitute for periplasmic MalE in the transport of maltose, allowing growth on minimal medium containing maltose as the sole carbon source. Thus, the MalE segment of the hybrid faces the periplasm, and the hybrid functions in the same way as the uncleaved preMalE described by Fikes & Bassford (1987). However, when D+2 was replaced by S+2, the protein was sorted to the outer membrane, as expected, and was severely handicapped in its ability to promote maltose utilisation. We are currently in the process of characterising mutations that restore this function.

What kinds of mutations might one expect to cause defective sorting of the MalE S+2 lipoprotein? One possibility is that increased non-specific proteolysis could separate a soluble, active MalE fragment from its outer membrane anchor. The fragment released into the periplasm would function just as a normal MalE polypeptide to restore maltose uptake. Mutations that affect the processing of the lipoprotein at any stage of the pathway (Fig. 1), would trap the precursor in the cytoplasmic membrane in the same orientation as the non-cleavable MalE variant constructed by Fikes & Bassford (1987). Mutations that convert S+2 codon into a D+2 codon could also arise, but such mutations should be rare because two bases must be changed to achieve this codon conversion. However, other mutations in and around codon +2 might define a hierarchy of amino acid sequences that are more- or less-efficiently recognised by different components of the sorting machinery. It is important to note that a systematic analysis of this region has not been performed.

The most interesting mutations would be those affecting the sorting machinery. *E. coli* might be unable to tolerate the complete misrouting of all of its outer membrane lipoproteins but mutations that partially reduce the function of any of the components of the outer membrane sorting pathway machinery, leading to a redistribution of outer membrane lipoproteins, should not be lethal and would be specifically selected in our procedure. A mutation that broadens the spectrum of lipoproteins recognised by the putative cytoplasmic membrane lipoprotein anchor protein to include S+2 lipoproteins would have the same effect. In order to obtain this kind of mutation, it might be advisable to use a fatty acylated MalE in which amino acid +2 is structurally similar to aspartate (e.g., glutamate or asparagine). This would minimise the structural changes required to reduce specificity, allowing redistribution of MalE but not that of other outer membrane lipoproteins (which generally have S+2).

A second series of novel lipoproteins was constructed for use in an alternative, biochemical approach to the study of lipoprotein sorting. This time, the reporter protein used was an extracellular cellulase, CelZ, from the Gram-negative bacterium *Erwinia chrysanthemi*. As with the MalE hybrids, the CelZ lipoprotein with S+2 was located in the outer membrane while the D+2 variant was located in the cytoplasmic membrane.

The fractionation data obtained with the CelZ and MalE hybrid proteins were different from those reported previously in studies with native PulA. These studies showed that PulA S+2 was located in the outer membrane whereas the wild-type (D+2) form of the protein appeared to be associated with membrane fragments whose density was higher than cytoplasmic membrane vesicles and lower than the outer membrane (Bouvier *et al.*, 1991; Ichihara *et al.*, 1981; Poquet *et al.*, 1993). PulA is much larger (120kDa) than any known lipoprotein, including the lipidated forms of MalE and CelZ constructed in this study. We noted previously that PulA tends to be released from the cell envelope as lipoprotein micelles either when secreting cells enter into stationary phase (d'Enfert *et al.*, 1987) or when the cells are disrupted (Poquet *et al.*, 1993). The presence of these protein micelles could create artefacts in standard fractionation procedures in which membrane fragments or vesicles are separated according to their density. We are currently testing this possibility by attempting to purify vesicles and membrane fragments by affinity chromatography. For example, we have shown that membrane fragments harbouring the lipidated forms of CelZ can be purified by chromatography on insoluble cellulose (Avicel; unpublished observation). In the future, the application of such an approach should allow us to distinguish between membrane-bound lipoproteins and those that have been released as lipoprotein micelles. This should enable us to identify more easily the region where lipoproteins are located.

A further application of the same technique might be to identify proteins that interact directly with lipoproteins. For example, by binding fatty acylated CelZ to Avicel in the presence of non-ionic detergents, we might be able to mimic the conditions necessary for its recognition by protein P20, its putative outer membrane receptor or (in the case of the D+2 variant) the hypothetical protein that retains lipoproteins in the cytoplasmic membrane. Clearly, the conditions necessary for this interactions to occur will have to be established empirically. For example, the choice of detergent, the requirement for specific classes of phospholipids and the involvement of protein complexes all have to be studied.

These techniques should provide access to a unique phenomenon, namely the simultaneous transport of a polypeptide chain and a lipid moiety through an aqueous environment with subsequent insertion of the lipids into a membrane different from that in which they originated. A analogous system is involved in the transport of lipids between organellar membranes in eukaryotes (Kearns *et al.*, 1996). Furthermore, the transport of phospholipids, and in particular lipopolysaccharide from the cytoplasmic membrane to the outer membrane, presents very similar mechanistic problems to those encountered in lipoprotein sorting. Thus, one might anticipate that a system homologous to that involving P20 might be involved in more general aspects of outer membrane biogenesis.

REFERENCES

Bishop, R. E., Penfold, S. S., Frost, L. S., Höltje, J.-V. & Weiner, J. H. (1995). Stationary phase expression of a novel *Escherichia coli* outer membrane lipoprotein and its relationship with mammalian apolipoprotein D. *J Biol Chem* **270**, 23097-23103.

Bouvier, J., Pugsley, A. P. & Stragier, P. (1991). A gene for a new lipoprotein in the *dapA-purC* interval of the *Escherichia coli* chromosome. *J Bacteriol* **173**, 5523-5531.

Braun, V. & Wu, H. C. (1994). Lipoprotein, structure, function, biosynthesis as model for protein export. In *Bacterial Cell Wall* (Ghuysen, J.-M. & Hakenbeck, R., eds.), pp. 319-341. Elsevier, Amsterdam, The Netherlands.

d'Enfert, C., Chapon, C. & Pugsley, A. P. (1987). Export and secretion of the lipoprotein pullulanase by *Klebsiella pneumoniae. Mol Microbiol* **1**, 107-116.

Demchick, P. & Koch, A. L. (1996). The permability of the wall fabric of *Escherichia coli* and *Bacillus subtilis. J Bacteriol* **178**, 768-773.

Dijkstra, A. J. & Keck, W. (1996). Peptidoglycan as a barrier to transenvelope transport. *J Bacteriol* **178**, 5555-5562.

Fikes, J. D. & Bassford, P. J. J. (1987). Export of unprocessed precursor maltose-binding protein to the periplasm of *Escherichia coli* cells. *J Bacteriol* **169**, 2352-2359.

Hayashi, S., Chang, S.-Y., Chang, S. & Wu, H. C. (1984). Modification and processing of *Bacillus licheniformis* prepenicillinase in *Escherichia coli*. Fate of mutant penicillinase lacking lipoprotein modification site. *J Biol Chem* **259**, 10448-10454.

Ichihara, S., Hussain, M. & Mizushima, M. (1981). Characterization of new lipoproteins and their precursors of *Escherichia coli. J Biol Chem* **256**, 3125-3129.

Inukai, M., Takeuchi, K., Shimizu, K. & Arai, M. (1978). Mechanism of action of globomycin. *J Antibiot* **31**, 1203-1205.

Inukai, M., Takeuci, M., Shimizu, K. & Arai, M. (1979). Existence of the bound form of prolipoprotein in *Escherichia coli* B cells treated with globomycin. *J Bacteriol* **140**, 1098-1101.

Kearns, B. G., Alb, J. G., Cartee, R. T. & Bankaitis, V. A. (1996). Functional analysis of phosphatidylinositol transfer proteins. In *Molecular dynamics of biomembranes* (Op den Kamp, J. A. F., ed.), pp. 327-338. Springer Verlag, Berlin, Germany.

Lange, R. & Hengge-Aronis, R. (1994). The *nlpD* gene is located in an operon with *rpoS* on the *Escherichia coli* chromosome and encodes a novel lipoprotein with a potential function in cell wall formation. *Mol Microbiol* **13**, 733-743.

Lugtenberg, E. J. J. & Peters, R. (1976). Distribution of lipids in cytoplasmic and outer membranes of *Escherichia coli* K12. *Biochim Biophys Acta* **441**, 38-47.

Ma, D., Cook, D. N., Alberti, M., Pon, N. G., Nikaido, H. & Hearst, J. E. (1993). Molecular cloning and characterization of *acrA* and *acrE* genes of *Escherichia coli*. *J Bacteriol* **175**, 6299-6313.

Matsuyama, S., Tajima, T. & Tokuda, H. (1995). A novel periplasmic carrier protein involved in the sorting and transport of *Escherichia coli* lipoproteins destined for the outer membrane. *EMBO J* **14**, 3365-3372.

Perego, M., Higgins, C. F., Pearce, S. R., Gallagher, M. P. & Hoch, J. A. (1991). The oligopeptide transport system of *Bacillus subtilis* plays a role in the initiation of sporulation. *Mol Microbiol* **5**, 173-185.

Poquet, I., Kornacker, M. G. & Pugsley, A. P. (1993). The role of the lipoprotein sorting signal (aspartate +2) in pullulanase secretion. *Mol Microbiol* **9**, 1061-1069.

Pugsley, A. P. (1993). The complete general secretory pathway in gram-negative bacteria. *Microbiol Rev* **57**, 50-108.

Pugsley, A. P. & Kornacker, M. G. (1991). Secretion of the cell surface lipoprotein pullulanase by *Escherichia coli*. collaboration or competition between the specific secretion pathway and the lipoprotein sorting pathway. *J Biol Chem* **266**, 13640-13645.

Pugsley, A. P., Kornacker, M. G. & Ryter, A. (1990a). Analysis of the subcellular location of pullulanase produced by *Escherichia coli* carrying the *pulA* gene from *Klebsiella pneumoniae* strain UNF5023. *Mol. Microbiol.* **4**, 59-72.

Pugsley, A. P., Poquet, I. & Kornacker, M. G. (1991). Two distinct steps in pullulanase secretion by *Escherichia coli* K12. *Mol. Microbiol.* **5**, 865-873.

Pugsley, A. P., d'Enfert, C., Reyss, I. & Kornacker, M. G. (1990b). Genetics of extracellular protein secretion by Gram-negative bacteria. *Ann Rev Genet* **24**, 67-90.

Sankaran, K. & Wu, H. C. (1994). Lipid modification of bacterial lipoprotein. Transfer of diacylglyceryl moiety from phosphatidylglycerol. *J Biol Chem* **269**, 19701-19706.

Snyder, W. B., Davis, L., Danese, P. N., Cosma, C. L. & Silhavy, T. (1995). Overproduction of NlpE, a new outer membrane lipoprotein, suppresses the toxity of periplasmic LacZ by activation of the Cpx signal transduction pathway. *J Bacteriol* **177**, 4216-4223.

von Heijne, G. (1989). The structure of signal peptides from bacterial lipoproteins. *Prot Eng* **2**, 531-534.

Yamaguchi, K., Yu, F. & Inouye, M. (1988). A single amino acid determinant of the membrane localization of lipoproteins in *E. coli*. *Cell* **53**, 423-432.

In Vitro Studies of the Interactions Between Signal Peptides and Signal Recognition Factors

Ning Zheng, Joanna L. Feltham & Lila M. Gierasch
Department of Chemistry, University of Massachusetts, Amherst, MA 01003, USA

1. Introduction

1.1 Cellular Factors Involved in Signal Sequence Recognition

Protein translocation across biological membranes is a fundamental cellular function of both prokaryotes and eukaryotes *(1-3)*. As the synthesis of almost all proteins is carried out in the cytoplasm, the proper export of a subset of these proteins across membranes requires cellular mechanisms which can specifically recognize the protein substrates and target them to the correct membrane site. The past two decades has witnessed extensive genetic and biochemical studies of two topologically equivalent protein translocation systems, namely, protein export through the *E. coli* cytoplasmic membrane and protein translocation across the endoplasmic reticulum (ER) of eukaryotic cells. These studies have identified a number of proteins involved in the translocation process, and multiple mechanisms for protein targeting and translocation have emerged which are both unique and common for both systems.

Conceptually, accurate and efficient protein translocation requires the following functions from cellular factors facilitating the process (Figure 1): (1) specific recognition of a wide range of signal sequences which lack consensus but which share very general characteristics (a

NATO ASI Series, Vol. H 106
Lipid and Protein Traffic
Pathways and Molecular Mechanisms
Edited by Jos A. F. Op den Kamp

charged N-terminus followed by a hydrophobic core and a more hydrophilic C-terminus); (2) maintenance of the translocation competency of the protein substrates; (3) delivery of the substrates to the correct translocation site; and (4) initiation of the transfer of polypeptide across the membrane. In mammalian cells, protein translocation across the ER membrane is predominantly co-translational *(1)*. The process is facilitated by the signal recognition particle (SRP), which can recognize the signal sequence as it is extruded from the ribosome, pause translation of the nascent chain, and finally deliver the nascent chain/ribosome complex to the ER membrane through its interaction with SRP receptor.

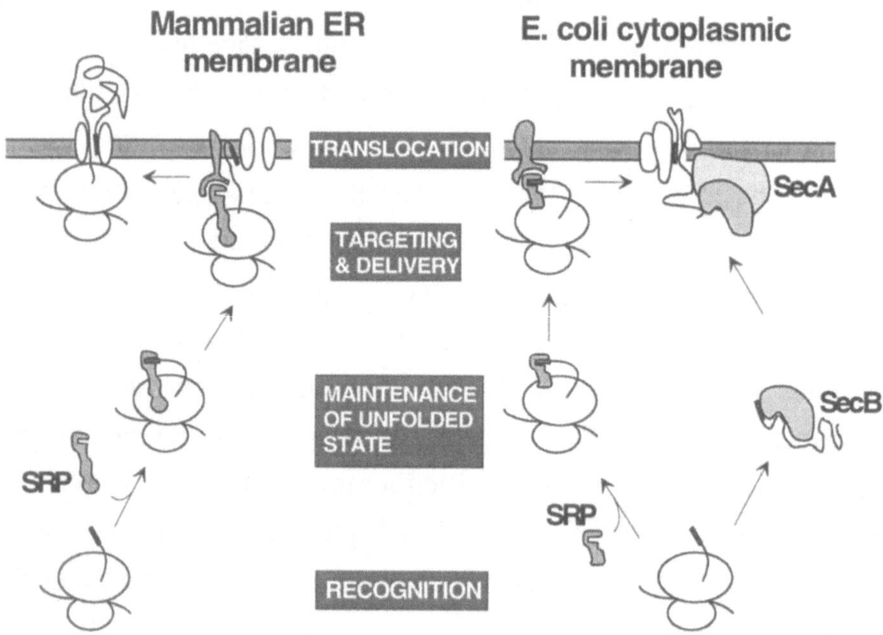

Figure 1. Protein targeting and translocation across membranes in eukaryotes and prokaryotes.

In contrast, extensive genetic screening in *E. coli* has identified a distinct post-translational protein export pathway, in which SecA and SecB proteins play the important roles of recognizing signal sequences, keeping substrates unfolded, and targeting them the membrane *(4, 5)*. Recent studies, however, have revealed that a eukaryotic SRP-like cellular factor also exists in *E. coli* *(6)* and is essential for the targeting and translocation of at least a subset of *E. coli* proteins (*(7)*, see below). The universality of SRP's role has been further confirmed by its identification in yeast *(8)*. In addition to SRP and SecA/B, a number of other cellular proteins, such as Sec62/63/71 in yeast, and Sec61p and TRAM from the mammalian translocon, have been suggested to be capable of recognizing signal sequences differentially and mediating the targeting and translocation process *(9)*.

As the protein translocation pathways have been delineated and more cellular factors involved have been identified, several interesting and important questions need to be answered: (1) How do these cellular factors specifically recognize widely divergent signal sequences? (2) Structurally, do different cellular factors recognize signal sequences through different strategies? (3) How do these cellular factors functionally interact in order to accomplish protein targeting and translocation? *In vitro* studies with isolated cellular factors and signal sequences are essential to address these important questions.

1.2 Use of Signal Peptides as Models for Signal Sequences

The fact that N-terminal signal sequences typically function independently from the rest of the polypeptide has invited researchers to use synthetic peptides as models for signal sequences in *in vitro* experiments. Our lab and others *(10-18)* have used synthetic signal peptides and their analogues to investigate the correlation between biological function and structural behavior, especially upon interacting with lipid bilayers. Our analyses using various biophysical methods including circular dichroism (CD), fluorescence and nuclear magnetic resonance (NMR) spectroscopies have indicated a direct correlation between the functional competency of *E. coli* LamB and OmpA signal sequence analogues *in vivo* and their ability to spontaneously insert deeply into lipid bilayers *in vitro.* This correlation has been extended by showing that functional signal peptides have a stronger tendency to adopt an α-helical conformation in membrane-mimetic environments than those which are nonfunctional, although it is apparent that helical propensity is not the only determinant of export-competency. NMR chemical shift analysis of these peptides in the same membrane-mimetic environments illustrates that the helical content is highest in the hydrophobic core. In order to gain a more comprehensive structural view of signal peptides when bound to lipid vesicles, trNOE NMR techniques were used to determine that a LamB peptide analogue inserted into a lipid bilayer is helical from the beginning of the hydrophobic core through the C-terminus. Monitoring the line-broadening of certain peptide resonances as a result of interaction with spin labeled lipids allowed development of a model in which the helical signal peptide inserts deeply into the acyl layer without completely traversing the membrane.

During these studies, we have not only characterized the structural behavior of purified synthetic signal peptides, but have also developed model signal peptide analogues that are suitable for further studies with other cellular factors. As more and more proteins have been identified from various biological systems which recognize signal sequences, it has become both necessary and feasible to examine the interaction of signal peptides with these different proteinaceous cellular factors.

2. Specific Case Study

2.1 An Overview of *E. coli* Signal Recognition Particle

The initial discovery of the *E. coli* SRP was facilitated by the cloning of the SRP54 protein subunit of mammalian SRP. When the sequence of the mammalian SRP54 protein became available, a previously identified *E. coli* protein of unknown function, p48, was found to be an SRP54 homologue (and renamed Ffh for fifty-four-homologue) *(6)*. Further studies have shown that Ffh forms a complex with a 4.5S RNA molecule which shares similarity with the 7S RNA component of mammalian SRP *(19)*. In addition, another *E. coli* protein called FtsY has been shown to exhibit sequence similarity to the SRα subunit of the mammalian SRP receptor *(6)* . Despite the fact that the *E. coli* SRP homologue was never identified by extensive genetic screening for secretion deficiencies, it has been proposed that the *E. coli* SRP has a role in protein export based on the above observations *(20)*. Although this proposal was once highly controversial *(21)*, new studies have provided convincing evidence indicating the indispensable role in protein secretion played by the *E. coli* SRP.

Using an *E. coli* strain with the *ffh* gene under control of the inducible arabinose promoter, Phillips and Silhavy have demonstrated that Ffh is essential for growth and viability of *E. coli* *(22)*. Various precursor proteins whose export is known to be SecB-independent were inefficiently exported following Ffh depletion. *In vitro* crosslinking studies by Luirink *et al.* have further confirmed that Ffh recognizes signal sequences *(23)*. Luirink *et al.* obtained a crosslinked product of Ffh and the nascent chain of preprolactin in a crude *E. coli* cell extract mixture. The fact that the crosslinking was specific to functional signal sequences and was competed by mammalian SRP54 strongly suggests that Ffh specifically recognizes signal sequences in a manner similar to mammalian SRP. In fact, when the SRP54 subunit of mammalian SRP was replaced by Ffh, the resultant complex was able to carry out both signal recognition and elongation arrest *(24)*. Moreover, Ffh/4.5S RNA and FtsY can substitute for their mammalian counterparts SRP and SRα *in vitro* to cotranslationally target and translocate nascent secretory proteins to microsomal membranes *(25)*.

The relationship between the SecA-dependent post-translational targeting pathway and the *E. coli* SRP-facilitated one is still unclear. By analogy with the mammalian SRP, it is tempting to propose that *E. coli* SRP, working in concert with its receptor on the cytoplasmic membrane, serves in a distinct targeting pathway that mediates the co-translational translocation of certain proteins for which post-translational translocation would be inefficient. Indeed, the newest studies by Ulbrandt *et al.* using a genome-wide screening approach have revealed that the membrane insertion of a subset of polytopic inner membrane proteins was severely blocked upon inhibition of the SRP pathway in *E. coli (7)*. Although several other inner membrane proteins (as well as many preproteins) showed only slight or no SRP dependence under the same experimental conditions, these results clearly indicate that the *E. coli* SRP is involved in protein secretion and its failure to be identified by previous genetic screens is most likely due to the limited number of protein substrates analyzed.

Based on sequence comparison and analysis, mammalian SRP54 and Ffh have been predicted to share a similar domain structure: an N-terminal GTP binding domain (NG domain) followed by a C-terminal methionine-rich domain (M domain). Limited protease digestion of mammalian SRP54 indeed yields two corresponding stable domains. Crystal structures of the NG domains of both Ffh and the related protein FtsY in the nucleotide-free conformation *(26, 27)* show them to be structurally similar to other members of the GTPase superfamily, but with a few novel modifications. First, the N-terminal N domain forms an antiparallel four helix bundle which packs against helices α6 and α7 of the G domain. Secondly, a helix-strand-helix insertion within the canonical GTPase "effector loop" is predicted to be the site of interaction of Ffh with its receptor, FtsY, and *vice versa*. Finally, the nucleotide-free conformation appears to be stabilized by a close network of interactions among those side chains which in other GTP-binding proteins are involved in nucleotide binding.

Crosslinking studies followed by protease digestion have revealed that the bound signal sequence is in proximity to the M domain of SRP54, which was therefore postulated to be the peptide-binding domain *(28-30)*. In order to explain how SRP54 and its homologues specifically bind widely divergent signal sequences, Bernstein *et al.* (1989) proposed that the M domains of mammalian SRP54 and Ffh bind signal sequences with so called "methionine bristles" *(6)*. The model predicts that a hydrophobic groove formed by amphipathic helices provides a peptide binding site where substrates with different sequences are accommodated by flexible methionine sidechains. A recent deletion analysis of the *Bacillus subtilis* Ffh protein *(31)*, however, argues against this proposal. In this study, only deletion of the relatively methionine-poor helix 1 of the M domain prevented binding of a precursor protein, whereas deletion of M domain helices 2 and 3, which are abundant in methionine residues, had little effect on precursor binding. Interestingly, the M domain of SRP54 has been identified as containing the RNA binding site as well *(32)*, and in the Bacillus subtilis Ffh protein a minimal fragment for RNA binding has been identified *(33)* which includes the predicted helices 2 and 3 and the positively charged sequence between them.

2.2 The Role of 4.5S RNA in Stabilizing the M Domain

In order to elucidate the role of the 4.5S RNA in *E. coli* SRP function, we investigated 4.5S RNA interactions with Ffh and an over-expressed His-tagged M domain. We were able to show that the M domain exists in a highly helical molten-globule-like state in the absence of the 4.5S RNA, as evidenced by high proteolytic sensitivity, noncooperative thermal denaturation, and poor chemical shift dispersion by NMR *(34)*. When 4.5S RNA is bound, on the other hand, the M domain exhibits much lower sensitivity to V8 digestion and melts cooperatively, indicating a well-packed structure.

Once crosslinking studies had revealed the signal recognition role played by SRP, it became essential to analyze the functional interaction between isolated signal sequences and purified SRP. This interaction was first demonstrated by Miller *et al.*, taking advantage of the regulation of signal recognition and targeting events by GTP *(35)*. Studies have shown that GTP binding and mutually-stimulated GTP hydrolysis by SRP54 (Ffh in *E. coli*) and the SRP receptor (FtsY in *E. coli*) may regulate many steps in the process of targeting nascent polypeptides *(36)*. Using purifed SRP54/SRP receptor and overexpressed *E. coli* Ffh/FtsY, Miller *et al. (35)* have shown that functional synthetic signal peptides can inhibit both the basal GTPase activity of SRP54 and its stimulated GTP hydrolysis by the SRP receptor.

Following this assay, we have examined the ability of a series of LamB synthetic peptide analogues to inhibit the basal GTPase activity of overexpressed *E. coli* Ffh or the Ffh/4.5S RNA complex. We found that the presence of the 4.5S RNA is crucial to maintain the promiscuous nature of Ffh. In the presence of 4.5S RNA, analogues of the LamB signal sequence with tryptophan substitutions at hydrophobic core positions 13, 14, 16 or 18 inhibit the Ffh GTPase activity with an IC_{50} similar to that of a LamB sequence without tryptophans (1 μM). Without the stabilizing influence of the RNA, however, the peptides with tryptophan at positions 13, 14 and 18 become less efficacious (IC_{50} ~20 μM), while the 16W mutant and the peptide without tryptophan still have a half-maximal effect at 1 μM. These results suggest that the RNA plays a crucial role in structuring the M domain to form a native signal sequence binding site which can accommodate a wide variety of ligands.

2.3 Domain Interactions and Conformational Changes of Ffh upon Signal Peptide Binding

The fact that signal peptides can inhibit the GTPase activity of Ffh invited us to test possible conformational changes of the protein upon peptide binding. One simple way to monitor conformational changes is to examine limited protease digestion of the protein in the absence and presence of peptides. Surprisingly, we found that in the absence of 4.5S RNA, signal peptide causes a drastic destabilization of the entire Ffh protein *(34)*. Whereas V8 protease cleaves the isolated Ffh protein into two somewhat stable domains, in the presence of a LamB signal peptide analogue the isolated protein is digested completely within a very short period. The dramatic effect of signal peptide on the proteolytic lability of Ffh implies a global destabilization of the protein upon peptide binding, which we confirmed by further structure analysis using CD spectroscopy. Perhaps the most surprising finding of all is that a purified isolated NG domain is similarly destabilized upon signal peptide binding, which suggests a more direct role of the NG domain in the signal recognition event. These results would be consistent with a report that mutations within a conserved N domain region of SRP54 result in reduced translation arrest and as such may be deficient in signal sequence binding *(37)*.

When the stabilizing 4.5S RNA is bound to Ffh, however, addition of signal sequence results in a very different conformational change such that the linker region between the M and NG domains becomes protected from proteolysis (Figure 2). Presumably it is this conformational change which results in functional regulation of the GTPase domain upon signal sequence binding.

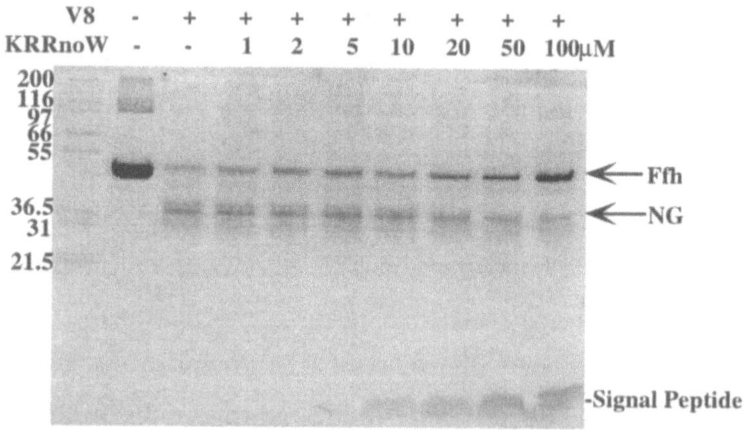

Figure 2. Titration of LamB signal peptide analogue KRRnoW (identical to wild-type LamB except for KRR insertion near N-terminus to enhance solubility characteristics) into V8 protease digestions of Ffh in the presence of 4.5S RNA.

2.4 A Model for Signal Sequence Recognition by *E. coli* SRP

Determination of the detailed structure of Ffh with both RNA and signal peptide bound will provide a picture of the signal recognition event. Yet speculation based on our results can give direction to future studies which will then be complementary to the awaited structure of the complex. Our biochemical and biophysical studies have revealed the following functional and structural relationships among Ffh, 4.5S RNA, and signal peptides: (1) 4.5S RNA provides structural stability to the M domain and helps create a native binding site which can accommodate many different signal sequences; (2) signal sequence binding to the isolated Ffh protein or even to the NG domain results in dramatic destabilization of these polypeptides, which suggests a role for the NG domain in formation of the peptide binding site; and (3) signal peptide binding to the Ffh/4.5S RNA complex causes a conformational change in which the flexible linker between the NG and M domains becomes protected from proteolysis. Based on our results, we speculate that, in contrast to what has been previously proposed, bound signal sequence might directly contact the NG domain of Ffh. This idea is supported by several lines of evidence. First, as discussed above, signal peptide destabilizes the isolated NG domain of Ffh *(34)*. Second, treatment of the NG domain of SRP54 with NEM blocks crosslinking of a signal sequence *(30)*, and finally, mutation of conserved residues within the N domain of SRP54 hinders translation arrest, presumably *via* reduced signal sequence binding*(37)*. However, since multiple crosslinking studies *(28-30)* as well as deletion analyses *(31)* point to the M domain being critically involved in signal sequence recognition, we propose that signal sequence in fact binds at an interface between NG and M. The consequence of signal sequence binding at this interface or cleft is an inter-domain conformational change which results in protection of the linker region from V8 digestion, as well as a reduction in NG domain GTPase activity. Indeed, our studies have suggested that the two domains of Ffh must work in concert for functional signal sequence recognition, possibly through simultaneous peptide binding.

Acknowledgment

This work was supported by a grant from the National Institutes of Health (GM34962).

References

1. Walter, P. and Johnson, A. E. Signal sequence recognition and protein targeting to the endoplasmic reticulum membrane. *Ann. Rev. of Cell Biol.* **10**, 87-119 (1994).
2. Landry, S. J. and Gierasch, L. M. Recognition of Nascent Polypeptides for Targeting and Folding. *Trends Biochem. Sci.* **16**, 159-163 (1991).
3. Althoff, S., Selinger, D. and Wise, J. A. Molecular evolution of SRP cycle components: functional implications. *Nucleic Acids Res.* **22**, 1933-1947 (1994).
4. den Blaauwen, T. and Driessen, A. J. M. Sec-Dependent Preprotein Translocation in Bacteria. *Arch. Microbiol.* **165**, 1-8 (1996).
5. Driessen, A. J. M., Dewit, J. G., Kuiper, W., Vanderwolk, J. P. W., et al. SecA, a novel ATPase that converts chemical energy into a mechanical force to drive precursor protein translocation. *Biochem. Soc. Trans.* **23**, 981-985 (1995).
6. Bernstein, H. D., Poritz, M. A., Strub, K., Hobben, P. J., et al. Model for signal sequence recognition from amino-acid sequence of 54K subunit of signal recognition particle. *Nature* **340**, 482-486 (1989).
7. Ulbrandt, N. D., Newitt, J. A. and Bernstein, H. D. The *E. coli* Signal Recognition Particle is Required for the Insertion of a Subset of Inner Membrane Proteins. *Cell* **88**, 187-196 (1997).
8. Hann, B. C., Poritz, M. A. and Walter, P. *Saccharomyces cerevisiae* and *Schizosaccharomyces pombe* contain a homologue to the 54-kD subunit of the signal recognition particle that in *S. cerevisiae* is essential for growth. *J. Cell Biol.* **109**, 3223-3230 (1989).
9. Zheng, N. and Gierasch, L. M. Signal Sequences: The Same Yet Different. *Cell* **86**, 849-852 (1996).
10. Batenburg, A. M., Brasseur, R., Ruysschaert, J. M., van Scharrenburg, G. J., et al. Characterization of the Interfacial Behavior and Structure of the Signal Sequence of *Escherichia coli* Outer Membrane Pore Protein PhoE. *J. Biol. Chem.* **263**, 4202-4207 (1988).

11. Batenburg, A. M., Demel, R. A., Verkleij, A. J. and de Kruijff, B. Penetration of the Signal Sequence of *Escherichia coli* PhoE Protein into Phospholipid Model Membranes Leads to Lipid-specific Changes in Signal Peptide Structure and Alterations of Lipid Organization. *Biochemistry* **27**, 5678-5685 (1988).
12. Briggs, M. S., Gierasch, L. M., Zlotnick, A., Lear, J. D., et al. In vivo function and membrane binding properties are correlated for Escherichia coli lamB signal. *Science* **228**, 1096-1099 (1985).
13. McKnight, C. J., Briggs, M. S. and Gierasch, L. M. Functional and Nonfunctional LamB Signal Sequences can be Distinguished by their Biophysical Properties. *J. Biol. Chem.* **264**, 17293-17297 (1989).
14. Hoyt, D. W. and Gierasch, L. M. Hydrophobic Content and Lipid Interactions of Wild-Type and Mutant OmpA Signal Peptides Correlate with Their *In vivo* Function. *Biochemistry* **30**, 10155-10163 (1991).
15. Jones, J. D. and Gierasch, L. M. Effect of Charged Residue Substitutions on the Membrane-Interactive Properties of Signal Sequences of the *Escherichia coli* LamB Protein. *Biophys. J.* **67**, 1534-1545 (1994).
16. Jones, J. D. and Gierasch, L. M. Effect of Charged Residue Substitutions on the Thermodynamics of Signal Peptide-Lipid Interactions for the *Escherichia coli* LamB Signal Sequence. *Biophys. J.* **67**, 1546-1561 (1994).
17. Rizo, J., Blanco, F. J., Kobe, B., Bruch, M. D., et al. Conformational Behavior of *Escherichia coli* OmpA Signal Peptides in Membrane Mimetic Environments. *Biochemistry* **32**, 4881-4894 (1993).
18. Wang, Z. L., Jones, J. D., Rizo, J. and Gierasch, L. M. Membrane-Bound Conformation of a Signal Peptide - A Transferred Nuclear Overhauser Effect Analysis. *Biochemistry* **32**, 13991-13999 (1993).
19. Poritz, M. A., Bernstein, H. D., Strub, K., Zopf, D., et al. An *E. coli* Ribonucleoprotein Containing 4.5S RNA Resembles Mammalian Signal Recognition Particle. *Science* **250**, 1111-1117 (1990).
20. Rapoport, T. A. Protein Translocation - A Bacterium Catches Up. *Nature* **349**, 107-108 (1991).

21. Bassford, P., Beckwith, J., Ito, K., Kumamoto, C., et al. The primary pathway of protein export in *E. coli*. *Cell* **65**, 367-368 (1991).
22. Phillips, G. J. and Silhavy, T. J. The *E. coli ffh* Gene Is Necessary for Viability and Efficient Protein Export. *Nature* **359**, 744-746 (1992).
23. Luirink, J., High, S., Wood, H., Giner, A., et al. Signal-Sequence Recognition by an *Escherichia coli* Ribonucleoprotein Complex. *Nature* **359**, 741-743 (1992).
24. Bernstein, H. D., Zopf, D., Freymann, D. M. and Walter, P. Functional Substitution of the Signal Recognition Particle 54-kDa Subunit by Its *Escherichia coli* Homolog. *Proc. Natl. Acad. Sci.* **90**, 5229-5233 (1993).
25. Powers, T. and Walter, P. Co-translational protein targeting catalyzed by the *Escherichia coli* signal recognition particle and its receptor. *EMBO J.* **16**, 4880-4886 (1997).
26. Montoya, G., Svensson, C., Luirink, J. and Sinning, I. Crystal structure of the NG domain from the signal-recognition particle receptor FtsY. *Nature* **385**, 365-368 (1997).
27. Freymann, D., Keenan, R. J., Stroud, R. M. and Walter, P. Structure of the conserved GTPase domain of the signal recognition particle. *Nature* **385**, 361-364 (1997).
28. Zopf, D., Bernstein, H. D., Johnson, A. E. and Walter, P. The Methionine-Rich Domain of the 54 kD Protein Subunit of the Signal Recognition Particle Contains an RNA Binding Site and Can Be Crosslinked to a Signal Sequence. *EMBO J.* **9**, 4511-4517 (1990).
29. High, S. and Dobberstein, B. The Signal Sequence Interacts with the Methionine-Rich Domain of the 54-kD Protein of Signal Recognition Particle. *J. Cell Biol.* **113**, 229-233 (1991).
30. Lutcke, H., High, S., Romisch, K., Ashford, A. J., et al. The Methionine-Rich Domain of the 54 kDa Subunit of Signal Recognition Particle Is Sufficient for the Interaction with Signal Sequences. *EMBO J.* **11**, 1543-1551 (1992).
31. Takamatsu, H., Bunai, K., Horinaka, T., Oguro, A., et al. Identification of a Region Required for Binding to Presecretory Protein in *Bacillus Subtilis* Ffh, a Homologue of the 54-kDa Subunit of Mammalian Signal Recognition Particle. *Eur. J. Biochem.* **248**, 575-582 (1997).

32. Romisch, K., Webb, J., Lingelbach, K., Gausepohl, H., et al. The 54-kD Protein of Signal Recognition Particle Contains a Methionine-Rich RNA Binding Domain. *J. Cell Biol.* **111**, 1793-1802 (1990).
33. Kurita, K., Honda, K., Suzuma, S., Takamatsu, H., et al. Identification of a region of *Bacillus subtilis* Ffh, a homologue of mammalian SRP54 protein, that is essential for binding to small cytoplasmic RNA. *J. Biol. Chem.* **271**, 13140-13146 (1996).
34. Zheng, N. and Gierasch, L. M. Domain Interaction in E. coli SRP: Stabilization of M Domain by RNA is Required for Effective Signal Sequence Modulation of NG Domain. *Molec. Cell* **1**, 000-000 (1997).
35. Miller, J. D., Bernstein, H. D. and Walter, P. Interaction of *E. coli* Ffh/4.5S ribonucleoprotein and FtsY mimics that of mammalian signal recognition particle and its receptor. *Nature* **367**, 657-659 (1994).
36. Millman, J. S. and Andrews, D. W. Switching the model: a concerted mechanism for GTPases in protein targeting. *Cell* **89**, 673-676 (1997).
37. Newitt, J. A. and Bernstein, H. D. The N-domain of the signal recognition particle 54-kDa subunit promotes efficient signal sequence binding. *Eur. J. Biochem.* **245**, 720-729 (1997).

A Pres1-specific Sequence Determines the Dual Topology of the Hepatitis B Virus Large Envelope Protein and Binds the Heat Shock Protein Hsc70

Heike Löffler-Mary, Margaret Werr and Reinhild Prange
Institute of Medical Microbiology, Augustusplatz, D-55101 Mainz

Keywords. Hepatitis B virus, envelope proteins, protein translocation, dual topology, posttranslational, Hsc70

1. Introduction

The hepatitis B virus (HBV) is a double-shelled sphere, 42 nm in diameter, with an inner nucleocapsid enclosed by the viral envelope. The envelope carrying the major surface antigen (HBsAg) is composed of cellular lipids and three related viral proteins, the large (L), middle (M), and small (S) envelope protein. They are encoded in a single open reading frame of the HBV genome and initiate at three separate in-phase start codons spaced at intervals of 108 (serotype ayw) and 55 codons. The segments downstream of the three initiation codons are called the preS1 and preS2 regions and the S gene, respectively. All three proteins are found in two forms, either glycosylated at Asn146 of the S sequence or unglycosylated at this site. The M protein is additionally glycosylated at Asn4 (reviewed in Heermann and Gerlich, 1991). While the S protein is the predominant constituent of the envelope, the L protein plays a pivotal function in the viral life cycle by mediating attachment of HBV to liver cells (Neurath *et al.*, 1986, 1992) and by interaction with viral cores (Bruss and Ganem, 1991; Ueda *et al.*, 1991). Envelopment of cytosolic capsids by transmembrane envelope proteins triggers the assembly of viral particles which are thought to bud into the lumen of a pre-Gogi compartment and to leave the cell via the constitutive pathway of secretion (Simon *et al.*, 1988; Huovila *et al.*, 1992).

NATO ASI Series, Vol. H 106
Lipid and Protein Traffic
Pathways and Molecular Mechanisms
Edited by Jos A. F. Op den Kamp

Upon biosynthesis, all three envelope proteins are cotranslationally inserted into the endoplasmic reticulum (ER) membrane which is mediated by signal-anchor and stop-transfer sequences of the S region (Eble *et al.*, 1987). These topogenic signals also govern the cotranslational translocation of the upstream preS2 region of the M protein into the ER lumen thereby leading to a S-like topology (Eble *et al.*, 1990; see Fig. 1). Conversely, the preS (preS1 plus preS2) region of the L protein fails to be cotranslationally translocated and remains on the cytosolic side of the ER membrane (i.e., internal in the mature particle; see Fig 1) (Bruss *et al.*, 1994; Ostapchuk *et al.*, 1994; Prange and Streeck, 1995).

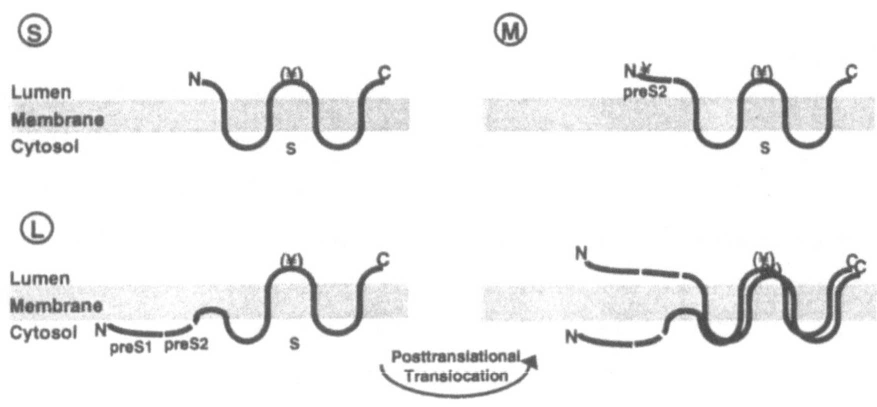

Fig. 1: Models of the topology of the HBV envelope proteins in the ER membrane.

S: The predicted four membrane spanning segments of S protein project the N- and C-termini into the ER lumen.

M: The M protein shows a similar topology like S. The preS2 and S regions are indicated.

L: The preS domain (preS1 + preS2 as indicated) of the L protein is initially exposed to the cytosol. During maturation, the topology of L changes, as its entire preS region is partially posttranslationally translocated.
The partial glycosylation at Asn146 in the S region is indicated by (¥); ¥ marks the glycosylation site Asn4 in the preS2 region of M.

This novel topology of the L protein exhibiting lumenal (external) and cytosolic (internal) preS domains may aid to accomplish its crucial functions for the outcome of a viral infection, as mediating receptor binding and encapsidation of viral cores, respectively. The unique topogenic properties of the L protein imply regulated mechanisms that control its dual topology. In this study we have therefore analyzed whether specific sequences of L repress the cotranslational translocation of the preS1 region and have investigated if the cognate heat shock protein Hsc70 plays a role in this novel process.

2. The preS1-specific sequence 70-94 of L determines cytosolic anchorage.

To examine the structural requirements for the unusual topology of L we constructed a series of preS1 deletion mutants using site-directed mutagenesis or restriction endonuclease reactions. Wild-type and mutant L genes were transiently expressed in COS-7 cells. The proteins were metabolically labeled with [^{35}S]methionine/cysteine and analyzed with polyclonal L-specific antiserum and SDS-PAGE. As shown in Fig. 2, the wild-type L protein was synthesized in non-glycosylated (p39) and glycosylated (gp42) form (lane 1) due to partial modification at Asn146 in the S domain. N-linked glycosylation was confirmed by treatment with PNGaseF (lane 2). Deletion of 38 residues from the C-terminus of the preS1 region (LΔ70-107) dramatically altered the glycosylation pattern of preS: in addition to the expected non- and single-glycosylated forms, double- and triple-glycosylated polypeptides were found modified within preS (Fig. 2, lane 3) as apparent from digestion with PNGaseF (Fig. 2, lane 4). Since N-linked glyans are added cotranslationally to polypeptide chains as the consensus sequence emerges on the luminal side of the ER membrane the glycosylation pattern of mutant LΔ70-107 indicates a cotranslational translocation of its preS region. This was confirmed by protease protection experiments. To map the critical sequence determining cytosolic anchorage of preS more precisely, further mutants were constructed. Mutant LΔ93-106 lacking the C-terminal part of the cytosolic anchorage determinant displayed a wild-type phenotype, occurring in non- and single-glycosylated form only (Fig. 2, lane 5). Conversely, mutant

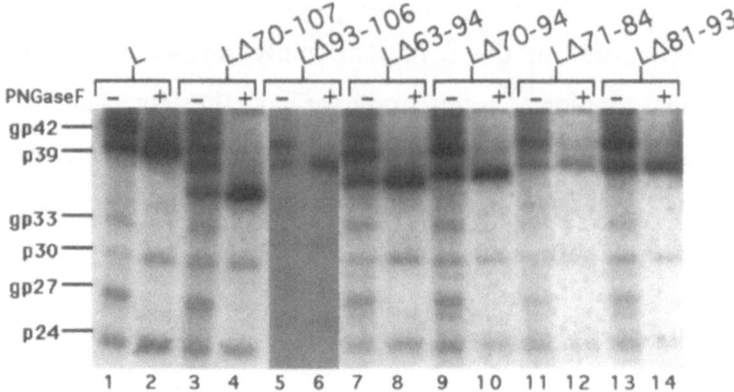

Fig. 2: Analysis of the topology of wild-type and mutant L proteins. After labeling of COS7 cells transfected with the indicated constructs, the proteins were immunoprecipitated and subjected to SDS-PAGE. Samples were either left untreated (-) or were digested with PNGaseF (+) for analysis of N-linked glycosylation. Non-glycosylated (p) and glycosylated (gp, ggp) forms of L (42 kDa, 39 kDa), M (33 kDa, 30 kDa) and S (27 kDa, 24 kDa) are indicated on the left.

LΔ63-94 devoid of the N-terminal part of the determinant, displayed the glycosylation pattern characteristic for cotranslational translocation (Fig. 2, lane 7). Therefore, the cytosolic anchorage determinant of L most likely resides within residues 70 to 94. This was confirmed by mutant LΔ70-94 exhibiting the translocated phenotype (Fig. 2, lane 9). L proteins with shorter deletions retaining part of aa sequence 70 to 94 (LΔ71-84 and LΔ81-93) were synthesized in non- and single-glycosylated form only (Fig. 2, lanes 11 and 13, respectively). We conclude from these data that the cytosolic anchorage determinant is completely located in the aa sequence 70 to 94 of L.

3. The cytosolic anchorage determinant is effective in another protein.

We transferred the cytosolic anchorage determinant into another protein to investigate if it is active independently of surrounding sequences. The HBV middle protein (M) was chosen as a marker because its preS2 region is cotranslationally translocated into the ER lumen (Eble *et al.*, 1990; see Fig. 1). In transfected cells the M protein therefore appeared predominantly in single-glycosylated form (gp33), modified at Asn4 in preS2, and, to a lesser extent, in twice-glycosylated form (ggp36), partially modified at Asn146 in S in addition (Fig. 3, lane 1). Removal of the glycans by PNGaseF converted both forms of M into its non-glycosylated version (p30; Fig. 3, lane 2).

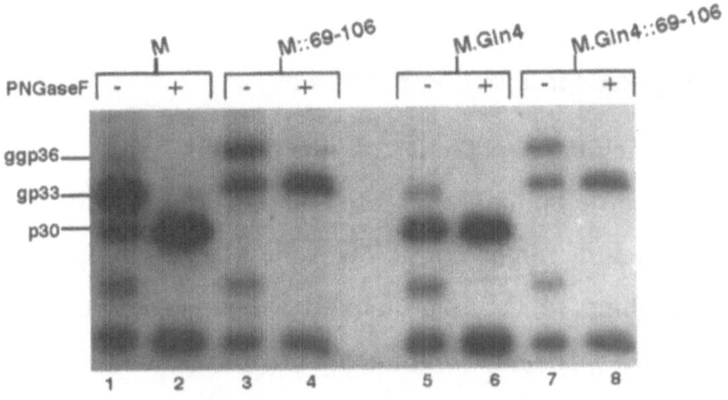

Fig. 3: Glycosylation and transmembrane topology of wild-type and mutant M proteins carrying the cytosolic anchorage determinant of L. Immonoprecipiteted lysates of labeled cells transfected with the indicated constructs were either left untreated (-) or were digested with PNGaseF. Numbers to the left show molecular masses of the non- (p) and the glycosylated (gp, ggp) forms of the M protein.

When the cytosolic anchorage determinant of L (i.e. aa 69 to 106 were used) was inserted into the very C-terminus of the preS2 region of M, its glycosylation pattern was altered: the mutant M::69-106 was synthesized in non- and single-glycosylated form only and lacks the double-glycosylated version (Fig. 3, lanes 3 and 4). This indicates that one of its two potential glycosylation sites (Asn4 in preS2, Asn146 in S) was inaccessible to cotranslational modification. To identify this residue involved, we mutated the carbohydrate acceptor site asparagine to glutamine at position 4 in M and M::69-106. As expected, the glycosylation-defective wild-type M protein (M.Gln4) was now obtained in non- and single-glycosylated form (Fig. 3, lanes 5 and 6). Importantly, mutant M.Gln4::69-106 displayed virtually the same pattern of glycosylation (Fig. 3, lanes 7 and 8) as M::69-106. We took the absence of preS2-linked glycosylation of mutant M::69-

106 as a proof that cotranslational translocation of its preS2 region was inhibited by the inserted cytosolic anchorage determinant, We could confirm this by protease protection experiments using trypsin. Taken together, the cytosolic disposition of the preS2 domain of mutant M::69-106 demonstrate that the repression of cotranslational translocation is an intrinsic feature of the cytosolic anchorage determinant of L, and it is even operative in a heterologous protein.

4. Hsc70 is a specific interacting partner of L.

The cytosolic anchorage potential of the preS1-specific sequence 70-94 suggests that specific binding of cytosolic cellular factors may be involved. A possible candidate might be the cytosolic Hsc70 chaperone, which has been shown to bind to the L protein of the related duck virus, DHBV (Swameye *et al.*, personal communication). To investigate a possible association between L and the chaperone, the L protein was synthesized in a coupled transcription/translation/translocation system using rabbit reticulocyte lysate and dog pancreas microsomes. The S gene was included as a control. Proteins were either left untreated or subjected to immunoprecipitation with a Hsc70-specific MAb prior to SDS-PAGE. As shown in Fig. 4, the S and L proteins both appear in non-glycosylated and glycosylated form, although glycosylation was less effcient (lanes 1 and 2, respectively). Importantly, the L protein (Fig. 4, lane 5), but not the S protein (Fig. 4, lane 4) was efficiently coimmunoprecipitated with the Hsc70-specific MAb, thus demonstrating a specific interaction between L and Hsc70. We were next interested to know whether the LΔ70-107 mutant, competent for cotranslational translocation of preS, interact with Hsc70. The LΔ70-107 protein was synthesized in the cell-free system with almost the same efficiency as wild-type L (Fig. 4, lanes 3 and 2, respectively). However, unlike L, the LΔ70-107 mutant failed to efficiently coprecipitate with Hsc70 (Fig. 4, lane 6). These data demonstrate Hsc70 as a specific interacting partner of L and suggest aa 70 to 107 of L as a major Hsc70 binding site.

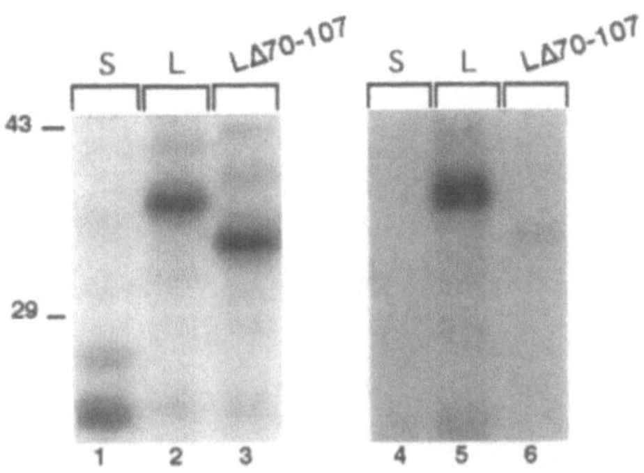

Fig. 4: Left: Analysis of *in vitro* synthesized S, L and LΔ70-107 proteins by SDS-PAGE. Right: Immunoprecipitation of *in vitro* synthesized S, L and LΔ70-107 proteins with a Hsc70-specific MAb.

To confirm the specifity of Hsc70 binding and to map the binding site(s) we assessed the ability of Hsc70 to bind to a series of GST fusion proteins containing sequences of the preS1 region of L. The fusion proteins were designed such as to cover either the N-terminal (GST::L1-42), the internal (GST::L42-69) or the C-terminal (GST::L69-106) third of preS1. For further mapping of the binding site(s), a GST::L42-62 construct was analyzed in addition. The recombinant GST::UAS carries a HBV-unrelated sequence of similar length and was included as a control. Equal amounts of immobilized fusion proteins were incubated with rabbit reticulocyte lysate and the fusion proteins along with bound protein(s) of the lysate were then eluted with glutathione and were subjected to immunoblotting with an anti-Hsp/Hsc70-specific antibody. Using this assay we could clearly identify Hsp/Hsc70 in the presence of the GST::L69-106 fusion protein (Fig. 5, lane 4), but neither in the constructs encoding the first 42 or 62 residues of L (Fig. 5, lanes 3 and 5, respectively) nor in the controls (Fig. 5, lanes 2 and 7). Unexpectedly, however, Hsp/Hsc70 was also found to bind to the GST::L42-69 fusion protein (Fig. 5, lane 6). In combining these data residues 63 to 106 of L interacted with Hsp/Hsc70. Whether aa 63 to 68 of L comprise an individual Hsp/Hsc70 binding site or are part of an overlapping site (i.e., aa 63-106) needs to be determined further.

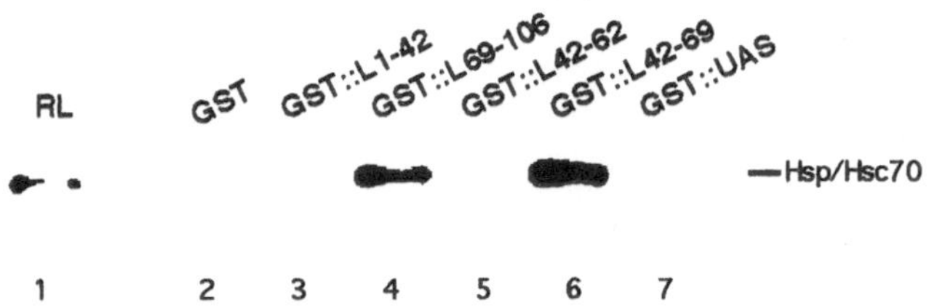

Fig. 5: ***In vitro*** **interaction of Hsp/Hsc70 with L.** GST and the indicated GST fusion proteins, immobilised on glutathione sepharose, were incubated with reticulocyte lysates. Proteins were eluted with glutathione and analyzed by ECL immunblotting with a Hsp/Hsc70-specific MAb.

5. Discussion

In this study we have identified and characterized a preS1-specific sequence which confers the repression of cotranslational translocation of the preS domain of L, a prerequiste for its unusual dual topology. Our mutational analysis revieled that solely aa 70 to 94 account for the (initial) cytoplasmic disposition of the preS domain of L and act as a compact and functional module independently of surrounding sequences. Moreover, we found that the cytosolic anchorage determinant is even active in another protein. In parallel, the cytosolic cognate Hsc70 chaperone was identified herein as a specific interacting partner of L. Hsc70 is known to keep nascent polypeptide chains in an unfolded state to ensure a translocation-competent conformation and to assist in posttranslational transmembrane transport processes (Deshaies *et al.*, 1988; Zimmermann *et al.*, 1988; Brodsky, 1996). Accordingly, the observed interaction between Hsc70 and L might dictate the (initial) cytosolic disposition of preS. In support of this view, we have mapped the interaction site(s) for Hsc70 to aa 63 to 107 of L which overlaps with the cytosolic anchorage determinant. Moreover, deletion mutant LΔ70-107, competent for cotranslocational translocation of preS, failed to bind Hsc70 efficiently. These data strongly suggest that interaction between L and the chaperone is involved in the unconventional translocation of L. Given the strict correlation between repression of cotranslational translocation and Hsc70 binding and vice versa we propose a role of Hsc70 in chaperoning the assembly of HBV. By controlling the translocation and topology of L, Hsc70 might stabilize the cytosolic configuration of preS required for virion formation (Bruss and Vieluf, 1995). Alternatively, bound Hsc70 might directly facilitate contacts between L and the viral nucleocapsid, as the Hsc70 binding site(s) of L (*i.e.*, 63 to 107) overlaps with C-terminal preS1 sequences proposed to be essential for virus assembly (Bruss and Thomssen, 1994; Dyson and Murray, 1995). Although further studies are needed to clarify the precise mechanism of Hsc70 action in the morphogenesis of HBV, this work provides evidence for a role of Hsc70 in the novel process of sequence-specific repression of cotranslational translocation of the HBV envelope proteins.

Acknowledgments

We thank Rolf E. Streeck for support throughout this work. This work was supported by the Deutsche Forschungsgemeinschaft (SFB 311).

References

Brodsky, J. L. (1996). Post-translational protein translocation: not all hsc70s are created equal. *Trends Biochem. Sci.* **21**, 122-126.

Bruss, V., and Ganem, D. (1991). The role of envelope proteins in hepatitis B virus assembly. *Proc. Natl. Acad. Sci. USA* **88**, 1059-1063.

Bruss, V., Lu, X., Thomssen, R., and Gerlich, W. H. (1994). Post-translational alterations in transmembrane topology of the hepatitis B large envelope protein. *EMBO J.* **13**, 2273-2279.

Bruss, V., and Thomssen, R. (1994). Mapping a region of the large envelope protein required for hepatitis B virion maturation. *J. Virol.* **68**, 1643-1650.

Bruss, V., and Vieluf, K. (1995). Functions of the internal pre-S domain of the large surface protein in hepatitis B virus particle morphogenesis. *J. Virol.* **69**, 6652-6657.
Deshaies, R. J., Koch, B. D., Werner-Washburne, M., Craig, E. A., and Schekman, R. (1988). A subfamily of stress proteins facilitates translocation of secretory and mitochondrial precursor polypeptides. *Nature (London)* **332**, 800-805.
Dyson, M. R., and Murray, K. (1995). Selection of peptide inhibitors of interaction involved in complex protein assemblies: association of the core and surface antigens of hepatitis B virus. *Proc. Natl. Acad. Sci. USA* **92**, 2194-2198.
Eble, B. E., Lingappa, V. R., and Ganem, D. (1990). The N-terminal (preS2) domain of a hepatitis B virus surface glycoprotein is translocated across membranes by downstream signal sequences. *J. Virol.* **64**, 1414-1419.
Eble, B. E., Macrae, D. R., Lingappa, V. R., and Ganem, D. (1987). Multiple topogenic sequences determine the transmembrane orientation of hepatitis B surface antigen. *Mol. Cell. Biol.* **7**, 3591-3601.
Heermann, K. H., and Gerlich, W. H. (1991). Surface proteins of hepatitis B viruses, *In* "*Molecul*ar biology of the hepatitis B vi*rus*" (A. McLachlan, Ed.), pp. 109-143. CRC Press, Inc., Boca Raton.
Huovila, A.-P. J., Eder, A. M., and Fuller, S. D. (1992). Hepatitis B surface antigen assembles in a post-ER, pre-Golgi compartment. *J. Cell Biol.* **118**, 1305-1320.
Neurath, A. R., Kent, S. B. H., Strick, N., and Parker, K. (1986). Identification and chemical synthesis of a host cell receptor binding site on hepatitis B virus. *Cell* **46,** 429-436.
Neurath, A. R., Strick, N. and Sproul, P. (1992). Search for Hepatitis B virus cell receptors reveals binding sites for interleukin 6 on the virus envelope protein. *J. Exp. Med.* **175**, 461 - 469.
Ostapchuk, P., Hearing, P., and Ganem, D. (1994). A dramatic shift in the transmembrane topology of a viral envelope glycoprotein accompanies hepatitis B viral morphogenesis. *EMBO J.* **13**, 1048-1057.
Prange, R., and Streeck, R. E. (1995). Novel transmembrane topology of the hepatitis B virus envelope proteins. *EMBO J.* **14**, 247-256.
Simon, K., Lingappa,V.R. and Ganem, D. (1988). Secreted hepatitis B surface antigen polypeptides are derived from a transmembrane precursor. *J. Cell Biol.* **107,** 2163-2168
Swameye, I., Kuhn, C., Hild, M., Klingmüller, U. and Schaller, H. Personal communication.
Ueda, K. L., Tsurimoto, T., and Matsubara, K. (1991). Three envelope proteins of hepatitis B virus: large S, middle S, and major S proteins needed for the formation of Dane particles. *J. Virol.* **65**, 3521-3529.
Zimmermann, R., Sagstetter, M., Lewis, M. J., and Pelham, H. R. B. (1988). Seventy-kilodalton heat shock proteins and an additional component from reticulocyte lysate stimulate import of M13 procoat protein into microsomes. *EMBO J.* **7**, 2875-2880.

Characterization of Components of the General Secretion Pathway of *Aeromonas hydrophila*

S. P. Howard, I. C. Schoenhofen, R. Jahagirdar and C. Stratilo
Department of Biology, University of Regina, SK, Canada S4S 0A2

1 Secretion across the Gram negative outer membrane.

One of the most widespread protein secretion mechanisms under study includes the *exe* genes in *Aeromonas* species (Jiang and Howard, 1992; Howard et al., 1993; Karlyshev and MacIntyre, 1995), and is termed the main terminal branch of the general secretory pathway or the type II secretion system (Salmond and Reeves, 1993; Pugsley et al., 1997). This system includes homologous genes such as the *pul* genes in *Klebsiella oxytoca,* where the system was first described (Pugsley et al., 1990), the *xcp* genes of *Pseudomonas* species (Bally et al., 1992), the *out* genes of *Erwinia* species (He et al., 1991), and the *eps* genes of *Vibrio cholerae* (Sandkvist et al., 1993), among others. Proteins which utilize this pathway are *sec* gene and signal sequence-dependent, and appear to transit the periplasm on their way out of the cell. The system is composed of an operon of 12 or 13 genes (i.e., *exeC-exeN* in *A. hydrophila* (Howard et al., 1993), *pulC-O* in *K. oxytoca* (Pugsley et al., 1990)), all of which encode proteins which are localized to the envelope, but only one of which (ExeD and its homologs) contains a classical signal sequence. In addition, the *pul* system (D'Enfert and Pugsley, 1989) as well as the *out* system of *E. chrysanthemi* (Condemine et al., 1992) include a lipoprotein (PulS/OutS) which has been shown to serve as a chaperone for the PulD protein (Hardie et al., 1996a). In addition to *exeC-N,* two further genes, *exeA* and *exeB* are required in *A. hydrophila* (Jahagirdar and Howard, 1994), as discussed further below (Figure 1).

NATO ASI Series, Vol. H 106
Lipid and Protein Traffic
Pathways and Molecular Mechanisms
Edited by Jos A. F. Op den Kamp

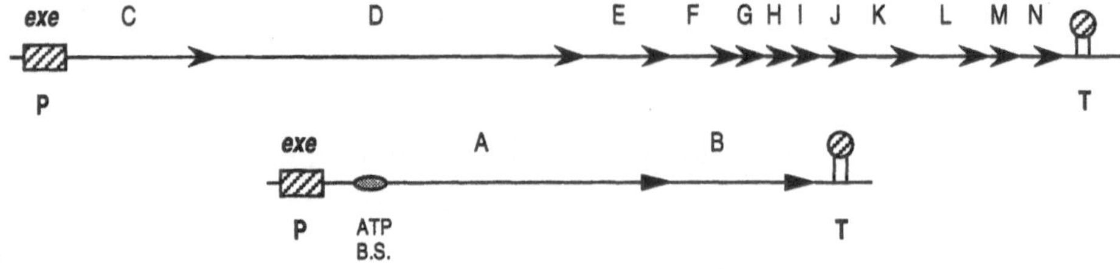

Fig. 1 Structure of the *exeC-N* and *exeAB* operons. P and T signify promoter and terminator sequences, respectively.

2 Properties of the ExeC-N proteins and their homologs.

It has been well established that the *exe* genes and their homologs are required for extracellular secretion, but little is known about how most of them function in this process. The most detailed information concerns a group of these gene products which bear strong homology to proteins involved in the assembly of the adhesive, type IV pili which many bacteria elaborate (including *Aeromonas* (Hokama and Iwanaga, 1991; Pepe et al., 1996)) and which have been well studied in *Ps. aeruginosa* (Paranchych and Frost, 1988). Each of the type II operons encodes a number of proteins which contain consensus type IV prepilin cleavage sequences at their amino termini, although the remainder of the proteins do not bear homology to the type IV pilin structural protein (Pugsley, 1993). Five of the *A. hydrophila* Exe proteins (ExeG-K) contain consensus prepilin cleavage sequences, and ExeG was shown to be processed when coexpressed with the *Ps. aeruginosa* prepilin peptidase (Howard et al., 1993). As might be expected, it has been shown that the prepilin peptidase itself is required for the functional assembly of this secretion pathway. In *Ps. aeruginosa*, the prepilin peptidase was isolated as the secretion gene *xcpA* (Bally et al., 1991) and as *pilD*, part of the type IV pilin assembly operon *pilBCD* (Nunn et al., 1990). The *pilB* and *pilC* genes encode homologs of ExeE and ExeF, which are represented by *xcpR* and *xcpS* in the *Ps. aeruginosa* type II secretion operons. A similar situation exists in *A. hydrophila,* in which the *tapBCD* genes represent homologs of *pilBCD,* and as for *Ps. aeruginosa,* the prepilin peptidase (TapD) was shown to be required for extracellular secretion (Pepe et al., 1996). Finally, the recently discovered *pilQ* gene, also required for pilin assembly, encodes a protein homologous to ExeD and its homologs (Martin et al., 1993). Thus in addition to the requirement for prepilin peptidase, a total of 7 of the 12 genes of the *exeC-N* operon have homologs that are pilin-like or involved in some way in pilin assembly. This has led to the hypothesis that the secretion apparatus forms some sort of pseudo-pilus through which or along which proteins are secreted from the cell, but such a pseudo-pilus has not been observed (Pugsley, 1996).

The ExeE protein and its homologs contain a consensus ATP binding site, which is required for the function of the protein in the secretion process (Turner et al., 1993; Possot and Pugsley, 1994; Howard et al., 1996). The first mutation isolated in the *A. hydrophila* type II secretion pathway was in the *exeE* gene, and we and others proposed that ExeE and its homologs might act to transduce the energy derived from ATP hydrolysis to other Exe proteins which were responsible for the actual translocation across the outer membrane (Jiang and Howard, 1992). Since the PulE homolog PilB is part of the pilin assembly operon, it has also been suggested that the hydrolysis of ATP by PulE may be involved not in secretion *per se* but in assembly of this (pilus-like?) secretion apparatus (Turner et al., 1993; Possot et al., 1992). Sandkvist et al. (1995) have shown that EpsE, the ExeE homolog of *V. cholerae,* is an autokinase, and that it associates with the inner membrane, but only if

EpsL is also being produced by the bacteria. Due to the high level of homology between these two proteins, it was possible to construct chimeras between ExeE and EpsE in order to determine which regions of the EpsE protein are involved in the interactions with EpsL. The results indicated that a region within the amino terminal 90 amino acids of EpsE is required for the interaction with EpsL which leads to stable interaction with the membrane and functional activity in the secretion of cholera toxin (Sandkvist et al., 1995).

An additional phenotype of *exeC-N* mutations (Jiang and Howard, 1991) is a substantial decrease in the quantities of major outer membrane proteins, espcially the OmpF homolog Protein II (Jeanteur et al., 1992). The cells lysed during osmotic shock procedures done to determine the localization of aerolysin entrapped in the nonsecretory mutant L1.97 (*exeE::Tn5-751*) indicating that they were also osmotically fragile, an effect probably caused by the altered outer membrane structure (Jiang and Howard, 1991; Jiang and Howard, 1992). The two phenotypes, altered membrane structure and non-secretion, were due to mutation in the same gene, as demonstrated by reconstruction of the original mutant via marker exchange mutagenesis (Howard et al., 1993). In addition, two other mutations in the *exeC-N* operon (*exeC* and *exeK*) were also altered in outer membrane structure, indicating that the entire operon is required for both protein secretion and normal assembly of the outer membrane. The *exe* mutant cells appear to be specifically downregulated with respect to Protein II synthesis since they were able to induce and properly assemble both the LamB and PhoE homologs of this bacteria under inducing conditions, and a strain containing a surface layer was unaffected in its ability to assemble it when it was mutated in *exeE* (Howard and Meiklejohn, 1995). Thus the defective membrane structure is not due to a general inability to make or assemble porin or surface proteins. Pulse chase analysis demonstrated downregulation at the level of synthesis of Protein II as opposed to mislocalization or rapid degradation in the mutants. The fact that mutations in the *exeC-N* genes cause this downregulation, whereas mutations in the *exeAB* genes do not, indicates that they affect different aspects or stages of the type II secretion pathway, as discussed further below.

3 ExeD and its homologs may form a secretion portal

The ExeD protein and its homologs contain standard signal peptidase I signal sequences, and the PulD protein was shown to be processed during its assembly into the outer membrane of *K. oxytoca* (Pugsley et al., 1990). It is now also clear that the type III secretion pathway contains an ExeD homolog, in addition to the type II pathway and the type IV pilin assembly pathway discussed above. Proteins related to ExeD are also involved in the assembly and release of the ssDNA bacteriophages f1 of *E. coli* and P13 of *Ps. aeruginosa.* The homology relationships of these proteins have been discussed in detail in the recent literature (Genin and Boucher, 1994; Hobbs and Mattick, 1993). It thus appears that an ExeD homolog is involved in a very wide variety of processes, all of which have in common translocation of a molecule across the outer membrane and onto the surface or free of the cell. The pIV protein of fI phage forms multimers which may comprise a pore through which assembling phage particles exit the cell, in a process which depends on interactions with pI, a 35 kDa inner membrane protein which contains a consensus ATP binding site (Kazmierczak et al., 1994; Russel, 1993). Studies on PulD have recently shown that it too forms a high molecular weight complex, and that PulS acts as a chaperone in the assembly of this complex in addition to protecting it from proteolytic degradation (Hardie et al., 1996a; 1996b).

4 Properties of ExeA and of ExeB and its homologs

The *exeAB* genes form an operon and like *exeC-N*, are required for protein secretion in *A. hydrophila* (Jahagirdar and Howard, 1994). ExeA is a 60 kDa protein containing a consensus ATP binding site, and is hydrophilic, but contains a highly hydrophobic stretch of 20 amino acids. The ExeB protein was deduced from the gene sequence to be a basic 25 kDa 226 residue protein with a very highly hydrophobic region of 24 amino acids from 37 - 60 and a region which would be less strongly hydrophobic at 193-218. An unusual feature of the protein is a region with numerous prolines which immediately follows the strongly hydrophobic region (13 in the following 47 residues).

A search of the sequence databanks with the *exeAB* genes retrieved two proteins with high homology to ExeB; the PulB protein (Pugsley et al., 1990) and the OutB protein of *E. chrysanthemi* (Condemine et al., 1992). The *pulB* gene, although reported to be not required for pullulanase secretion, is found associated with the other *pul* genes, and an *outB* mutant of *E. chrysanthemi* displayed a decreased efficiency of secretion of pectinases (30% of normal levels). Both proteins are significantly smaller (174 and 157 aa respectively) than the 226 residue ExeB protein, and the missing region is that which contains the highly hydrophobic region at the amino terminus. There are a number of possible explanations for the difference in ExeB structure in these bacteria. In the case of the *pul* operon, the type II pathway in *Klebsiella* may have become highly specialized (specific to the secretion of pullulanase) and therefore dispensed with or completely modified functions such as ExeB that are still required in *A. hydrophila,* which is a prolific secretor of many different proteins. Supporting arguments for this specialization-of-function hypothesis are that the *pul* genes are part of the maltose regulon in *Klebsiella* and that the prepilin peptidase gene is found associated with the *pul* genes, whereas in the *Vibrionaceae* and

Pseudomonaceae it is found with the type IV pilin assembly genes ((Nunn et al., 1990; Pepe et al., 1996). Such a hypothesis might also apply to the *out* genes of *E. chrysanthemi*, another member of the *Enterobacteriaceae*, but in this case as stated above the *outB* gene is still required for efficient secretion. A decision as to which if any of these explanations is correct awaits further information, especially concerning the structure and function of possible ExeB homologs of *Ps. aeruginosa* and *V. cholerae*.

5 Current research

Although we do not yet know how the type II pathway transports proteins across the outer membrane, it seems clear that: (1), There is a genuine periplasmic intermediate in the secretion pathway, in which proteins are free to fold into their tertiary conformations; (2), Secretion across the outer membrane *per se* via this pathway is an energy dependent process; (3), Many of the ExeC-N proteins and their homologs (i.e. ExeE,F,L, and the prepilin peptidase TapD) may be involved in the assembly of some sort of structure which is composed of the others (i.e Exe D, G,H,I, J, and K); and (4), The ExeA and B proteins are required for secretion across the outer membrane, but are most likely not involved in the assembly of the apparatus discussed above.

In our current research on the ExeA and B proteins, we are studying the role they play in the functioning of the secreton formed by products of the *exeC-N* operon (Howard et al., 1996). ExeA and B were purified from overproducing *E. coli* cells and antisera were used to demonstrate that both fractionate with the inner membrane in sucrose density gradient centrifugation of *A. hydrophila* membranes. Fusions to the secretion reporter alkaline phosphatase indicated that although neither contains a signal sequence, both proteins cross the inner membrane once and have a major domain on the periplasmic side of the membrane (Figure 2).

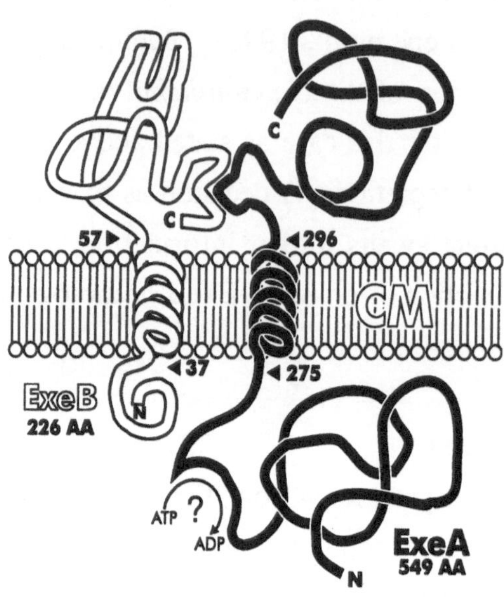

Fig. 2 Schematic diagram of the topology of the ExeA and B proteins. The location of the amino and carboxy termini as well as the boundaries of the regions which traverse the cytoplasmic membrane (CM) are shown.

When ExeA and ExeB were overproduced, ExeB was essentially completely stable, whereas if ExeB was overproduced alone, it was unstable, with a half-life of approximately 35 minutes. The simplest way to interpret the dependence of ExeB on ExeA for its stability is that the two proteins form a complex, and similar stabilizing interactions have been observed for other complex-forming proteins such as SecY/SecE and TonB/ExbB (Fischer et al., 1989; Nishiyama et al., 1992). *In vitro* mutagenesis was used to modify a highly conserved glycine residue in the consensus ATP binding site of ExeA, and each of three different substitutions of this residue (Val, Ala,and Asp) essentially completely prevented the secretion of aerolysin from the cells. Finally, searches of the sequence databases indicated that in addition to the homologies between ExeB and PulB/OutB described above, weak homology was found between ExeB and TonB, a protein involved in the energy dependent inward translocation of ligands such as siderophores and vitamin B12 across the outer membrane of *E. coli* (Braun, 1995). Although the similarity between the two proteins is low (18% identical, 39% similar amino acids), they are both highly basic and share a similar hydrophobicity profile in addition to being enriched for proline in the region immediately periplasmic to the membrane anchor.

These studies indicate that ExeA contains an essential ATP-binding site and that it forms a complex in the inner membrane with the TonB-like ExeB protein, which is required for passage of secreted proteins across the outer membrane. From these conclusions, we have proposed that ExeA and ExeB together act as the turnkeys of a secretion port formed by products of the *exeC-N* operon. It is possible that ExeA, through the hydrolysis of ATP, interacts with ExeB to cause it to directly or indirectly (i.e. perhaps through interactions with other Exe proteins) open a gated port in the outer membrane formed by the ExeD protein. One prediction of this hypothesis is that ATP will be required as an energy source for secretion. It had previously been shown by Wong and Buckley that the addition of the protonophore carbonyl cyanide m-chlorophenyl hydrazone (CCCP) blocks secretion of an accumulated pool of aerolysin from the periplasm (Wong and Buckley, 1989). This energy requirement was recently reexamined by Letellier et al. (1997), who extended the study by quantitating both the electrochemical potential ($\Delta\Psi$) across the inner membrane and ATP levels when cells were treated with CCCP or Arsenate. It was found that both forms of cellular energy were independently required for secretion, since CCCP collapsed $\Delta\Psi$ without strongly affecting ATP levels, whereas arsenate rapidly depleted

ATP levels without significantly modifying $\Delta\Psi$, but both poisons prevented release of the accumulated periplasmic pool of aerolysin from the cells. It should be noted that only $\Delta\Psi$ was required in similar studies of pullulanase secretion, which as described above is independent of PulB, the ExeB homolog of *K. oxytoca* (Possot et al., 1997).

In other current studies we are using crosslinking reagants to examine the structure of the putative ExeAB complex as well as interactions between it and other components of the secretion apparatus, purifying ExeA in order to determine whether or not it posesses ATPase activity, and examining the energy requirements for secretion in cells containing ExeA proteins with various point mutations in its ATP-binding site. These studies are aimed at increasing our understanding of how the ExeA and B proteins regulate the passage of secreted proteins through the apparatus formed by the products of the *exeC-N* operon.

Acknowledgements

Research in our laboratory is supported by the Medical Research Council of Canada. We are grateful to David Shivak for the artwork in Figure 2.

References

Bally, M., Ball, G., Badere, A. and Lazdunski, A. (1991) Protein secretion in *Pseudomonas aeruginosa*: The *xcpA* gene encodes an integral inner membrane protein homologous to *Klebsiella pneumoniae* secretion function protein PulO. J. Bacteriol. 173: 479-486.

Bally, M., Filloux, A., Akrim, M., Ball, G., Lazdunski, A. and Tommassen, J. (1992) Protein secretion in *Pseudomonas aeruginosa*: Characterization of seven xcp genes and processing of secretory apparatus components by prepilin peptidase. Mol. Microbiol. 6: 1121-1131.

Braun, V. (1995) Energy-coupled transport and signal transduction through the gram-negative outer membrane via TonB-ExbB-ExbD-dependent receptor proteins. FEMS Microbiol. Rev. 16: 295-307.

Condemine, G., Dorel, C., Hugouvieux-Cotte-Pattat, N. and Robert-Baudouy, J. (1992) Some of the *out* genes involved in the secretion of pectate lyases in *Erwinia chrysanthemi* are regulated by *kdgR*. Mol. Microbiol. 6: 3199-3211.

D'Enfert, C. and Pugsley, A.P. (1989) *Klebsiella pneumoniae pulS* gene encodes an outer membrane lipoprotein required for pullulanase secretion. J. Bacteriol. 171: 3673-3679.

Fischer, E., Günter, K. and Braun, V. (1989) Involvement of ExbB and TonB in transport across the outer membrane of *Escherichia coli*: Phenotypic complementation of *exb* mutants by overexpressed *tonB* and physical stabilization of TonB by ExbB. J. Bacteriol. 171: 5127-5134.

Genin, S. and Boucher, C.A. (1994) A Superfamily of Proteins Involved in Different Secretion Pathways in Gram-Negative Bacteria - Modular Structure and Specificity of the N-Terminal Domain. Mol. Gen. Genet. 243: 112-118.

Hardie, K.R., Lory, S. and Pugsley, A.P. (1996a) Insertion of an outer membrane protein in Escherichia coli requires a chaperone-like protein. EMBO J. 15: 978-988.

Hardie, K.R., Seydel, A., Guilvout, I. and Pugsley, A.P. (1996b) The secretin-specific, chaperone-like protein of the general secretory pathway: separation of proteolytic protection and piloting functions. Mol. Microbiol. 22: 967-976.

He, S.Y., Lindeberg, M., Chatterjee, A.K. and Collmer, A. (1991) Cloned *Erwinia chrysanthemi out* genes enable *Escherichia coli* to selectively secrete a diverse family of heterologous proteins into its milieu. Proc. Natl. Acad. Sci. USA 88: 1079-1083.

Hobbs, M. and Mattick, J.S. (1993) Common components in the assembly of type 4 fimbriae, DNA transfer systems, filamentous phage and protein-secretion apparatus: A general system for the formation of surface-associated protein complexes. Mol. Microbiol. 10: 233-243.

Hokama, A. and Iwanaga, M. (1991) Purification and characterization of Aeromonas sobria pili, a possible colonization factor. Infect. Immun. 59: 3478-3483.

Howard, S.P., Critch, J. and Bedi, A. (1993) Isolation and analysis of eight *exe* genes and their involvement in extracellular protein secretion and outer membrane assembly in *Aeromonas hydrophila*. J. Bacteriol. 175: 6695-6703.

Howard, S.P. and Meiklejohn, H.G. (1995) Effect of mutations in the general secretory pathway on outer membrane protein and surface layer assembly in *Aeromonas spp.*. Can. J. Microbiol. 41: 525-532.

Howard, S.P., Meiklejohn, H.G., Shivak, D. and Jahagirdar, R. (1996) A tonB-like protein and a novel membrane protein containing an ATP-binding cassette function together in exotoxin secretion. Mol. Microbiol. 22: 595-604.

Jahagirdar, R. and Howard, S.P. (1994) Isolation and Characterization of a 2nd Exe Operon Required for Extracellular Protein Secretion in Aeromonas-Hydrophila. J. Bacteriol. 176: 6819-6826.

Jeanteur, D., Gletsu, N., Pattus, F. and Buckley, J.T. (1992) Purification of *Aeromonas hydrophila* major outer-membrane proteins: *N*-terminal sequence analysis and channel-forming properties. Mol. Microbiol. 6: 3355-3363.

Jiang, B. and Howard, S.P. (1991) Mutagenesis and isolation of *Aeromonas hydrophila* genes which are required for extracellular secretion. J. Bacteriol. 173: 1241-1249.

Jiang, B. and Howard, S.P. (1992) The Aeromonas hydrophila exeE gene, required both for protein secretion and normal outer membrane biogenesis, is a member of a general secretion pathway. Mol. Microbiol. 6: 1351-1361.

Karlyshev, A.V. and MacIntyre, S. (1995) Cloning and study of the genetic organization of the exe gene cluster of Aeromonas salmonicida. Gene 158: 77-82

Kazmierczak, B.I., Mielke, D.L., Russel, M. and Model, P. (1994) Piv, a Filamentous Phage Protein That Mediates Phage Export Across the Bacterial-Cell Envelope, Forms a Multimer. J. Mol. Biol. 238: 187-198.

Letellier, L., Howard, S.P. and Buckley, J.T. (1997) Studies on the energetics of proaerolysin secretion across the outer membrane of Aeromonas species. Evidence for a requirement for both the protonmotive force and ATP. J. Biol. Chem. 272: 11109-11113.

Martin, P.R., Hobbs, M., Free, P.D., Jeske, Y. and Mattick, J.S. (1993) Characterization of *pilQ*, a new gene required for the biogenesis of type 4 fimbriae in *Pseudomonas aeruginosa*. Mol. Microbiol. 9: 857-868.

Nishiyama, K., Mizushima, S. and Tokuda, H. (1992) The carboxyl-terminal region of SecE interacts with SecY and is functional in the reconstitution of protein translocation activity in *Escherichia coli*. J. Biol. Chem. 267: 7170-7176.

Nunn, D., Bergman, S. and Lory, S. (1990) Products of three accessory genes, *pilB, pilC* and *pilD*, are required for biogenesis of *Pseudomonas aeruginoae* pili. J. Bacteriol. 172: 2911-2919.

Paranchych, W. and Frost, L.S. (1988) The physiology and biochemistry of pili. Adv. Microb. Physiol. 29: 53-114.

Pepe, C.M., Eklund, M.W. and Strom, M.S. (1996) Cloning of an Aeromonas hydrophila type IV pilus biogenesis gene cluster: Complementation of pilus assembly functions and characterization of a type IV leader peptidase/N-methyltransferase required for extracellular protein secretion. Mol. Microbiol. 19: 857-869.

Possot, O., D'Enfert, C., Reyss, I. and Pugsley, A.P. (1992) Pullulanase secretion in *Escherichia coli* K-12 requires a cytoplasmic protein and a putative polytopic cytoplasmic membrane protein. Mol. Microbiol. 6: 95-105.

Possot, O. and Pugsley, A.P. (1994) Molecular characterization of PulE, a protein required for pullulanase secretion. Mol. Microbiol. 12: 287-299.

Possot, O.M., Letellier, L. and Pugsley, A.P. (1997) Energy requirement for pullulanase secretion by the main terminal branch of the general secretory pathway. Mol. Microbiol. 24: 457-464.

Pugsley, A.P., D'Enfert, C., Reyss, I. and Kornacker, M.G. (1990) Genetics of extracellular protein secretion by gram-negative bacteria. Annu. Rev. Genet. 24: 67-90.

Pugsley, A.P. (1993) The complete general secretory pathway in gram-negative bacteria. Microbiol. Rev. 57: 50-108.

Pugsley, A.P. (1996) Multimers of the precursor of a type IV pilin-like component of the general secretory pathway are unrelated to pili. Mol. Microbiol. 20: 1235-1245.

Pugsley, A.P., Francetic, O., Possot, O.M., Sauvonnet, N. and Hardie, K.R. (1997) Recent progress and future directions in studies of the main terminal branch of the general secretory pathway in Gram-negative bacteria--a review. Gene 192: 13-19.

Russel, M. (1993) Protein-protein interactions during filamentous phage assembly. J. Mol. Biol. 231: 689-697.

Salmond, G.P.C. and Reeves, P.J. (1993) Membrane traffic wardens and protein secretion in Gram-negative bacteria. Trends Biochem. Sci. 18: 7-12.

Sandkvist, M., Morales, V. and Bagdasarian, M. (1993) A protein required for secretion of cholera toxin through the outer membrane of *Vibrio cholerae*. Gene 123: 81-86.

Sandkvist, M., Bagdasarian, M., Howard, S.P. and DiRita, V.J. (1995) Interaction between the autokinase EpsE and EpsL in the cytoplasmic membrane is required for extracellular secretion in *Vibrio cholerae*. EMBO J. 14: 1664-1673.

Turner, L.R., Lara, J.C., Nunn, D.N. and Lory, S. (1993) Mutations in the consensus ATP-binding sites of XcpR and PilB eliminate extracellular protein secretion and pilus biogenesis in *Pseudomonas aeruginosa*. J. Bacteriol. 175: 4962-4969.

Wong, K.R. and Buckley, J.T. (1989) Proton motive force involved in protein transport across the outer membrane of Aeromonas salmonicida. Science 246: 654-656.

Do the Type I Signal Peptidases of *Bacillus subtilis* Compete for Binding and Cleavage of Secretory Precursor Proteins?

Albert Bolhuis*, Harold Tjalsma*, Gerard Venema, Sierd Bron, and Jan Maarten van Dijl

* These authors contributed equally to the present work

Department of Genetics, Groningen Biomolecular Sciences and Biotechnology Institute, Kerklaan 30, 9751 NN Haren, The Netherlands

Abstract. *Bacillus subtilis* contains four closely related, chromosomally-encoded type I signal peptidases (SipS, SipT, SipU and SipV), which remove signal peptides from secretory precursor proteins. In the present studies, the role of SipS in protein secretion in *B. subtilis* was analysed. Interestingly, the absence of SipS had opposite effects on the secretion of different mature proteins into the growth medium. For example, the neutral protease NprE was secreted at reduced levels, whereas levansucrase was secreted at increased levels. Similarly, the processing of certain secretory precursor proteins was reduced, whereas processing of other precursors was improved. The latter observation indicates that the presence of SipS can interfere with efficient processing of certain precursor proteins, which raises the question whether the type I signal peptidases of *B. subtilis* compete for binding and cleavage of secretory precursor proteins.

Key words. *Bacillus subtilis*, leader peptidase, protein secretion, signal peptidase

1. Introduction

Bacillus subtilis and closely related bacilli secrete proteins directly into the growth medium, naturally and to high concentrations (Nagarajan, 1993; Simonen and Palva, 1993). In recent years, several components of the protein secretion machinery of *B. subtilis* have been identified. These include chaperones, components of the pre-protein translocase complex and signal peptidases. Although the general secretion pathway of *B. subtilis* is similar to that of *Escherichia coli*, there are also clear differences. A good example of a difference in the secretion machinery of bacilli and other organisms is the secretory pre-protein processing machinery which, in *B. subtilis*, seems to consist of at least four closely related type I signal peptidases (SPase I), also known as leader peptidases (Lep).

NATO ASI Series, Vol. H 106
Lipid and Protein Traffic
Pathways and Molecular Mechanisms
Edited by Jos A. F. Op den Kamp

2. Type I Signal Peptidases

Most bacterial proteins that are transported from the cytoplasm to other cellular compartments (*e.g.* export to the periplasm and outer membrane of Gram-negative bacteria), or to the growth medium (secretion) are synthesised with an amino-terminal signal peptide. The signal peptide is required for the targeting of proteins to the membrane and the initiation of protein translocation across this membrane (for reviews, see: von Heijne, 1990; Wickner *et al.*, 1991; Pugsley, 1993; Driessen, 1994). During or shortly after the translocation of a protein across the membrane, the signal peptide is removed by SPases. This is a prerequisite for the release of the secretory protein from the membrane and, in some cases, the post-translational modification of its amino-terminus (for reviews, see: Dalbey, 1994; Lory, 1994; Sankaran and Wu, 1994).

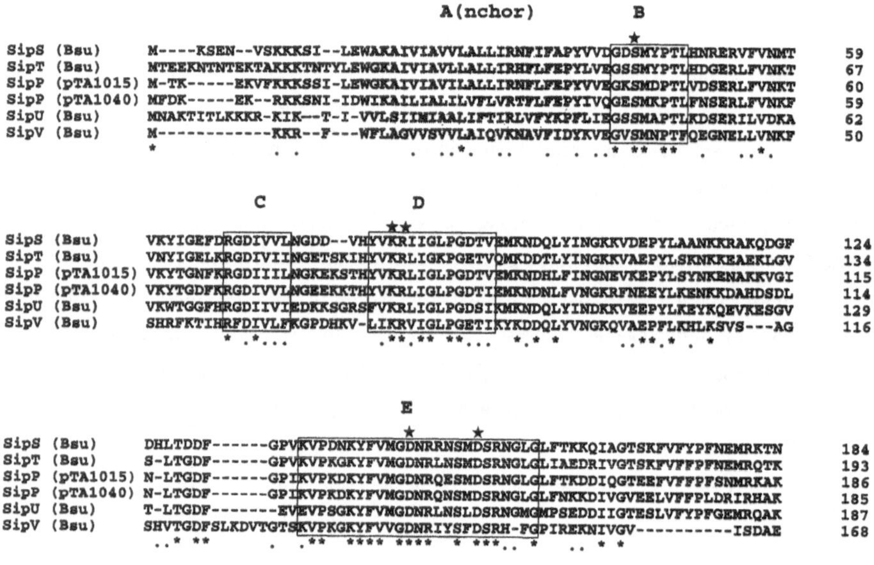

```
                                            A(nchor)          B
                                                              ★
SipS (Bsu)       M----KSEN--VSKKKSI--LEWAKAIVIAVVLALLIRNFIFAPYVVDGDSMYPTLHNRERVFVNMT     59
SipT (Bsu)       MTEEKNTNTEKTAKKKTNTYLEWGKAIVIAVLLALLIRHFLFEPYLVEGSSMYPTLHDGERLFVNKT     67
SipP (pTA1015)   M-TK-----EKVFKKKSSI-LEWGKAIVIAVILALLIRNFLFEPYVVEGKSMDPTLVDSERLFVNKT     60
SipP (pTA1040)   MFDK-----EK--RKKSNI-IDWIKAILIALILVFLVRTFLFEPYIVQGESMKPTLFNSERLFVNKF     59
SipU (Bsu)       MNAKTITLKKKR-KIK--T-I-VVLSIIMIAALIFTIRLVFYKPFLIEGSSMAPTLKDSERILVDKA     62
SipV (Bsu)       M------------KKR--F---WFLAGVVSVVLAIQVKNAVFIDYKVEGVSMNPTFQEGNELLVNKF     50
                 *     . .          .  ..*.  ..   .  .  ..* ** **  .   ..*.

                         C             D
                                      ★★
SipS (Bsu)       VKYIGEFDRGDIVVLNGDD--VHYVKRIIGLPGDTVEMKNDQLYINGKKVDEPYLAANKKRAKQDGF    124
SipT (Bsu)       VNYIGELKRGDIVIINGETSKIHYVKRLIGKPGETVQMKDDTLYINGKKVAEPYLSKNKKEAEKLGV    134
SipP (pTA1015)   VKYTGNFKRGDIIILNGKEKSTHYVKRLIGLPGDTVEMKNDHLFINGNEVKEPYLSYNKENAKKVGI    115
SipP (pTA1040)   VKYTGDFKRGDIVVLNGEEKKTHYVKRLIGLPGDTIEMKNDNLFVNGKRFNEEYLKENKKDAHDSDL    114
SipU (Bsu)       VKWTGGFHRGDIIVIEDKKSGRSFVKRLIGLPGDSIKMKNDQLYINDKKVEEPYLKEYKQEVKESGV    129
SipV (Bsu)       SHRFKTIHRFDIVLFKGPDHKV-LIKRVIGLPGETIKYKDDQLYVNGKQVAEPFLKHLKSVS---AG    116
                         * .*...         .**.** **...  *.* *      ..**.* . *

                                         E
                                      ★     ★
SipS (Bsu)       DHLTDDF------GPVKVPDNKYFVMGDNRRNSMDSRNGLGLFTKKQIAGTSKFVFYPFNEMRKTN     184
SipT (Bsu)       S-LTGDF------GPVKVPKGKYFVMGDNRLNSMDSRNGLGLIAEDRIVGTSKFVFFPFNEMRQTK     193
SipP (pTA1015)   N-LTGDF------GPIKVPKDKYFVMGDNRQESMDSRNGLGLFTKDDIQGTEEFVFFPFSNMRKAK     186
SipP (pTA1040)   N-LTGDF------GPIKVPKDKYFVMGDNRQNSMDSRNGLGLFNKKDIVGVEELVFFPLDRIRHAK     185
SipU (Bsu)       T-LTGDF-------EVEVPSGKYFVMGDNRLNSLDSRNGMGMPSEDDIIGTESLVFYPFGEMRQAK     187
SipV (Bsu)       SHVTGDFSLKDVTGTSKVPKGKYFVVGDNRIYSFDSRH-FGPIREKNIVGV----------ISDAE     168
                  ..* **     .  .**  **** ****  * ***.. *   .. * *
```

Figure 1. Type I SPases of *B. subtilis*. Comparison of the deduced amino acid sequences of type I SPases from *B. subtilis*. Identical amino acids [•], or conservative replacements [.], are marked. Putative transmembrane segments, indicated with A(nchor), were predicted as described by von Heijne *et al.* (1992). The conserved domains B-E, which are present in all known type I SPases of prokaryotic and eukaryotic organisms (van Dijl *et al.*, 1992), are indicated. Conserved residues which are critical for the activity and/or stability of SipS (Bsu) are marked (★) (van Dijl *et al.*, 1995).

Several homologous type I SPases have been identified in Gram-positive and Gram-negative bacteria, yeast mitochondria and the endoplasmic reticulum of yeast and higher eukaryotes. Considerable similarities can be observed between the known type I SPases, but only a few residues are strictly conserved in all enzymes. In the prokaryotic and mitochondrial enzymes these include serine and lysine residues, which are essential for enzymatic activity, possibly by forming a catalytic dyad (see: Dalbey and von Heijne, 1992; van Dijl, 1992; Dalbey *et al.*, 1997).

Thus far, four chromosomally-encoded type I SPases from *B. subtilis* have been identified (with EMBL, GenBank, DDBJ accession numbers in parentheses): SipS (Z11847), SipT (U45883), SipU (D38161), and SipV (M. A. Noback and S. Bron, unpublished results). Moreover, some strains of *B. subtilis* also contain cryptic plasmids specifying type I SPases (*sipP* [pTA1015; Z27459] and *sipP* [pTA1040; Z36269]; Meijer *et al.*, 1995). The known type I SPases from *B. subtilis* share a high degree of sequence similarity (30.5 % identical or conserved residues; Fig. 1). Furthermore, all these enzymes are 21 kDa proteins (on the avarage 186 residues) with one putative membrane spanning domain, a small amino-terminal domain with a predicted cytoplasmic localisation, and a large carboxyl-terminal domain with a predicted extra-cytoplasmic localisation. Finally, none of the SPases of *B. subtilis* is, by itself, essential for cell viability (van Dijl *et al.* 1992; Bolhuis *et al.*, 1996; our unpublished results).

In contrast to *B. subtilis*, *E. coli* contains only one type I SPase (Lep), which is essential for viability. Moreover, Lep of *E. coli* (323 residues) is much larger than the SPases of *B. subtilis*, and it has two amino-terminal membrane-spanning domains; the amino-terminus being located in the periplasm.

3. Effects of the absence of SipS on protein secretion in *B. subtilis degU32*(Hy) strains

Thus far, SipS (signal peptidase of *Bacillus subtilis*) is the best-studied SPase from a Gram-positive bacterium (van Dijl *et al*, 1992; van Dijl *et al.*, 1995; Bolhuis *et al.*, 1996). We have recently shown that SipS is required for the efficient secretion of various (unidentified) proteins into the growth medium. Unexpectedly, however in the same studies, we observed that the secretion of at least two other proteins of *B. subtilis* was increased in the absence of SipS. Unfortunately, none of these secreted proteins could be identified readily (Bolhuis *et al.*, 1996).

The *degU32*(Hy) mutation results in the overproduction of several degradative enzymes, such as levansucrase, subtilisin E, and the neutral protease NprE, which can be detected in the growth medium upon SDS-PAGE and Western blotting (see: Ferrari *et al.* 1993). To determine whether, and to what extent, the secretion of the latter proteins is affected in the absence of SipS, we introduced the *degU32*(Hy) mutation into *B. subtilis* 8G5 (parental strain) and *B. subtilis* 8G5 *sipS* (lacking a functional *sipS* gene; Bolhuis *et al.*, 1996). First, the effects of the *degU32*(Hy) mutation on protein secretion in the presence or absence of SipS were examined by SDS-PAGE.

The results confirmed our previous findings obtained with strains containing a wild-type *degU* gene: compared to the growth medium of the parental *degU32*(Hy) strain, the growth medium of the *degU32*(Hy) strain lacking SipS contained some secreted proteins in strongly reduced, similar, or increased quantities (Fig. 2A). Next, the secretion of levansucrase, subtilisin and NprE into the growth medium was examined by Western blotting. The results showed that the secretion of levansucrase was significantly increased in the absence of SipS (Fig. 2B). In contrast, the secretion of subtilisin was not affected (Fig. 2C), and that of NprE was slightly reduced (Fig. 2D). These findings show that the efficient secretion of only a subset of the secreted proteins, such as NprE, requires the presence of SipS, suggesting that the corresponding precursors are preferred substrates for SipS. Because none of the secreted proteins appeared to be completely absent from the growth medium of cells lacking SipS, it seems that their precursors can also be processed by other SPases of *B. subtilis*, such as SipT, SipU and SipV, albeit at a reduced efficiency. Opposite, the presence of SipS interferes with the secretion of other proteins, like levansucrase, suggesting that their precursors are preferred substrates for SipT, SipU or SipV.

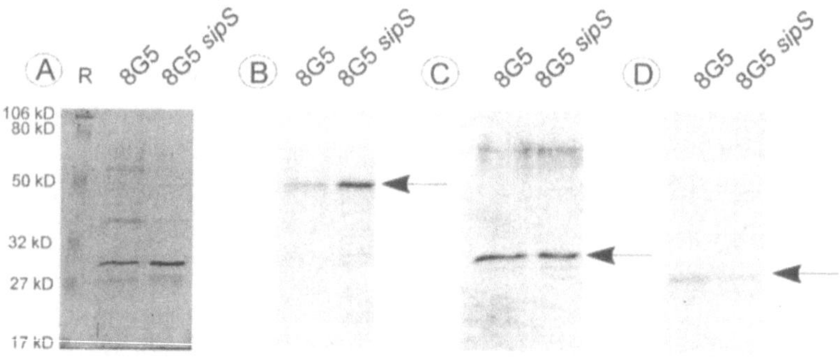

Figure 2. Protein secretion by *B. subtilis* 8G5 *sipS*. Proteins secreted by *B. subtilis* 8G5 *degU(32)*Hy (first lane) and *B. subtilis* 8G5 *degU(32)*Hy *sipS* (second lane) were compared. After overnight growth on glucose minimal medium, cells were removed from the culture by centrifugation. Secreted proteins were concentrated ten-fold and separated by SDS-PAGE. All samples were prepared from equal amounts of medium. Secreted proteins were visualised by staining with Coomassie Brilliant Blue (**A**), or Western blotting, using specific antibodies against levansucrase (**B**); subtilisin (**C**); and the neutral protease NprE (**D**). Positions of the latter three proteins and the molecular masses of the reference markers (R) are indicated.

4. Processing of secretory precursor proteins in the absence of SipS

For technical reasons, we were unable to study the rates of processing of the precursors of levansucrase, subtilisin and NprE by pulse-chase labeling. However, this type of analysis was possible with other precursor proteins, such as the hybrid precursor pre(A13i)-ß-lactamase (van Dijl *et al.*, 1992) and the precursor of the α-amylase AmyQ of *Bacillus amyloliquefaciens* (Palva, 1982; Kontinen and Sarvas, 1988).

Pre(A13i)-ß-lactamase, specified by plasmid pGDL42 (van Dijl *et al.*, 1992) was processed at a reduced rate in *B. subtilis* 8G5 *sipS* as compared to the parental strain (Fig. 3). In the presence of wild-type levels of SipS, about 35% of the total (A13i)-β-lactamase was mature after a chase of 10 minutes, whereas in the absence of SipS, only about 18% of the total (A13i)-β-lactamase was in its mature form under the same conditions. Efficient processing of pre(A13i)-β-lactamase could be restored in *B. subtilis* 8G5 *sipS* by transforming the strain with plasmid pGDL41 (van Dijl *et al.*, 1992), a derivative of pGDL42 containing the *sipS* gene (Fig. 3). This resulted in pre(A13i)-β-lactamase processing at a high rate: after 10 minutes of chase, about 86% of the total (A13i)-β-lactamase was mature. In fact, the rate of processing was even significantly increased compared to the parental strain, which can be attributed to the fact that SipS is overproduced in strains containing pGDL41 (Meijer *et al.* 1995; Bolhuis *et al.* 1996).

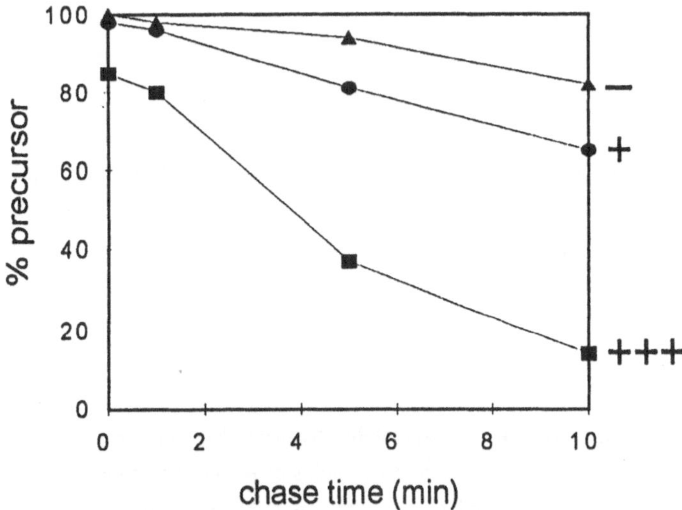

Figure 3. Kinetics of processing of pre(A13i)-β-lactamase in *B. subtilis* 8G5 *sipS* (pGDL42; no *sipS*, indicated by '-'), *B. subtilis* 8G5 (pGDL42; single copy of *sipS*, indicated by '+'), and *B. subtilis* 8G5 *sipS* (pGDL41; multiple copies of *sipS*, indicated by '+++'). Cells were labeled with [^{35}S]-methionine for 60 s prior to chase with excess non-radioactive methionine, and samples were withdrawn at the times indicated. The relative amounts of the precursor and mature forms of (A13i)-β-lactamase in each sample were determined by immunoprecipitation, SDS-PAGE and fluorography. The kinetics of processing are plotted as the percentage of the total (A13i)-β-lactamase (precursor + mature) in the precursor form at the time of sampling.

In contrast to pre(A13i)-ß-lactamase, pre-AmyQ, specified by plasmid pKTH10 (Palva, 1982; Kontinen and Sarvas, 1988), was processed at an increased rate in the absence of SipS. Pulse-chase labeling experiments revealed that, after a chase of 4 min, almost all AmyQ was mature in *B. subtilis* 8G5 *sipS*, whereas, under the same conditions, the parental strain still contained about 28% of the labeled AmyQ in the precursor form (Fig. 4; Bolhuis *et al.* 1996).

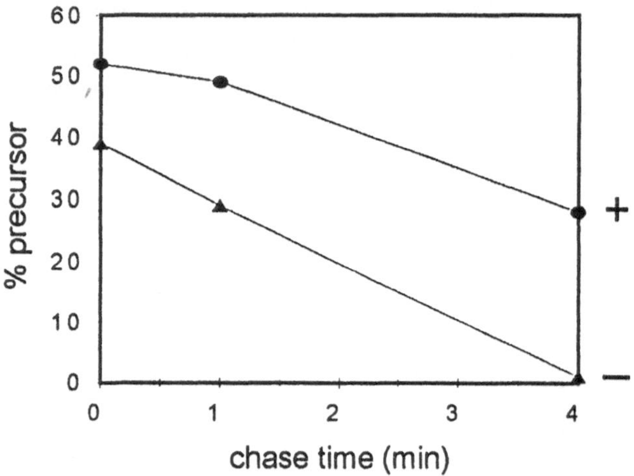

Figure 4. Kinetics of processing of *B. amyloliquefaciens* α-amylase in *B. subtilis* 8G5 (pKTH10; one copy of *sipS*, indicated by '+') and *B. subtilis* 8G5 *sipS* (pKTH10; no *sipS*, indicated by '-'). Cells were labeled with [^{35}S]-methionine for 60 s prior to chase with excess non-radioactive methionine, and samples were withdrawn at the times indicated. The kinetics of processing were determined as in Fig. 3.

5. Discussion

In paragraphs 3 and 4, we have described that the absence of SipS has very different effects on the processing and secretion of various secretory proteins of *B. subtilis*. What can we conclude from these observations and, in particular, how can we explain the unexpected finding that the processing and secretion of some proteins is improved in the absence of SipS?

At least three conclusions can be drawn from the observation that certain precursor proteins, such as pre(A13i)-ß-lactamase and at least one other precursor (pre-staphylokinase; data not shown; Bolhuis *et al.* 1996), are processed at a reduced rate in *B. subtilis* strains lacking SipS. First, SipS is involved in the processing of these precursors in the parental strain; second, in the absence of SipS, these precursors are processed by other SPases; and third, these other SPases can not fully compensate for the absence of SipS, indicating that the availability of SPases is limiting in the absence of SipS. The latter conclusion is corroborated by the fact that the rate of processing of pre(A13i)-ß-lactamase was not only restored but, in fact, strongly increased when SipS was overproduced in a strain lacking an intact chromosomal copy of *sipS*. The observation that cells lacking SipS secrete various proteins, such as NprE, at reduced levels into the growth medium is in accord with the observation that processing of secretory precursor proteins by SPase is a prerequisite for the release of the translocated mature protein from the cytoplasmic membrane (see: Dalbey *et al.*, 1997).

The observations that, in the absence of SipS, the rate of pre-AmyQ processing was increased and that certain proteins, such as levansucrase, accumulated at increased levels in the medium, can not be explained in a simple way. However, at least three explanations for these phenomena are conceivable. First, by analogy to the situation with Lep of *E. coli* (Wolfe *et al.* 1985), it seems probable that SipS requires the general protein-secretion pathway for its insertion into the membrane of *B. subtilis*. Thus, SipS could compete with secreted proteins for export sites, which would result in reduced levels of precursor processing, and subsequent, secretion of the mature protein into the medium. However, we consider this explanation unlikely, because SipS is produced in very low quantities as compared to AmyQ and levansucrase. Second, the expression of other SPase-encoding genes, such as *sipT, sipU*, or *sipV*, might be increased in the absence of SipS. However, preliminary results indicate that this is not the case (our unpublished results). In fact, we have shown previously that the transcription from the *sipS* promoter is not altered in the absence of a functional SipS protein (Bolhuis *et al.*, 1996). Third, as the presence of SipS seems to interfere with efficient processing of pre-AmyQ, SipS might compete with SipT, SipU, and SipV for binding to certain precursors, such as pre-AmyQ, that emerge at the outer surface of the membrane. If so, improved processing in the absence of SipS would be expected if SipS would also bind precursors, of which the signal peptides are removed more efficiently by SipT, SipU and/or SipV. A model in which these ideas are illustrated is shown in Fig. 5. Thus, pre-AmyQ seems to be a preferred substrate for one, or more, of the other SPases of *B. subtilis*, and a poor substrate for SipS (Bolhuis *et al.*, 1996).

Improved processing of pre-AmyQ in cells lacking SipS did not result in increased amounts of AmyQ in the medium (data not shown). This indicates that processing by SPase I is not a rate limiting step in the secretion of AmyQ. Indeed, it was shown that the lipoprotein PrsA, which is believed to promote the folding of secreted α-amylase into a protease-resistent conformation, sets a limit for high-level secretion of this enzyme (Kontinen and Sarvas, 1993). In contrast, in the absence of SipS, the cells secreted increased amounts of levansucrase into the medium. The latter finding shows that the presence of SipS sets, either directly or indirectly, a limit for to the secretion of levansucrase.

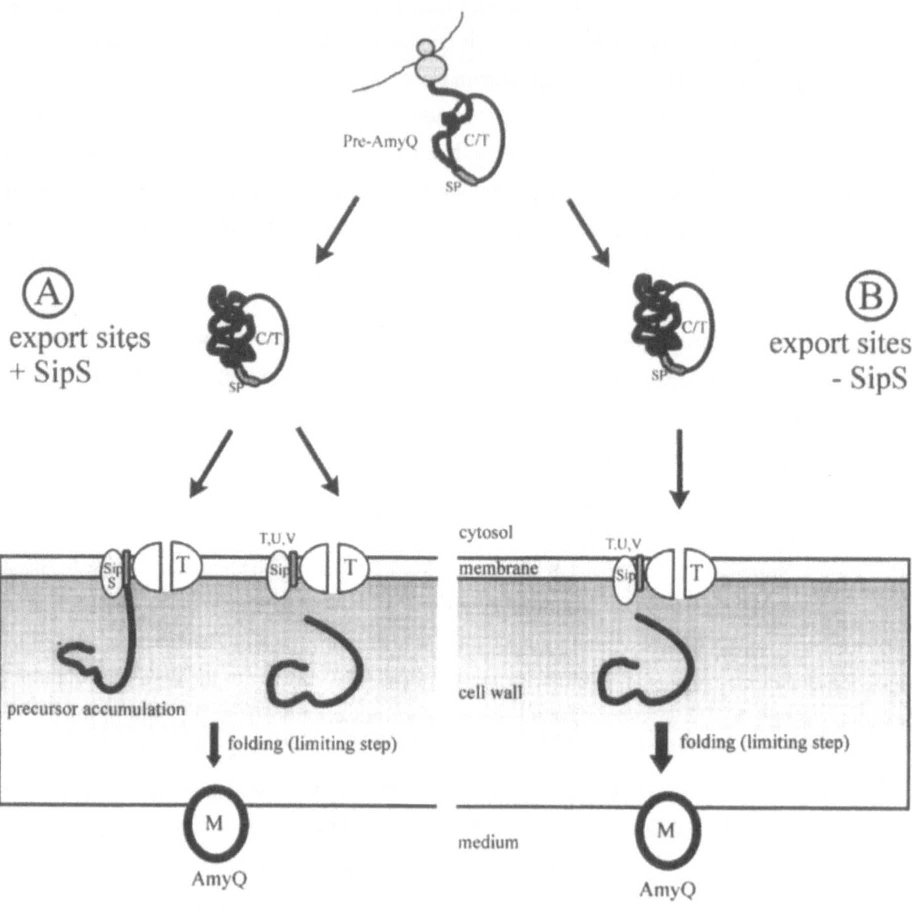

Figure 5. Competition model for binding and cleavage of pre-AmyQ by type I SPases. Like most exported proteins, pre-AmyQ is synthesised in the cytoplasm with an amino-terminal signal peptide (SP), which is essential for targeting of the precursor to the translocase (T) and initiation of protein translocation across the membrane. Prior to translocation, certain cytoplasmic factors are involved in maintaining the precursor in an export-competent state and targeting the precursor to the membrane (indicated with C/T for chaperones and targeting factors). During or shortly after translocation, the signal peptide is bound and removed by one of the SPases of *B. subtilis*. **A,** Pre-AmyQ bound to SipS is inefficiently cleaved, resulting in the accumulation of membrane-bound pre-AmyQ. Binding to one of the other SPases results in efficient processing of pre-AmyQ to the mature form. **B,** In the absence of SipS, pre-AmyQ is efficiently cleaved by SipT, SipU and/or SipV. No pre-AmyQ is "trapped" by SipS. Nevertheless, cells lacking SipS do not secrete increased amounts of AmyQ into the medium because the folding of the mature protein into its native, protease-resistant conformation sets a limit for high-level secretion of AmyQ (Kontinen and Sarvas, 1993).

6. Perspectives

Recently, we were able to show that, like SipS, SipT, SipU and SipV are able to cleave pre(A13i)-ß-lactamase, indicating that these four SPases of *B. subtilis* have a similar substrate specificity. Nevertheless, our present results indicate that the substrate specificities of at least some of these enzymes are not identical, as they seem to prefer different precursor proteins. Whether this preference is based on structural properties of the precursor proteins, or reflects more indirect effects, such as the way the precursor is delivered to the signal peptidase, is presently not known. To address this question, we are currently purifying the SPases of *B. subtilis*, which should allow us to compare the substrate specificities of these enzymes *in vitro*. Furthermore, we are combining various *sip* gene mutations to determine which SPase(s) are most important for precursor processing in *B. subtilis*. Finally, we are investigating the effects of (multiple) *sip* gene disruptions on the expression of the remaining *sip* genes. These investigations may provide answers to two intriguing questions: why has *B. subtilis* acquired so many SPases during its evolution, and how are the activities of all these enzymes concerted?

Acknowledgements. We thank J. Jongbloed, M. van Roosmalen and M. Noback for valuable discussions. A. B. was supported by Biotechnology Grant (BIO2-CT93-0254) from the European Union. H.T. was supported by Genencor International (Delft, the Netherlands) and Gist-brocades (Delft, the Netherlands). J.M.v.D. and S.B. were supported in part by Biotechnology Grants (BIO2-CT93-0254 and BIO4-CT96-0097) from the European Union.

References

Bolhuis, A., Sorokin, A., Azevedo, V., Ehrlich, S.D., Braun, P.G., de Jong A., Venema G., Bron, S., and van Dijl, J.M. (1996) *Bacillus subtilis* can modulate its capacity and specificity for protein secretion through temporally controlled expression of the *sipS* gene for signal peptidase I. Mol. Microbiol. 22: 605-618.

Dalbey, R.E. (1994) Bacterial signal peptidase I. In: Signal peptidases. von Heijne, G. (ed.). R.G. Landes Company, Austin, TX, USA, 5-15.

Dalbey, R.E., and von Heijne, G. (1992) Signal peptidases in prokaryotes and eukaryotes -a new protease family. Trends. Biochem. Sci. 17: 474-478.

Dalbey, R.E., Lively, M.O., Bron, S., and van Dijl, J.M. (1997) The chemistry and enzymology of the type I signal peptidases. Protein Sci. 6:1129-1138.

Driessen, A.J.M. (1994) How proteins cross the bacterial cytoplasmic membrane. J. Membr.Biol. 142: 145-159.

Ferrari, E., Jarnagin, A.S, and Schmidt, B.F. (1993) Commercial production of extracellular enzymes. In *Bacillus* subtilis and other Gram-positive bacteria. Sonenshein, A.L., Hoch, J.A., and Losick, R., (eds.). American Society for Microbiology, Washington, DC. 713-726

Kontinen, V.P., and Sarvas, M. (1988) Mutants of *Bacillus subtilis* defective in protein export. J. Gen. Microbiol. 134: 2333-2344.

Lory, S. (1994) Leader peptidases of type IV prepilins and related proteins. In: Signal peptidases. von Heijne,G. (ed.). R.G. Landes Company, Austin, TX, USA. 31-48.

Meijer, W.J.J., de Jong, A., Wisman, G.B.A., Tjalsma, H., Venema, G., Bron, S., and van Dijl, J.M. (1995) The endogenous *Bacillus subtilis (natto)* plasmids pTA1015 and pTA1040 contain signal peptidase-encoding genes: identification of a new structural module on cryptic plasmids. Mol. Microbiol. 17: 621-631

Nagarajan, V., (1993) Protein secretion. *In Bacillus subtilis* and other Gram-positive bacteria. Sonenshein, A.L., Hoch, J.A., and Losick, R., (eds.). American Society for Microbiology, Washington, DC. 713-726

Palva, I. (1982) Molecular cloning of α-amylase gene from *Bacillus amyloliquefaciens* and its expression in *Bacillus subtilis*. Gene 19: 81-87.

Pugsley, A.P. (1993) The complete general secretory pathway in Gram-negative bacteria. Microbiol. Rev. 57: 50-108.

Sankaran, K., and Wu, H.C. (1994) Signal peptidase II. In: Signal peptidases. von Heijne, G.(ed.). R.G. Landes Company, Austin, TX, USA, 17-29.

Simonen, M., and Palva, I. (1993) Protein secretion in *Bacillus* species. Microbiol. Rev. 57: 109-137.

Van Dijl, J.M., de Jong, A., Vehmaanperä, J., Venema, G., and Bron, S. (1992) Signal peptidase I of *Bacillus subtilis*: patterns of conserved amino acids in prokaryotic and eukaryotic type I signal peptidases. EMBO J. 11: 2819-2828.

Van Dijl, J.M., de Jong, A., Venema, G., and Bron, S (1995). Identification of the potential active site of the signal peptidase SipS of *Bacillus subtilis*: structural and functional similarities with LexA-like proteases. J. Biol. Chem. 270: 3611-3618.

Von Heijne, G. (1990) The signal peptide. J. Membrane Biol. 115: 195-201.

Von Heijne, G. (1992) Membrane protein structure prediction. Hydrophobicity analysis and the positive-inside rule. J.Mol.Biol. 225:487-494.

Wickner, W., Driessen, A.J.M., and Hartl, F-U. (1991) The enzymology of protein translocation across the *Escherichia coli* plasma membrane. Annu. Rev. Biochem. 60, 101-124.

Wolfe, P.B., Rice, M., and Wickner, W. (1985) Effects of two *sec* genes on protein assembly into the plasma membrane of *Escherichia coli*. J. Biol. Chem. 260: 1836-1841.

Studies on N-Glycosylation of Proteins in the Endoplasmic Reticulum

Qi Yan, Hangil Park, Glenn D. Prestwich[2] and William J. Lennarz[*1]
[*1]Department of Biochemistry and Cell Biology and the Institute for Cell and Developmental Biology, State University of New York at Stony Brook, Stony Brook, NY 11794, USA
[2]Department of Medicinal Chemistry, University of Utah, Salt Lake City, UT 84112, USA

Introduction

The N-glycosylation of proteins at asparagine residues in polypeptide chains is a ubiquitous modification in eukaryotic organisms that occurs at the lumenal face of the rough endoplasmic reticulum (ER). Proteins that are destined to become N-linked glycoproteins are synthesized on membrane bound ribosomes, and glycosylated during the process of translocation into the ER (cotranslational translocation). Subsequently these glycoproteins traverse the endomembrane system where they may be further processed, and then either delivered to various organelles, the plasma membrane, or secreted. In many cases oligosaccharide chains of these proteins are essential because if they are absent or altered, several processes can be affected. These processes are: (1) protein stability; (2) subcellular localization; (3) normal folding; or (4) protein secretion. The diverse biological roles for N-linked oligosaccharides have been reviewed (Varki, 1993).

The membrane associated enzyme that plays an essential role in the glycosylation process is OST[1]. As shown in Figure 1 below, it catalyzes the transfer of a preassembled high-mannose oligosaccharide from a lipid-linked oligosaccharide donor (Dol-PP-$GlcNAc_2Man_9Glc_3$) onto asparagine acceptor sites on nascent polypeptide chains being translocated into the lumen of the rough ER. The consensus sequence of the acceptor is -Asn-X-Thr/Ser-, where X is any amino acid other than proline (Gavel & von Heijne, 1990). This acceptor sequence specificity for N-linked glycosylation was confirmed using synthetic peptides as substrates in OST assays (Hart *et al.*, 1979; Bause & Legler, 1981; Bause, 1983). A simple tripeptide sequence N-X-T/S is the minimal sequence necessary for glycosylation, providing that the NH_2 and COOH termini are blocked by reagents that mimic peptide bonds (Welply *et al.*, 1983). Although the primary consensus sequence for N-linked glycosylation is required for the glycosylation reaction *in vitro*, not all potential sequences are modified *in vivo*. In addition, it has been shown that synthetic peptides containing a proline after the Thr or Ser of N-X-T/S are not glycosylated *in vitro* by OST

NATO ASI Series, Vol. H 106
Lipid and Protein Traffic
Pathways and Molecular Mechanisms
Edited by Jos A. F. Op den Kamp

(Bause, 1983). Peptides that contain N-X-T are 40-fold better substrates than peptides with N-X-S sequence when assayed as acceptors *in vitro*. It has been shown that factors other than primary and or secondary structures, also affect the glycosylation efficiency. Consensus sites that reside within 12-14 residues from a transmembrane domain of a protein are not modified *in vivo* (Nilsson & von Heijne, 1993). Furthermore, it is clear that a number of unglycosylated N-X-T/S sites can be posttranslationally glycosylated *in vitro* after unfolding, thereby demonstrating that certain secondary or tertiary features may interfere with N-linked glycosylation (Pless & Lennarz, 1977; Kronquist & Lennarz, 1978).

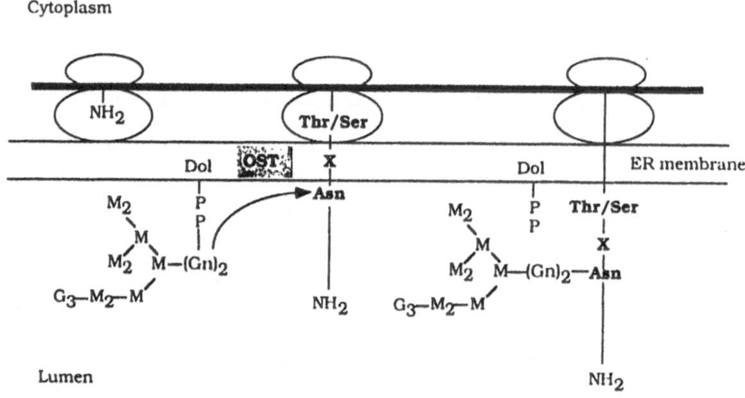

Figure 1. Diagram of the reaction catalyzed by OST.

The preferred oligosaccharide donor for glycosylation is Dol-PP-$GlcNAc_2Man_9Glc_3$, which is synthesized sequentially by a series of glycosyltransferases in the membrane of the RER. The exact topology of this synthesis is still a matter of debate (Lennarz, 1987; Abeijon & Hirschberg, 1992), although the final lipid linked oligosaccharide product is believed to be lumenally oriented. Incompletely assembled dolichol-linked oligosaccharides can also be transferred by OST *in vivo* and *in vitro* with lower efficiency (Sharma *et al.*, 1981; Huffaker & Robbins, 1983).

---

[1]**Abbreviations:**

Am, Amide; bh, Bolton-Hunter reagent; BP, benzophenone; Dol, dolichol; OST, oligosaccharyl transferase

Many studies have been done characterizing the substrates of OST, but only recently has information about the enzyme itself become available. This enzyme functions at a very unique junction of three RER-associated biosynthetic pathways: (1) formation of the lipid-linked oligosaccharide donor, (2) translocation of the nascent polypeptide acceptor into the ER lumen, and (3) chaperone-mediated folding and oligomeric-assembly reactions that are required to produce functional glycoproteins which are competent for subsequent intracellular transport. The yeast OST was initially purified as a hexameric complex consisting of six polypeptides (α, β, γ, δ, ε, ζ subunits) with enzymatic activity (Kelleher & Gilmore, 1994). Subsequently, two other groups (Knauer & Lehle, 1994; Pathak *et al.*, 1995) have described a tetrameric OST complex that lacks ε and ζ subunits. So far eight subunits have been cloned that may be components of this enzyme complex. Five of them are essential genes (Ost1, Ost2, Wbp1, Swp1 and Stt3). Based on the amino acid sequences of the characterized subunits, the yeast enzyme complex shows significant homology to the OST complexes purified from higher eukaryotes.

The OST initially purified from canine ER is a trimeric protein complex consisting ribophorin I, ribophorin II and a 48-kD protein named OST48 (Kelleher *et al.*, 1992; Silberstein *et al.*, 1992). The α subunit of yeast OST complex (Ost1p) is an essential type I membrane protein, which is heterogeneously glycosylated. Ost1p is 28% identical in sequence to human and rat ribophorin I (Silberstein *et al.*, 1995b). The β subunit of OST is the essential Wbp1 protein, and it is homologous to the canine OST 48 subunit (Silberstein *et al.*, 1992). Inactivation of the yeast OST with a cysteine-directed protein modification reagent, methyl methanethiosulfonate, correlates with the labeling of Wbp1 protein. Addition of Dol-PP-GlcNAc$_2$ protects the enzyme from inactivation. This suggests that Wbp1p may contain a site that influences the binding of the lipid-linked oligosaccharide substrate (Pathak *et al.*, 1995). More than a decade ago, Bause's group found that pig liver OST is inactivated irreversibly by a hexapeptide in which threonine has been substituted with epoxyethylglycine in the N-X-T consensus sequence. Very recently, the same group observed labeling of two peptides when they incubated the partially purified pig liver OST complex with Dol-PP-linked [^{14}C] oligosaccharides and the N-3,5-dinitrobenzoylated epoxy peptide. Although, the MW's of these two labeled proteins are very close to those of two of the OST subunits, ribophorin I and Ost48p (66 and 48 kDa, respectively), the evidence showing that these two labeled peptides are the two OST subunits is indirect since the only antibodies used to identify them were directed against the dinitrobenzene group (Bause *et al.*, 1997).

The δ subunit, Swp1p originally was identified as an allele-specific high copy suppressor of the *wbp1-2* mutant. *In vivo* depletion experiments indicate that Swp1p is also required for OST activity in yeast. A protein sequence comparison revealed that Swp1 protein is related to the carboxy-terminal half of mammalian ribophorin II. Consistent with the genetic interactions between these two subunits, it was found that Wbp1p and Swp1p could be chemically cross-linked in the yeast membrane (te Heesen *et al.*, 1993).

The OST2 gene that encodes the ε subunit of the OST was found to be essential for vegetative growth of yeast (Silberstein *et al.*, 1995a). This protein was found to be 40% identical in sequence to the human, hamster, and *X. laevis* Dad1 (defender against apoptotic cell death) protein. It has been proposed that Dad1 protein is a negative regulator of apoptosis. Recently there is evidence which shows that Dad1p is tightly associated with subunits of the mammalian OST both in the intact membrane and in the purified enzyme (Kelleher & Gilmore, 1997). Induction of a cell death pathway upon loss of DAD1 may reflect the essential nature of N-linked glycosylation.

All ER sugar transferases and synthases identified thus far are membrane-associated proteins that are composed of a single polypeptide chain. However, the yeast OST complex obviously is much more complicated because it is composed of at least eight different polypeptide subunits. To understand the biochemical functions of these subunits, we are developing photoprobes containing BP derivatives. BP probes have several advantages compared to other photoprobes: (1) they are chemically more stable than diazo esters, aryl azides, and diazirines; (2) they can be manipulated in ambient light and can be activated at 350-360nm, avoiding protein-damaging wavelengths, (3) BPs react preferentially with normally unreactive C-H bonds, even in the presence of aqueous solvent and bulk nucleophiles. These three properties combine to allow highly efficient covalent modifications of macromolecules, frequently with high site specificity. Based on modeling and experimental data, the reactive volume of the BP moiety has been approximated to be a sphere with a radius of 3.1Å centered on the ketone oxygen. In the presence of UV light, the ketone group of BP will form a diradical. This excited state (triplet state) may last 80-120 μs in the absence of an abstractable hydrogen, but will react rapidly in the presence of a suitably oriented H-donor. However, the triplet state will relax to the ground state if there is no C-H bond in close proximity to the activated probe (Dorman & Prestwich, 1994).

Although progress has been made in cloning the subunits and defining which protein subunits of the yeast enzyme are essential for viability, relatively little progress has been made in defining the role of individual subunits in catalysis. Our objective is to identify the subunit(s) of yeast OST that interact with the peptide substrate, and to study the possible interaction between subunit(s), that interact with the peptide substrate and other subunits.

Results and Discussion

BPA probe acts as a substrate for OST. To search for the peptide recognition subunit(s) of oligosaccharyl transferase, ^{125}I-bh-Asn-BPA-Thr-Am and ^{3}H-Ac-Asn-BPA-Thr-Am, two tripeptides containing the Asn-X-Thr/Ser acceptor sequence, plus a benzophenone modification on the side chain of phenylalanine in the X position, have been synthesized. It is assumed that binding of these two tripeptides to one or more subunits of OST would occur because they are potential substrates. Initially, both tripeptides were tested as substrates for N-glycosylation in the absence of photolysis. Microsomes were incubated with radiolabeled probe, and quantitation of the formation of labeled glycopeptide was carried out by the use of ConA-agarose beads, which bind α-mannosyl residues. The results indicate that when the concentration of labeled ^{3}H-Ac-Asn-BPA-Thr-Am was increased, the amount of glycosylated product also increased. Furthermore, when the concentration of enzyme was increased by increasing the amount of microsomes, the glycosylated peptide increased proportionally. However, when the concentration of an unlabeled peptide, Ac-Asn-Phe-Thr-Am, which is also a substrate for N-glycosylation, was increased, formation of ^{3}H-glycopeptide decreased, as expected if the cold acceptor peptide is competing with the labeled one. These findings established that ^{3}H-Ac-Asn-BPA-Thr-Am is a substrate for OST. Similar experiments with the ^{125}I-bh-Asn-BPA-Thr-Am probe gave similar results.

BPA probe is photoactivatable. Having established that ^{125}I-bh-Asn-BPA-Thr-Am and ^{3}H-Ac-Asn-BPA-Thr-Am are substrates of OST, their ability to serve as photoprobes was studied by determining if, following photolysis, ^{125}I-bh-Asn-BPA-Thr-Am labeled one or more microsomal proteins. It was found that this peptide radiolabeled several ER components in a photolysis-dependent manner. Irradiation of microsomes in the presence of ^{125}I-bh-Asn-BPA-Thr-Am produced three radiolabeled bands detected upon analysis by SDS-PAGE, with apparent molecular weights of approximately 55, 12 and 3 kDa. When the microsomes were incubated with ^{125}I-bh-Asn-BPA-Thr-Am in the dark without irradiation, a band was still detected at approximately 55 kDa; thus, the radiolabeling of this band is photolysis-independent. However, when microsomes were inactivated by either heating or by adding 4-fold molar excess of EDTA (to deplete the divalent cations which are necessary for OST activity), the radiolabeling of both the 55 kDa and 12 kDa bands was abolished. These results indicate that labeling of 55 kDa and 12 kDa bands requires active microsomes. Since the labeling of the 3 kDa band still occurred, it clearly is not enzymatic. Taken together, these data suggest that the labeling of 12 kDa band is photolysis-dependent, and is the result of an enzyme dependent reaction.

Radiolabeling of the 12 kDa band is due to the reactive ketone group of BPA. As described above, formation of the 55 kDa band does not require photolysis, but does require enzymatically active microsomes. It is assumed that the reactive group in bh-Asn-BPA-Thr-Am is benzophenone. To investigate the possibility that labeling of 55 kDa band is due to a chemical reaction with BP in the absence of UV light, and to further confirm that the 12 kDa labeling is due to the crosslinking with ketone group of BPA, the ketone group of BPA was reduced, and then the reduced photoprobe was tested for its ability to label the 55, 12 and 3 kDa bands. When reduced probe was incubated with microsomes, the 55 kDa band was still formed. This finding demonstrated that labeling of the 55 kDa band is not due to its reaction of the ketone group on the photoprobe. In contrast, the 12 kDa and 3 kDa bands were not labeled, indicating that the appearance of the 12 kDa and 3 kDa bands was related to the reaction of the ketone group of benzophenone.

Thermal denaturation of OST. Since earlier experiments established that boiled microsomes resulted in the loss of ability to radiolabel the 55 kDa and 12 kDa bands, and it was known that OST is heat-labile, the effect of temperature pretreatment of microsomes on both enzyme activity and photoaffinity labeling of OST was compared. It was found that formation of labeled glycopeptide progressively decreased when microsomes were preincubated at increasing temperatures. Compared to the unheated control, preincubation of microsomes at 55°C destroyed about 55% of the glycosylation activity and incubation at 70°C reduced the activity to background level. Similarly, the radiolabeling of the 12 kDa band decreased after preincubation of microsomes at elevated temperatures. Thus, the loss of glycosylation activity upon heating of microsomes parallels the loss of the ability of ^{3}H-Ac-Asn-BPA-Thr-Am to interact and radiolabel the 12 kDa component.

Photolabeling is inhibited by an OST substrate, Ac-N-F-T-Am, but is unaffected by a peptide that is not an OST substrate, Ac-Asp(NHCH$_3$)-L-T-Am. To establish that photolabeling is specific for the OST substrate, a competition assay was performed using an unlabeled peptide that was either a known OST substrate or a peptide known to not be an OST substrate. When microsomes were incubated with ^{125}I-bh-Asn-BPA-Thr-Am and an increasing amount of a competing acceptor peptide, Ac-Asn-Phe-Thr-Am, labeling of the 55 kDa and the 12 kDa bands decreased. Conversely, when microsomes were incubated with the labeled probe and an increasing amount of a nonacceptor peptide, Ac-Asp(NHCH$_3$)-Leu-Thr-Am, in which the amido group on the Asn side chain was methylated, the labeling of these two bands did not decrease. Formation of the 3 kDa band was not affected and appeared to be nonspecific, since its labeling did not decrease with an increasing amount of competitor.

The 12 kDa photolabeled component is membrane-associated. To further understand the connection between the formation of radiolabeled bands and the OST enzyme complex, the localization of the photolabeled 12 kDa product in the microsomes was investigated. Microsomes were irradiated in the presence of ^{125}I-bh-Asn-BPA-Thr-Am, and then sonicated for various time periods to rupture the membranes, thereby releasing lumenal contents. The released lumenal contents and membranes of the microsomes were separated by centrifugation and then analyzed by SDS-PAGE. It was found that the 12 kDa radiolabeled band was recovered in the membrane fraction, while the 55 kDa band was found in the lumenal contents. The separation of lumenal contents and membranes was confirmed by Western Blot with anti-BiP and anti-Wbp1p antibodies. Wbp1p is an ER membrane protein and it was found to be exclusively in the membrane fraction after centrifugation, while BiP is an ER lumenal protein and it was found to be in the soluble lumenal fraction. Since all of the OST subunits have been shown to be membrane-bound, these observations lend support to the idea that the 12 kDa band may be a component of OST.

<u>Photoaffinity labeling of microsomal proteins inhibits OST activity</u>. To further demonstrate that photolabeling correlates with OST activity, photolysis of microsomes was carried out with unlabeled photoprobe bh-Asn-BPA-Thr-Am. These microsomes were then tested for OST activity. If the photoprobe irreversibly binds to the peptide binding site, one would expect OST activity to decrease after photolysis. The results showed that photolysis with bh-Asn-BPA-Thr-Am inactivated OST activity in a time-dependent manner. After 5 minutes of photolysis, inactivation reached a maximum. In a control experiment, microsomes that were photolyzed without probe were found to retain most of their OST activity.

Summary

The current studies have identified a 12 kDa band in microsomes that is labeled in a manner consistent with it being a protein component of OST. Future studies will concentrate on testing this idea. If the 12 kDa protein is shown to be a subunit of OST it will be characterized by cloning and sequencing, and then its function will be studied.

Acknowledgment

This work was supported by grants GM33184 and GM33185 to WJL from the National Institutes of Health.

References

Abeijon, C. & Hirschberg, C. B. (1992). Topography of glycosylation reactions in the endoplasmic reticulum. *Trends Biochem. Sci.* **17**(1), 32-6.

Bause, E. (1983). Active-site-directed inhibition of asparagine N-glycosyltransferases with epoxy-peptide derivatives. *Biochem. J.* **209**(2), 323-30.

Bause, E. & Legler, G. (1981). The role of the hydroxy amino acid in the triplet sequence Asn-Xaa-Thr(Ser) for the N-glycosylation step during glycoprotein biosynthesis. *Biochem. J.* **195**(3), 639-44.

Bause, E., Wesemann, M., Bartoschek, A. & Breuer, W. (1997). Epoxyethylglycyl peptides as inhibitors of oligosaccharyltransferase: double-labelling of the active site. *Biochem. J.* **322**(Pt 1), 95-102.

Dorman, G. & Prestwich, G. D. (1994). Benzophenone photophores in biochemistry. *Biochemistry* **33**(19), 5661-73.

Gavel, Y. & von Heijne, G. (1990). Sequence differences between glycosylated and non-glycosylated Asn-X-Thr/Ser acceptor sites: implications for protein engineering. *Protein Eng.* **3**(5), 433-42.

Hart, G. W., Brew, K., Grant, G. A., Bradshaw, R. A. & Lennarz, W. J. (1979). Primary structural requirements for the enzymatic formation of the N-glycosidic bond in glycoproteins. Studies with natural and synthetic peptides. *J. Biol. Chem.* **254**(19), 9747-53.

Huffaker, T. C. & Robbins, P. W. (1983). Yeast mutants deficient in protein glycosylation. *Proc. Natl. Acad. Sci. U. S. A.* **80**(24), 7466-70.

Kelleher, D. J. & Gilmore, R. (1994). The Saccharomyces cerevisiae oligosaccharyltransferase is a protein complex composed of Wbp1p, Swp1p, and four additional polypeptides. *J. Biol. Chem.* **269**(17), 12908-17.

Kelleher, D. J. & Gilmore, R. (1997). DAD1, the defender against apoptotic cell death, is a subunit of the mammalian oligosaccharyltransferase. *Proc. Natl. Acad. Sci. U. S. A.* **94**(10), 4994-9.

Kelleher, D. J., Kreibich, G. & Gilmore, R. (1992). Oligosaccharyltransferase activity is associated with a protein complex composed of ribophorins I and II and a 48 kd protein. *Cell* **69**(1), 55-65.

Knauer, R. & Lehle, L. (1994). The N-oligosaccharyltransferase complex from yeast. *FEBS Lett.* **344**(1), 83-6.

Kronquist, K. E. & Lennarz, W. J. (1978). Enzymatic conversion of proteins to glycoproteins by lipid-linked saccharides: a study of potential exogenous acceptor proteins. *J. Supramol. Struct.* **8**(1), 51-65.

Lennarz, W. J. (1987). Protein glycosylation in the endoplasmic reticulum: current topological issues. *Biochemistry* **26**(23), 7205-10.

Nilsson, I. M. & von Heijne, G. (1993). Determination of the distance between the oligosaccharyltransferase active site and the endoplasmic reticulum membrane. *J. Biol. Chem.* **268**(8), 5798-801.

Pathak, R., Hendrickson, T. L. & Imperiali, B. (1995). Sulfhydryl modification of the yeast Wbp1p inhibits oligosaccharyl transferase activity. *Biochemistry* **34**(13), 4179-85.

Pless, D. D. & Lennarz, W. J. (1977). Enzymatic conversion of proteins to glycoproteins. *Proc. Natl. Acad. Sci. U. S. A.* **74**(1), 134-8.

Sharma, C. B., Lehle, L. & Tanner, W. (1981). N-Glycosylation of yeast proteins. Characterization of the solubilized oligosaccharyl transferase. *Eur. J. Biochem.* **116**(1), 101-8.

Silberstein, S., Collins, P. G., Kelleher, D. J. & Gilmore, R. (1995a). The essential OST2 gene encodes the 16-kD subunit of the yeast oligosaccharyltransferase, a highly conserved protein expressed in diverse eukaryotic organisms. *J. Cell Biol.* **131**(2), 371-83.

Silberstein, S., Collins, P. G., Kelleher, D. J., Rapiejko, P. J. & Gilmore, R. (1995b). The alpha subunit of the Saccharomyces cerevisiae oligosaccharyltransferase complex is essential for vegetative growth of yeast and is homologous to mammalian ribophorin I. *J. Cell Biol.* **128**(4), 525-36.

Silberstein, S., Kelleher, D. J. & Gilmore, R. (1992). The 48-kDa subunit of the mammalian oligosaccharyltransferase complex is homologous to the essential yeast protein WBP1. *J. Biol. Chem.* **267**(33), 23658-63.

te Heesen, S., Knauer, R., Lehle, L. & Aebi, M. (1993). Yeast Wbp1p and Swp1p form a protein complex essential for oligosaccharyl transferase activity. *EMBO J.* **12**(1), 279-84.

Varki, A. (1993). Biological roles of oligosaccharides: all of the theories are correct. *Glycobiology* **3**(2), 97-130.

Welply, J. K., Shenbagamurthi, P., Lennarz, W. J. & Naider, F. (1983). Substrate recognition by oligosaccharyltransferase. Studies on glycosylation of modified Asn-X-Thr/Ser tripeptides. *J. Biol. Chem.* **258**(19), 11856-63.

Role of Cop Coats and GTPases in Transport of Cargo Through the Early Secretory Pathway

William E. Balch
The Scripps Research Institute
Department of Cell Biology
10550 N. Torrey Pines Road
La Jolla, CA 92130

The secretory pathway of eukaryotic cells comprises a series of sequential compartments whose compartmental boundaries are bridged through the activity of small 60-80 nm transport vesicles. Cargo selection and vesicle formation is now recognized to be mediated by export signals and cytosolic coat component. Vesicle formation involves the recruitment of cytosolic proteins which form in combination with membrane-associated components the vesicle coat which drives membrane deformation and which are likely to play a critical role in cargo selection (2). The assembly and disassembly of these coat complexes are regulated by the activity of small GTPases belonging to Ras superfamily and include Sar1, ARF and Rab proteins. There are now three well-characterized and structurally unrelated coats participating in vesicle formation. These include: (1) clathrin-coated vesicles which facilitate protein transfer between the plasma membrane, endocytic and trans-Golgi compartments (49); (2) COPI (coatomer) coated vesicles which have an important role in retrograde transport from the Golgi to the ER (24, 28, 30) and COPII vesicles which participate in the anterograde transport of cargo from the endoplasmic reticulum (ER) to the Golgi (10, 48). The activity of the COPII and COPI coats and their respective GTPases are briefly examined below.

NATO ASI Series, Vol. H 106
Lipid and Protein Traffic
Pathways and Molecular Mechanisms
Edited by Jos A. F. Op den Kamp

Budding from the endoplasmic reticulum (ER) is mediated by COPII.

Transport through the exocytic pathway is initiated by budding of COPII carriers from the ER. ER export has been dissected genetically in yeast (27) and studied biochemically using a variety of cell-free assays which support ER to Golgi transport by yeast and mammalian semi-intact cells, and microsomes prepared from cell homogenates (5, 12, 18, 40, 42). These studies have led to the identification and functional characterization of a number of gene products specifically involved in ER export. The soluble components necessary for ER export comprise the COPII coat complex. This consists of the small GTPase Sar1, and the heterodimeric complexes Sec23/Sec24, and Sec13/Sec31 found in both yeast and mammalian cells. The assembly and disassembly of the coat is regulated by the cyclical activation and inactivation of the Sar1 GTPase, respectively (4, 9, 10, 29, 35, 42). These cytosolic components are sufficient to promote vesicle budding from yeast (10) and mammalian ER membranes in vitro (3). Activation of the Sar1 GTPase to the GTP-bound form requires the function of the membrane-associated guanine nucleotide exchange factor (GEF) (11). Coat disassembly is currently believed to be initiated by the activity of Sec23, a Sar1 specific GTPase activating protein (GAP) which promotes the hydrolysis of Sar1-GTP to the GDP-bound form (62). GTP hydrolysis is essential for the disassembly of the coat lattice and for the subsequent fusion of COPII vesicles to the target Golgi membrane (10, 42). Although previous studies have led to a rudimentary identification and characterization of some of the components involved in export from the ER, neither the overall mechanism nor the individual roles of COPII coat components in cargo selection and vesicle budding are known.

Insight into the mechanism of COPII function in mammalian cells has come from the use of an in vitro budding assay which utilizes microsomal membranes prepared from cells infected with vesicular stomatitis virus (VSV) (42). VSV inhibits host protein synthesis while synthesizing a single type 1 transmembrane glycoprotein, the vesicular stomatitis virus glycoprotein or VSV-G. Therefore, in infected cells, VSV-G is the major cargo traversing the secretory pathway. Incubation of microsomes prepared from infected cells leads to the efficient budding of VSV-G containing ER-derived 60-80 nm COPII coated vesicular carriers. Export in vitro is energy- and cytosol-dependent and requires the activation of Sar1 to the GTP-bound form to promote budding. Mammalian homologues to yeast Sar1, Sec13/31 and Sec23/24 found in rat liver cytosol are necessary and sufficient for the export of VSV-G from mammalian ER microsomes.

Pleomorphic elements function in ER to Golgi transport.

Pre-Golgi intermediates harboring ER-derived cargo are found distributed throughout the cytoplasm (7, 25). A detailed stereological characterization of the structure of these export sites (8) has revealed that intermediates were found associated with one or more bud bearing ER cisternae which face towards a central cavity (Fig. 1). The central cavity is filled with a collection of closely opposed vesicles and convoluted tubules which are discontinuous with the ER. These appear to be inter-connected between one another when examined using serial thin-section. These central compact structures are referred to as 'vesicular-tubular clusters' or VTCs (6). The close juxtaposition of ER-derived buds and a central VTC comprises a morphological unit of functional organization referred to as an 'export complex'(Fig. 1). ER-derived buds cannot be detected outside export complexes. This suggests a local specialization of the cytoplasm which is likely to be enriched in COPII transport factors promoting vesicle budding. VTCs can also be readily detected in the perinuclear region where they are frequently in continuity with one another to form an array of tubulo-cisternal elements referred to as the cis Golgi network (CGN) (32)

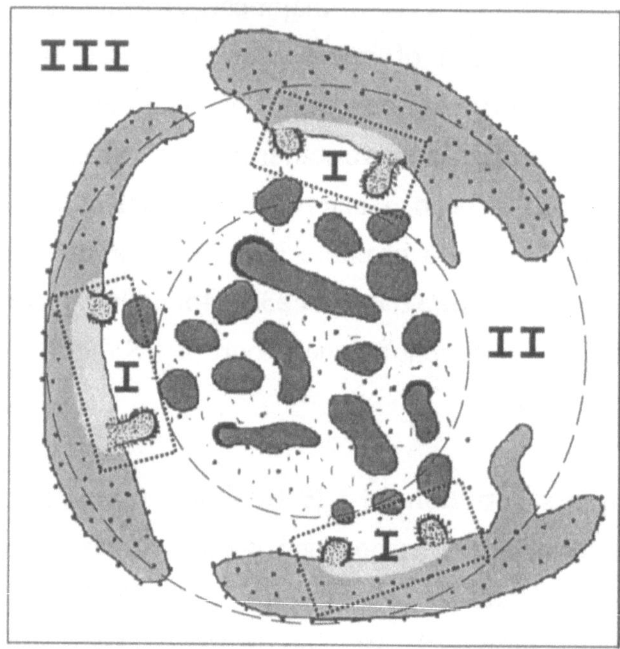

Fig. 1. Export complexes are defined as structures in which Individual ER buds (I) surround a central cavity containing VTCs (II). The entire regional specialization (III) found in the cytoplasm defines an export complex. Figure obtained from (8).

Biochemical composition of export complexes. The separation of a COPII vesicular carrier from ER budding elements defines the first boundary between the ER and downstream organelles of the secretory pathway. This is consistent with the fact that VTCs lack continuity with the ER (7, 8, 43). Although VTCs define a unique compartment, their composition is still under investigation. During ER export, ribophorin, components of the protein translocation machinery such as Sec61, as well as resident ER proteins including the folding chaperones calnexin and BiP, are efficiently excluded from COPII vesicles (42). ER to Golgi intermediates can be readily identified by the presence of two closely related (90% identity) VTC marker proteins p53 and p58 (45, 50) which are type 1 transmembrane proteins which continuously recycle between the ER and pre/cis-Golgi elements, but are concentrated in VTCs at steady-state. Other proteins now recognized to be enriched in VTCs are the small GTPases Rab1 (39, 44) and Rab2 (14), and Gos28 (55), Sly1 (16) and syntaxin 5 (Syn5) (17). Each of these proteins are believed to be involved in vesicle fusion (20) and actively recycle between VTCs and the ER. Whether VTCs contain post-ER processing enzymes involved in palmitoylation, addition of O-linked sugars, trimming of mannose residues from N-linked core oligosaccharides or contain the enzymes for the addition GlcNAc-1-P to lysosomal hydrolyases and the corresponding uncovering enzyme (phosphodiester glycosidase) remains to be determined. Because no markers can be defined as 'resident' proteins, it is readily apparent that VTCs are highly dynamic structures.

VTCs undergo maturation in a COPI-dependent fashion.

Following release from the ER, COPII vesicles rapidly lose their coats in response to hydrolysis of GTP to GDP by the Sar1 GTPase (4). A property which highlights the unique function of intermediates from the ER is the presence of buds on tubular elements of VTCs which contain the COPI coat complex (28, 41). The COPI coat consists of a heptameric complex whose recruitment to membranes is regulated by the ARF family of GTPases. COPI, originally discovered for its potential role in intra-Golgi transport by Rothman and colleagues (60), is now recognized to mediate the formation of vesicular carriers which promote the retrograde transport of proteins containing terminal di-lysine/di-phenylalanine motifs (21, 24, 30, 53). The mechanism by which VTCs acquire

COPI coats has been investigated biochemically using an assay which reconstitutes budding from ER microsomes in vitro (42). In these experiments it was found that uncoated COPII vesicles can initiate the recruitment of COPI components prior to their fusion to VTCs. This 'coupled exchange' between COPII and COPI coats was proposed to serve as a tagging mechanism to mark components for rapid retrieval to the ER from VTCs. Therefore, segregation of retrograde and anterograde transported proteins represents a key activity of the tubular elements comprising VTCs. Morphological evidence for segregation stems from the observation that p58, which is initially present in COPII vesicles together with other cargo molecules destined for the Golgi and cell surface (42), can be physically separated from cargo within the tubular elements of VTCs (4, 56). In this regard, the function of VTCs may be similar to that of endosomes which segregate components derived from the cell surface.

The mechanism by which anterograde cargo present in these peripheral export sites is mobilized to the central Golgi region involves microtubules (MTs) (31, 33, 46). Studies now suggest that when VTCs reach the central Golgi region they appear to fuse to form the more elaborate labyrinth of tubular elements comprising the CGN. Combined with the fact that ER-budding elements are exclusively associated with VTCs at steady-state (8), we are left with a view that export complexes are mobile collecting/recycling structures.

In general, the combined evidence leads us to conclude that COPII performs an exclusive role in sorting and concentration of cargo during ER export, whereas COPI functions in retrograde recycling from VTCs and possibly downstream Golgi compartments. Thus, VTCs are likely to be highly dynamic structures which undergo continuous maturation through input from COPII vesicles and output through COPI directed recycling. Because COPII vesicles and VTCs are compositionally related, they may use similar machineries to target their anterograde cargo to the cis Golgi compartment. However, the mechanism by which anterograde cargo is delivered to Golgi compartments remains elusive. Although COPI coated vesicles have been proposed to mediate this event (36), the strong evidence supporting its essential if not exclusive role in retrograde transport, leaves this an important direction of investigation for future studies.

Rab GTPases direct vesicle targeting and fusion

The Rab family of small GTPases includes more than 30 members which which share 30% to 95% amino acid identity and regulate vesicular traffic between specific compartments of the endocytic and exocytic pathways (52). Rab proteins, like other members of the Ras superfamily, contain four highly conserved sequence motifs required for guanine nucleotide binding (13), Rab proteins function as molecular switches which undergo conformational changes pending the status of bound GDP and GTP to regulate the assembly/disassembly of protein complexes involved in vesicle targeting and fusion. The Rab cycle involves the sequential action of proteins which escort Rab in the cytosol (guanine nucleotide dissociation inhibitors (GDIs) and those which facilitate GDP-exchange (guanine nucleotide exchange proteins (GEPs)) and promote GTP hydrolysis (guanine nucleotide activating proteins GAPs)). In this cycle, Rab proteins are delivered to a membrane-associated receptor(s) which promote dissociation from GDI (guanine-nucleotide displacement factors (GDFs)) and promote activation to the GTP-bound form on the emerging vesicle through the activity of GEF. Subsequently, during or following vesicle targeting/fusion, Rab function is terminated and recycled by GDI. The ancient origins of the structural motifs comprising the guanine nucleotide binding site of Rab GTPases suggests a common mode of biological function for the Rab family which remains to be deduced.

GDI functions as a recycling for factor Rab GTPase function in vesicular transport

Rab proteins bind to an 'escort' referred to as guanine nucleotide dissociation inhibitor or GDI. This class of proteins regulates the membrane association and recycling of Rab GTPases. They are now recognized to be a family of isozymes which are differentially expressed during development (38, 61). Complex formation between Rab proteins and GDI requires post-translational modification of carboxyl-terminal cysteine residue(s) with geranylgeranyl lipids (19, 23, 34, 54, 58). The cytosolic pools of GDI complexed with Rab1, Rab3, Rab5 and Rab9 have been shown to be essential for fusion between specific subsets of exocytic and endocytic compartments (37, 54, 57). The delivery of Rab proteins to membranes via their respective GDI complexes has been shown to be accompanied by the release of GDI and exchange of GDP for GTP to form the activated Rab (54, 57). Following GTP hydrolysis and fusion, GDI re-extracts the inactive GDP-bound form of Rab from the membrane. In general, the physiological role of GDI *in vivo* in vesicular traffic is to bind and stabilize Rab in its GDP-bound form, thereby providing a cytosolic reservoir recycle Rab for reuse in multiple rounds of targeting and fusion.

GDI and the choroideraemia (CHM) form a superfamily of GDI-like proteins which bind and regulate Rab GTPase function

GDI is strongly related to the mammalian *choroideremia* gene product CHM. CHM is identical to the component A subunit (referred to Rab escort protein (REP1) of the Rab geranylgeranyl transferase II, a protein complex that is responsible for prenylation of Rab proteins terminating in di-cysteine containing motifs (1, 15, 51). The role of component A is to escort newly synthesized Rab proteins to the catalytic subunits of the transferase for prenylation (1). A homologue in yeast, MRS6p (MS14p), is essential and has identical biochemical properties to CHM/CHML in geranylgeranylation of yeast Rab homologues (22, 26). Following prenylation, CHM-REP delivers Rab to the membranes in an analogous fashion to GDI. Two isoforms of CHM-REP (90% identity) have now been identified and appear to have differences in the spectrum of Rab GTPases which they target to the catalytic subunits for efficient prenylation.

Alignment of CHM-REP and GDI shows four sequence conserved regions (referred to as SCRs) located in the N-terminal and central regions with an insert of variable length separating the less conserved SCRs 1 and 2. Residues in SCR1A and 1B, and SCR3B are particularly diagnostic of GDI/CHM-REP related proteins due to the presence of invariant di- and tripeptides (59). The structure of a-GDI has been recently been solved (47). Examination of the structure of GDI reveals that the SCR1, which includes the entire N-teminus and SCR3B, being distant from SCR1 by nearly 140 residues in sequence, structurally winds back to contact the SCR1 motif to form a compact subdomain found at the apex of GDI (referred to as the GDI-CHM consensus domain (GCD)). Tyr39,

Glu233, and Arg240 are strongly conserved residues in the SCRs 1B and 3B and are found in surface exposed strands and helices forming GCD with their side-chains exposed into the upper cleft. When these particular residues are mutated, GDI completely losses its ability to bind Rab (47). Therefore, GCD is a region which is critical for Rab-binding. From the structure of GDI (47), it is now apparent that a highly conserved face contains SCR1, 2, and 3 and the opposite face shows little evolutionary conservation between members of the GDI and CHM-REP families. The conserved face is likely to play a critical role in its interaction with Rab and/or membrane associated receptors involved in Rab recycling.

GTPases and the assembly of protein complexes directing coat formation and vesicle targeting and fusion

In general, it is now apparent that protein-protein interactions dictate the biochemical events dictating vesicle formation, targeting and fusion. The assembly of these complexes is dictated by the regulatory activity of small GTPases which belong to the Rab, Sar1 and ARF gene families. Future work will focus on the specific biochemical roles of these GTPases in mediating movement of cargo through the secretory pathway of eukaryotic cells.

References cited:

1. Anders, D. A., M. C. Seabra, M. S. Brown, S. A. Armstrong, T. E. Semland, F. P. M. Cremers, and J. L. Goldstein. 1993. cDNA cloning of component A of rab geranylgeranyl transferase and demonstration of its role as a Rab escort protein. *Cell.* 73:1091-1099.

2. Aridor, M., and W. E. Balch. 1996. Principles of selective transport: coat complexes hold the key. *Trends in Cell Biol.* 6:315-320.

3. Aridor, M., S. Bannyhkh, T. Rowe, H. Plutner, C. N. Nuoffer, and W. E. Balch. 1997. Functional coupling of cargo to the COPII budding machinery. *J. Cell Biol.*

4. Aridor, M., S. I. Bannykh, T. Rowe, and W. E. Balch. 1995. Sequential coupling between COPII and COPI vesicle coats in endoplasmic reticulum to Golgi transport. *J. Cell Biol.* 131:875-893.

5. Baker, D., L. Hicke, M. Rexach, M. Schleyer, and R. Schekman. 1988. Reconstitution of SEC gene product-dependent intercompartmental protein transport. *Cell.* 54:335-344.

6. Balch, W. E., J. M. McCaffery, H. Plutner, and M. G. Farquhar. 1994. Vesicular stomatitis virus glycoprotein is sorted and concentrated during export from the endoplasmic reticulum. *Cell.* 76:841-852.

7. Bannykh, S. I., and W. E. Balch. 1997. Membrane dynamics at the ER/Golgi interface. *J. Cell Biol.* In press.

8. Bannykh, S. I., T. Rowe, and W. E. Balch. 1996. Organization of endoplasmic reticulum export complexes. *J. Cell Biol.* 135:19-35.

9. Barlowe, C., C. d'Enfert, and R. Schekman. 1993. Purification and characterization of SAR1p, a small GTP-binding protein required for transport vesicle formation from the endoplasmic reticulum. *J. Biol. Chem.* 268:873-879.

10. Barlowe, C., L. Orci, T. Yeung, M. Hosobuchi, S. Hamamoto, N. Salama, M. F. Rexach, M. Ravazzola, M. Amherdt, and R. Schekman. 1994. COPII: a membrane coat formed by sec proteins that drive vesicle budding from the endoplasmic reticulum. *Cell.* 77:895-907.

11. Barlowe, C., and R. Schekman. 1993. SEC12 encodes a guanine-nucleotide-exchange factor essential for transport vesicle budding from the ER. *Nature.* 365:347-349.

12. Beckers, C. J. M., D. S. Keller, and W. E. Balch. 1987. Semi-intact cells permeable to macromolecules: use in reconstitution of protein transport from the endoplasmic reticulum to the Golgi complex. *Cell.* 50:523-534.

13. Bourne, H. R., D. A. Sanders, and F. McCormick. 1991. The GTPase superfamily: conserved structure and molecular mechanism. *Nature.* 349:117-127.

14. Chavrier, P., R. G. Parton, H.-P. Hauri, K. Simons, and M. Zerial. 1990. Localization of low molecular weight GTP binding proteins to exocytic and endocytic compartments. *Cell.* 62:317-329.

15. Cremers, F. P. M., S. A. Armstrong, M. C. Seabra, M. S. Brown, and J. L. Goldstein. 1994. REP-2, a rab escort protein encoded by the choroideremia-like gene. *J. Biol. Chem.* 269:2111-2117.

16. Dascher, C., and W. E. Balch. 1996. Mammalian sly1 regulates syntaxin 5 function in endoplasmic reticulum to Golgi transport. *J. Biol. Chem.* 271:15866-15869.

17. Dascher, C., J. Matteson, and W. E. Balch. 1994. Syntaxin 5 regulates endoplasmic reticulum (ER) to Golgi transport. *J. Biol. Chem.* 269:29363-29366.

18. Davidson, H. W., and W. E. Balch. 1993. Differential inhibition of multiple vesicular transport steps between the endoplasmic reticulum and trans Golgi network. *J. Biol. Chem.* 268:4216-4226.

19. Dirac-Svejstrup, A. B., T. Soldati, A. D. Shapiro, and S. R. Pfeffer. 1994. Rab-GDI presents functional rab9 to the intracellular transport machinery and contributes selectivity to rab9 membrane recruitment. *J. Biol. Chem.* 269:15427-15430.

20. Ferro-Novick, S., and R. Jahn. 1994. Vesicle fusion from yeast to man. *Nature.* 370:191-193.

21. Fiedler, K., M. Veit, M. A. Stamnes, and J. E. Rothman. 1996. Bimodal interaction of coatomer with a family of putative cargo and vesicle coat protein receptors, the p24 proteins. *Science.* In press.

22. Fujimura, K., K. Tanaka, A. Nakano, and A. Toh-e. 1994. The saccharomyces cerevisiae MS14 gene encodes the yeast counterpart of component A of rab geranylgeranyltransferase. *J. Biol. Chem.* 269:9205-9212.

23. Garrett, M. D., J. E. Zahner, C. M. Cheney, and P. J. Novick. 1994. GDI1 encodes a GDP dissociation inhibitor that plays an essential role in the yeast secretory pathway. *EMBO J.* 13:1718-1728.
24. Gaynor, E., and S. D. Emr. 1997. COPI-dependent anterograde transport: cargo-selective ER-to-Golgi protein transport in yeast COPI mutants. *J. Cell Biol.* 24:789-802.
25. Hauri, H.-P., and A. Schweizer. 1992. The endoplasmic reticulum-Golgi intermediate compartment. *Curr. Opin. Cell Biol.* 4:600-608.
26. Jiang, Y., and S. Ferro-Novick. 1994. Identification of yeast component A: reconstitution of the geranylgeranyltransferase that modifies Ypt1p and Sec4p. *Proc. Natl. Acad. Sci. USA.* 91:4377-4381.
27. Kaiser, C. A., and R. Schekman. 1990. Distinct sets of SEC genes govern transport vesicle formation and fusion early in the secretory pathway. *Cell.* 61:723-733.
28. Kreis, T. E., M. Lowe, and R. Pepperkok. 1995. COPs regulating membrane traffic. *Ann. Rev. Cell Dev. Biol.* 11:677-706.
29. Kuge, O., C. Dascher, L. Orci, T. Rowe, M. Amherdt, H. Plutner, M. Ravazzola, G. Tanigawa, J. E. Rothman, and W. E. Balch. 1994. Sar1 promotes vesicle budding from the endoplasmic reticulum but not Golgi compartments. *J. Cell Biol.* 125:51-65.
30. Letourneur, F., E. C. Gaynor, S. Hennecke, C. Demolliere, R. Duden, S. D. Emr, H. Riezman, and P. Cosson. 1995. Coatomer is essential for retrieval of dilysine-tagged proteins to the endoplasmic reticulum. *Cell.* 79:1199-1207.
31. Lippincott-Schwartz, J., N. B. Cole, A. Marotta, and P. A. Conrad. 1995. Kinesin is the motor for microtubule-mediated Golgi-to-ER membrane traffic. *J. Cell Biol.* 128:293-306.
32. Mellman, I., and K. Simon. 1992. The Golgi complex: in vitro veritas? *Cell.* 68:829-840.
33. Mizuno, M., and S. J. Singer. 1994. A possible role for stable microtubules in intracellular transport from the endoplasmic reticulum to the Golgi apparatus. *J. of Cell Sci.* 107:1321-1331.
34. Musha, T., M. Kawata, and Y. Takai. 1992. The geranylgeranyl moiety but not the methyl moiety of the smg-25A/rab3A protein is essential for the interactions with membrane and its inhibitory GDP/GTP exchange protein. *J. Biol. Chem.* 267:9821-9825.
35. Oka, T., and A. Nakano. 1994. Inhibition of GTP hydrolysis by Sar1p causes accumulation of vesicles that are a functional intermediate of the ER-to-Golgi transport in yeast. *J. Cell Biol.* 124:425-434.
36. Ostermann, J., L. Orci, K. Tani, M. Amherdt, M. Ravazzola, Z. Elazar, and J. E. Rothman. 1993. Stepwise assembly of functionally active transport vesicles. *Cell.* 75:1015-1025.
37. Peter, F., C. Nuoffer, S. N. Pind, and W. E. Balch. 1994. Guanine nucleotide dissociation inhibitor (GDI) is essential for rab1 function in budding from the ER and transport through the Golgi stack. *J. Cell Biol.* 126:1393-1406.
38. Pfeffer, S. R., B. Dirac-Svejstrug, and T. Soldati. 1995. Rab GDP dissociation inhibitor: putting Rab GTPases in the right place. *J. Biol. Chem.* 270:17057-17059.

39. Pind, S., C. Nuoffer, J. M. McCaffery, H. Plutner, H. W. Davidson, M. G. Farquhar, and W. E. Balch. 1994. Rab1 and Ca^{2+} are required for the fusion of carrier vesicles mediating endoplasmic reticulum to Golgi transport. *J. Cell Biol.* 125:239-252.
40. Plutner, H., H. W. Davidson, J. Saraste, and W. E. Balch. 1992. Morphological analysis of protein transport from the endoplasmic reticulum to Golgi membranes in digitonin permeabilized cells: role of the p58 containing compartment. *J. Cell Biol.* 119:1097-1116.
41. Rothman, J. E., and F. Wieland. 1996. Protein sorting by transport vesicles. *Science*. 272:272-234.
42. Rowe, T., M. Aridor, J. M. McCaffery, H. Plutner, and W. E. Balch. 1996. COPII vesicles derived from mammalian endoplasmic reticulum (ER) microsomes recruit COPI. *J. Cell Biol.* 135:895-911.
43. Saraste, J., and E. Kuismanen. 1984. Pre- and post-Golgi vacuoles operate in the transport of Semliki Forest virus membrane glycoproteins to the cell surface. *Cell*. 38:535-549.
44. Saraste, J., U. Lahtinen, and B. Goud. 1995. Localization of the small GTP-binding protein rab1p to early compartments of the secretory pathway. *J. Cell Sci.* 108:1541-1552.
45. Saraste, J., G. E. Palade, and M. G. Farquhar. 1987. Antibodies to rat pancreas Golgi subfractions- identification of a 58 kD cis Golgi protein. *J. Cell Biol.* 105:2021-2029.
46. Saraste, J., and K. Svensson. 1991. Distribution of the intermediate elements operating in ER to Golgi transport. *J. Cell Sci.* 100:415-430.
47. Schalk, I., K. Zeng, S.-K. Wu, E. A. Stura, J. Matteson, M. Huang, A. Tandon, I. A. Wilson, and W. E. Balch. 1996. Structure and mutational analysis of Rab GDP dissociation inhibitor. *Nature*. 381:42-48.
48. Schekman, R., and L. Orci. 1996. Coat proteins and vesicle budding. *Science*. 271:1526-1533.
49. Schmid, S. L. 1997. *Ann. Rev. Biochem.* In press.
50. Schweizer, A., J. A. M. Fransen, T. Bachi, L. Ginsel, and H.-P. Hauri. 1988. Identification, by a monoclonal antibody, of a 53-kD protein associated with a tubulo-vesicular compartment at the cis-side of the Golgi apparatus. *J. Cell Biol.* 107:1643-1653.
51. Seabra, M. C., M. S. Brown, C. A. Slaughter, T. C. Sudhof, and J. L. Goldstein. 1992. Purification of component A of Rab geranylgeranyl transferase: Possible identity with the choroideremia gene product. *Cell*. 70:1049-1057.
52. Simons, K., and M. Zerial. 1993. Rab proteins and the road maps for intracellular transport. *Neuron*. 5:613-620.
53. Sohn, K., L. Orci, M. Ravazzola, M. Bremser, F. Lottspeich, K. Fiedler, J. B. Helms, and F. T. Wieland. 1996. A major transmembrane protein of Golgi-derived COPI-caoted vesicles involved in coatomer binding. *J. Cell Biol.* 135:12239-1248.
54. Soldati, T., M. A. Riederer, and S. R. Pfeffer. 1993. Rab GDI: a solubilizing and recycling factor for rab9 protein. *Mol. Biol. Cell*. 4:425-434.
55. Subramaniam, V. N., J. Krijnse-Locker, B. L. Tang, M. Ericsson, A. R. bin Mohd Yusoff, G. Griffiths, and W. Hong. 1995. Monoclonal antibody

HFD9 identifies a novel 28 kDa integral membrane protein on the cis-Golgi. *J. Cell Sci.* 108:2405-2414.
56. Tang, B. L., S. H. Low, H.-P. Hauri, and W. Hong. 1995. Segregation of ERGIC53 and the mammalian KDEL receptor upon exit from the 15°C compartment. *Eur. J. Cell Biol.* 68:397-410.
57. Ullrich, O., H. Horiuchi, C. Bucci, and M. Zerial. 1994. Membrane association of Rab5 mediated by GDP-dissociation inhibitor and accompanied by GDP/GTP exchange. *Nature.* 368:157-160.
58. Ullrich, O., H. Stenmark, K. Alexandrov, L. Hubert, K. Kaibuchi, T. Sasaki, Y. Takai, and M. Zerial. 1993. Rab GDP dissociation inhibitor as a general regulator for the membrane association of Rab proteins. *J. Biol. Chem.* 268:18143-18150.
59. Waldherr, M., A. Ragnini, R. J. Schweyen, and M. S. Boguski. 1993. MRS6-yeast homolog of the choroideremia gene. *Nature Genetics.* 3:193-194.
60. Waters, M. G., T. Serafini, and J. E. Rothman. 1991. 'Coatomer': a cytosolic protein complex containing subunits of non-clathrin-coated Golgi transport vesicles. *Nature.* 349:248-251.
61. Wu, S. K., K. Zeng, I. Wilson, and W. E. Balch. 1996. The GDI-CHM/REP connection: structural insights into the Rab GTPase cycle. *Trends in Biochem.* 21: 472-476..
62. Yoshihisa, T., C. Barlowe, and R. Schekman. 1993. Requirement for a GTPase-activating protein in vesicle budding from the endoplasmic reticulum. *Nature.* 259:1466-1468.

Mapping the Golgi Retention Signal in the Cytoplasmic Tail of the Uukuniemi Virus G1 Glycoprotein

Agneta M. Andersson and Ralf F. Pettersson

Ludwig Institute for Cancer Research
Box 240
171 77 Stockholm, Sweden.

1 Introduction

The majority of newly synthezised secretory and membrane proteins are transported along the secretory pathway from the endoplasmic reticulum (ER) to the plasma-membrane (PM), where they are secreted or remain inserted into the membrane. A minority of the transported proteins contain sorting or retention signals which instead direct them to or retain them at specific sites within the cell.

Enveloped viruses acquire their lipoprotein coat by budding through one of the cellular membranes. Most viruses mature at the PM whereas others bud at intracellular membranes. The viral membrane proteins, which form the spikes on the virus particle are transported to and retained in the budding compartment where they accumulate, indicating that these proteins contain retention signals. Viral membrane proteins are synthesized, processed, modified and transported according to cellular mechanisms and are therefore useful tools when studying intracellular membrane trafficking and protein compartmentalization (Pettersson, 1991).

We are using the spike proteins G1 and G2 of Uukuniemi virus (UUK), a *Phlebovirus* within the *Bunyaviridae* family, as a model

NATO ASI Series, Vol. H 106
Lipid and Protein Traffic
Pathways and Molecular Mechanisms
Edited by Jos A. F. Op den Kamp

system to study intracellular transport and retention of proteins in the Golgi complex (GC) (Pettersson and Melin, 1996). UUK virus is a spherical enveloped virus with a genome consisting of three circular, single-stranded, negative-sense RNA segments named L, M and S according to their sizes. The M RNA segment encodes a 110 kDa precursor protein which is cotranslationally cleaved into G1 and G2. These two glycoproteins are associated with the envelope forming the spikes on the viral surface. In the ER, G1 folds rapidly, whereas G2 folds slowly and less efficiently (Persson and Pettersson, 1991). They form heterodimers in the ER which are transported to and retained in the GC, where they accumulate. The L RNA segment encodes the RNA polymerase and the S RNA segment encodes the nucleocapsid protein N and a non-structural protein with an as yet unknown function.

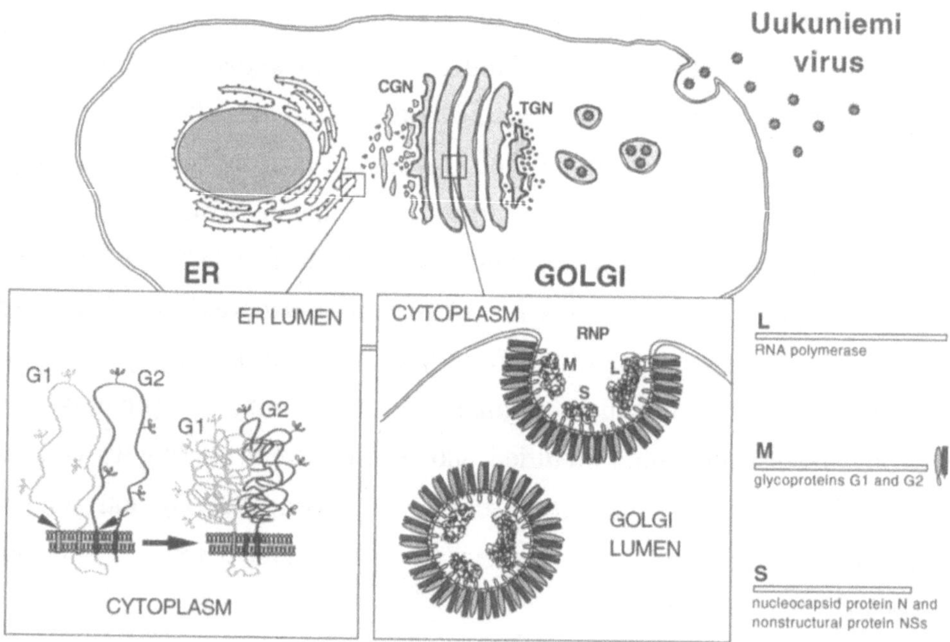

Fig. 1. Model summerizing the genome structure of Uukuniemi virus, the transport of the glycoproteins G1 and G2, virus assembly, the budding in the GC and export of virus particles.

The RNA polymerase and the nucleocapsid protein are both associated with the RNA genome. The interaction between the ribonucleoproteins and the spike proteins are thought to trigger the budding. Intracellularly maturing viruses are in most cases transported by large vesicles to the PM, where they fuse, releasing the virus into the extracellular space (Fig. 1).

The aim of this project is to define the region in G1 which specifies the retention of the protein in the GC and to determine the structure(s) to which G1 might be binding. We also wish to elucidate the mechanism underlying the budding of UUK virus in the GC.

2 Results

Our previous results have shown that G1 expressed alone can exit the ER, be transported to and retain in the Golgi, whereas G2 is dependent on G1 for ER to Golgi transport (Melin et al., 1995). We therefore concluded that G1 and G2 must interact with each other in the ER, and that G1 contains the signal for targeting the heterodimer to the GC.

2.1 The cytoplasmic tail of G1 contains the Golgi retention signal.

To localize the domain in G1 containing the Golgi retention signal, we constructed several chimeric proteins by replacing different domains of G1 with the corresponding domains of vesicular stomatitis virus (VSV) G protein, chicken lysozyme or CD4 (CD4 chimeric constructs are shown in Fig. 2) (Andersson et al., 1997b). The results, shown by immunofluorescence (IF) microscopy in Fig. 3 and summerized in Fig. 2, indicated that neither the ectodomain nor the transmembrane domain (TMD) of G1 are essential for Golgi localization. Instead, the cytoplasmic tail of G1 (Andersson et al., 1997a) was found to be both necessary and sufficient for Golgi retention. A conclusion which was further supported by the fact that the soluble green fluorescent protein (GFP) could be targeted to the GC by attaching the cytoplasmic tail of G1 to either the N- or the C-terminus of GFP (Fig. 2).

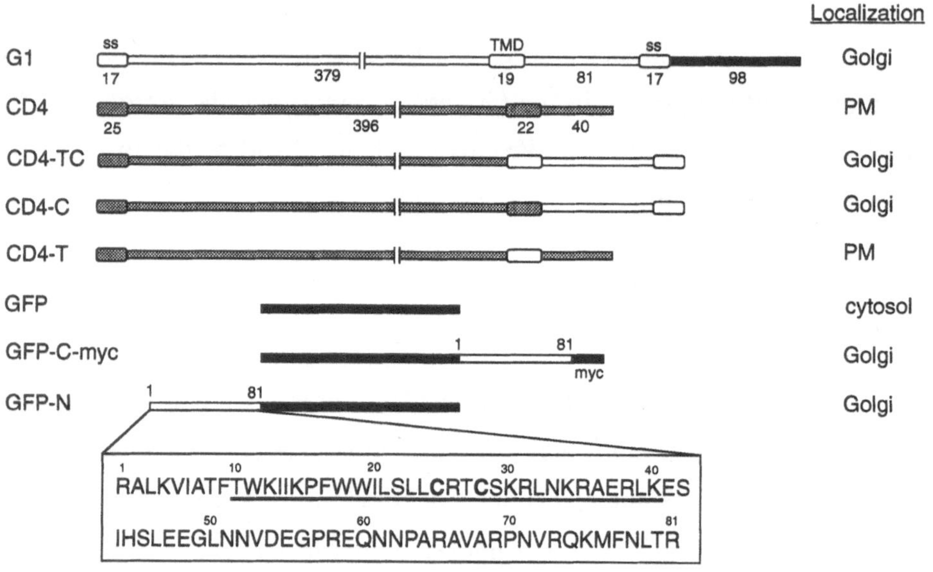

Fig. 2. Schematic representation of chimeras between wild type G1 (open bar), wild type CD4 (light-grey shaded bar) and GFP (dark-grey shaded bar). The amino acid sequence of the cytoplasmic tail of G1 is numbered and shown below, with the cysteines marked in bold.

2.2 The Golgi retention signal is narrowed down to residues 4 to 50 in the cytoplasmic tail of G1.

To map the Golgi retention signal in more detail, we made CD4 chimeric proteins with progressive deletions of the G1 tail, starting from the C-terminus (Andersson et al., 1997b). The CD4-G1 tail chimeras contained 98, 81, 49, 19 or 4 residues of the G1 tail. Deletion of the hydrophobic signal sequence of G2 did not affect the Golgi localization. When half of the tail was deleted it had a minor effect on the Golgi retention, allowing a portion of the chimeric protein to leak out to the cell surface whereas deleting 79 residues caused a greatly increased cell surface staining. Finally, the CD4 chimera containing 4 residues of the G1 tail was expressed on the surface with nearly the same efficiency as wild type CD4. The results suggest that the Golgi retention signal is located approximately between residues 4 to 50 in the cytoplasmic tail of G1, counting from the TMD border. In this short region, there are two cysteines which are palmitylated. A replacement of either one or both of them with alanine(s), did not influence retention of the chimeric proteins in the Golgi. At present, the role(s) of the palmitylation in the cytoplasmic tail of G1 is unknown, but they probably anchor the cytoplasmic tail to the lipid bilayer, thereby affecting the conformation of the tail.

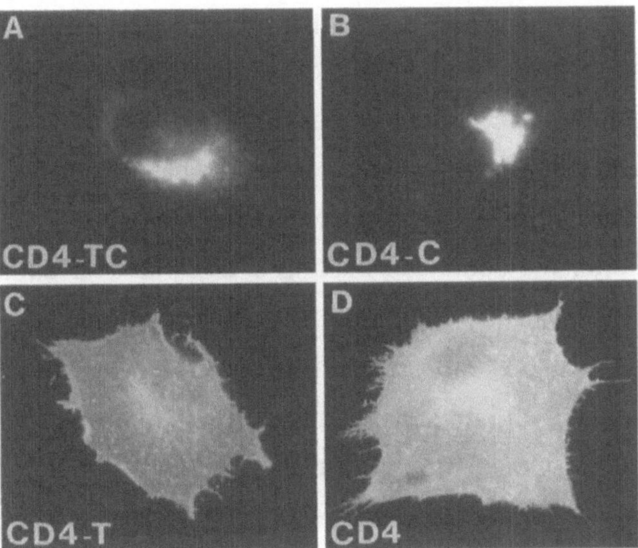

Fig. 3. Localization of CD4 chimeric proteins by indirect IF microscopy using a monoclonal antibody to CD4.

2.3 Determination of the shortest region needed for GC localization.

To determine the shortest region needed for Golgi localization, we expressed the cytoplasmic tail of G1 alone as a myc-tagged 81-residue peptide (Fig. 4). We found that the cytoplasmic tail of G1 was localized to the GC shown by costaining with mannosidase II, a medial GC marker protein, but the G1 tail also localized to as yet unidentified vesicles (Fig. 5). We have tried to determine the origin of these vesicles by IF microscopy using different antibodies against marker proteins known to be localized in the ER, intermediate compartment, medial Golgi, TGN, endosomes and lysosomes. None of these marker proteins overlap in localization with the vesicles. Thus, the origin of the vesicles remain unknown.

We continued narrowing down the Golgi retention signal by making progressive deletions of the cytoplasmic tail of G1, starting at the N-terminus. The G1 tail peptide lacking the first 10 membrane-proximal residues was localized to the Golgi, but not to the vesicular structures. When the following 10 residues were deleted, the tail lost its association with the Golgi membranes. These results indicate that residues 10 to 20 are needed for Golgi localization when the cytoplasmic tail of G1 is expressed alone. We continued making C-terminal truncations of the G1 tail peptide 10-81, and found that the G1 tail peptide deleted down to residues 10 to 40 was still targeted to the Golgi membranes. We thus conclude that the Golgi retention signal is located between residues 10 to 40 (underlined sequence in Fig. 2) in the cytoplasmic tail of G1, a result which is in agreement with the results from the expression of the membrane-spanning CD4-G1 tail chimeric proteins.

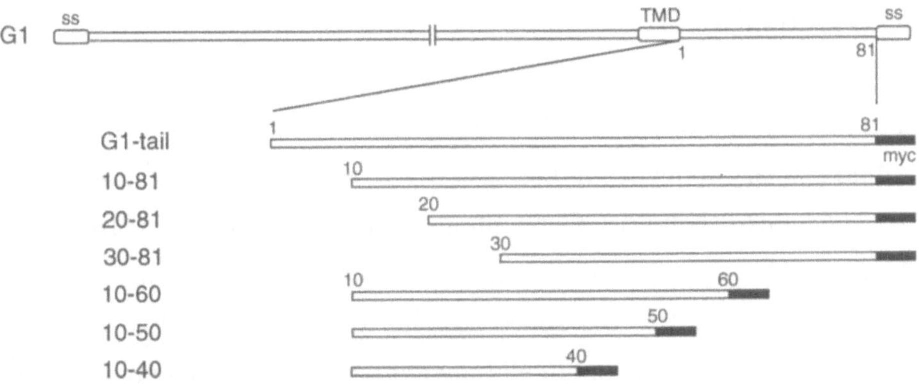

Fig. 4. Schematic representation of the cytoplasmic tail of G1 expressed as a myc-tagged 81-residue peptide or N- and C-terminus truncated G1 tail peptides.

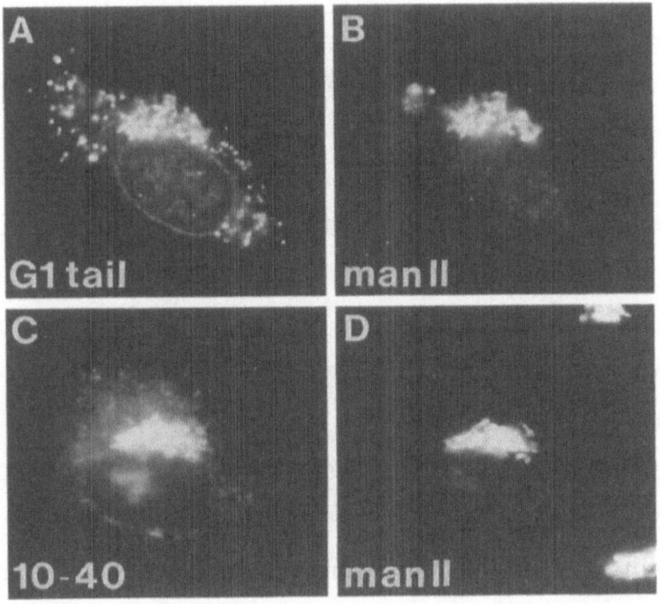

Fig. 5. Localization of the cytoplasmic tail of G1 (residues 1-81) and the peptide 10-40 by indirect IF by using a monoclonal antibody to myc (A, C) in combination with a polyclonal antibody to mannosidase II (B, D), a GC marker protein.

3 Discussion

It is known that members of the *Bunyaviridae* family bud into the GC, but no primary sequence homologies have been found which may explain the mechanism behind GC retention of the glycoproteins G1/G2. It is therefore possible that the signal consists of a three-dimensional structure, common to the spike proteins of viruses in this family. At present, the data show that in all bunyaviruses, the GC retention signal is situated in the N-terminally located glycoprotein (GN) of the precursor, whereas the C-terminal glycoprotein (GC) accumulates in the GC by interacting with GN containing the Golgi retention signal. Our results have shown that the cytoplasmic tail of G1 is both necessary and sufficient for Golgi retention. This is clearly different from previously reported results where the TMD and flanking sequences of some glycosyltransferases are essential for Golgi retention. One model for GC retention suggests that the length of the TMD determine to what membrane the protein will be transported to and retained at (Munro, 1991, 1995). A second model suggests that hetero-oligomerization mediated by the TMD and a flanking region of enzymes located in the same cisternae would result in the formation of large aggregates (kin recognition), which would be excluded from the transport vesicles (Nilsson et al., 1994, 1996).

The first model does not seem to apply to the retention of G1. What could be the mechanism behind the retention of G1/G2 in the GC? One possibility is that the cytoplasmic tails of G1 molecules interact with each other, forming large complexes, thereby excluding the packaging of the heteodimers into the transport vesicles. Another possibility is that the cytoplasmic tail of G1 contains a retrieval signal which mediates the recycling of escaped proteins back to the GC. However, we favour a model according to which the cytoplasmic tail of G1 interacts with a

Golgi membrane or intercisternal matrix protein(s). Experiments to identify interacting partners are in progress.

4 Summary

In this study we have shown that the cytoplasmic tail of G1 contains the Golgi retention signal, indicating no or only a minor role of the ectodomain and/or the TMD for GC localization. By making CD4 chimeric proteins with progressive deletions of the cytoplasmic tail of G1, we could further narrow down the region containing the Golgi retention signal to be located approximately between residues 4 to 50, counting from the TMD border. To determine the shortest region needed for Golgi localization, we expressed the cytoplasmic tail of G1 as a 81-residue peptide and made progressive deletions of both the N- and C-terminus. We found that the shortest peptide, ranging from residues 10 to 40, was still able to be targeted to the Golgi membranes, a result which is in agreement with recent results with the CD4 membrane-spanning chimeras (Andersson et al., 1997b). Thus, we conclude that the motif responsible for retaining the G1/G2 heterodimer in the GC, is located between residues 10 to 40 in the cytoplasmic tail of G1.

The primary amino acid sequence does not give a clue as to which residues are important for GC retention. A more detailed mapping of this region may help in defining critical residues needed for Golgi retention.

5 Acknowledgement

We thank Anita Bergström and Elisabeth Raschperger for excellent technical assistance.

6 References

Andersson AM, Melin L, Persson R, Raschperger E, Wikström L and Pettersson RF (1997a) Processing and membrane topology of the spike proteins G1 and G2 of Uukuniemi virus. J. Virol. **71**:218-225

Andersson AM, Melin L, Bean A and Pettersson RF (1997b) A retention signal necessary and sufficient for Golgi localization maps to the cytoplasmic tail of a *Bunyaviridae* (Uukuniemi virus) membrane glycoprotein. J. Virol. **71**:4717-4727

Melin L, Persson R, Andersson A, Bergström A, Rönnholm R and Pettersson RF (1995) The membrane glycoprotein G1 of Uukuniemi virus contains a signal for localization to the Golgi complex. Virus Res. **36**:49-66

Munro S (1991) Sequences within and adjacent to the transmembrane segment of α-2,6-sialyltransferase specify Golgi retention. EMBO J. **10**:3577-3588

Munro S (1995) An investigation of the role of the transmembrane domains in Golgi protein retention. EMBO J. **14**:4695-4704

Nilsson T, Hoe MH, Slusarewicz P, Rabouille C, Watson R, Hunte F, Watzele G, Berger EG and Warren G (1994) Kin recognition between medial Golgi enzymes in HeLa cells. EMBO J. **13**:562-574

Nilsson T, Rabouille C, Hui N, Watson R and Warren G (1996) The role of the membrane-spanning domain and stalk region of N-acetylglucosaminyltransferase I in retention, kin recognition and structural maintenance of the Golgi apparatus in HeLa cells. J. Cell Sci. **109**:1975-1989

Persson R and Pettersson RF (1991) Formation and intracellular transport of a heterodimeric viral spike protein complex. J. Cell biol. **112**:257-266.

Pettersson RF (1991) Protein localization and virus assembly at intracellular membranes. p. 67-98. In RW Compans (ed.), Protein traffic in eucaryotic cells. Springer-Verlag, Berlin Heidelberg, Germany.

Pettersson RF and Melin L (1996) Synthesis, assembly and intracellular transport of *Bunyaviridae* membrane proteins. p. 159-188. In RM Eliott (ed.), The *Bunyaviridae*. Plenum Press, New York, N.Y.

REP-Mediated Protein Prenylation

U. Bialek, B. E. Bauer, M. Miaczynska, S. Lorenzetti, R. J. Schweyen and A. Ragnini*

University of Vienna, Vienna Biocenter, Institute of Microbiology and Genetics, Dr. Bohrgasse 9, A-1030 Vienna, Austria. E-mail: antonell@gem.univie.ac.at

Keywords: REP mutants, Ypt binding, membrane binding, choroideremia

Introduction

Addition of isoprenoids to polypeptides is a common post-translational modification occurring in yeast, animal and plant cells. Isoprenoid precursors are derived from the mevalonate pathway and are incorporated into many cellular products such as sterols, heme A, ubiquinone, dolichols, carotenoids and prenylated proteins (Goldstein and Brown, 1990).

Protein prenylation involves addition of 15- or 20- carbon chains (farnesyl or geranylgeranyl moieties, respectively) via thioether bonds to particular C-terminal cysteine motifs of proteins (Zhang and Casey, 1996). Due to the high hydrophobicity of the attached lipids, prenylated proteins tend to associate spontaneously with intracellular membranes. Thermodynamic studies on the binding of prenylated peptides to phospholipid vesicles have shown that, while the geranylgeranyl modification ensures stable membrane association of the modified polypeptides, farnesylation *per se* is not enough to give a stable association without additional lipidic modifications, for example myristoylation (Shahinian and Silvius, 1995; Bhatnagar and Gordon, 1997).

Different classes of prenylated polypeptides are known, such as lamins, fungal mating factors, subunits of trimeric G proteins, kinases and Ras-like small G proteins (Casey and Seabra, 1996). Among the Ras-superfamily, the Ras proteins are farnesylated (C_{15}) while Rho/Rac and Rab/Ypt proteins are geranylgeranylated (C_{20}). The farnesyltransferase (FTase) and the geranylgeranyltransferase I (GGTase I) recognise proteins terminating in CAAX (where A is an aliphatic amino acid). The terminal amino acid of the recognition motif discriminates between the two enzymes: in the case of serine, methionine or glutamine the protein is recognised by the FTase, while leucine is recognised by GGTase I. These two heterodimeric enzymes share one subunit (alpha) while the other (beta) is specific for each of

1

[1] *to whom correspondence should be addressed

NATO ASI Series, Vol. H 106
Lipid and Protein Traffic
Pathways and Molecular Mechanisms
Edited by Jos A. F. Op den Kamp

them (Casey and Seabra, 1996; Zhang and Casey, 1996). Although substrate specificity of FTase and GGTase I is quite high, cross-specificity has been observed (James et al., 1995; Armstrong et al., 1995). Single amino acid changes at either position 159 or 362 in the beta subunit of the yeast FTase result in a conversion of the protein substrate specificity from FTase I to GGTase I (Del Villar et al., 1997).

The Rab/Ypt-type of small G proteins have different C-terminal motifs, CC, CXC or CCXX, which are recognised by the GGTase II (Peter et al., 1992). In contrast to other prenyltransferases, the presence of the C-terminal cysteine motif is not sufficent *per se* for substrate recognition by the GGTase II (Casey and Seabra, 1996) since other features located in the N-terminal part of the protein are also required (Stenmark et al., 1994; Beranger et al., 1994). GGTase II substrate recognition depends on the presence of an additional subunit termed the Rab escort protein (REP; Andres et al., 1993). The possible mechanism of GGTase II-mediated protein prenylation in mammals has been described as follows (Fig. 1): REP binds to a newly synthetised Rab protein in its inactive GDP-bound form (Seabra, 1996). The resulting heterodimer is recognised as a substrate by the catalytic subunits of the GGTase II with subsequent lipid transfer (Shen and Seabra, 1996). Following prenylation, REP and Rab stay associated; meanwhile the GGTase II catalytic subunits are released. The prenylated Rab is then delivered to its target membrane. Upon membrane association of Rab, REP is released to the cytosol and can take part in the next round of prenylation (Alexandrov et al., 1994; Shen and Seabra, 1996).

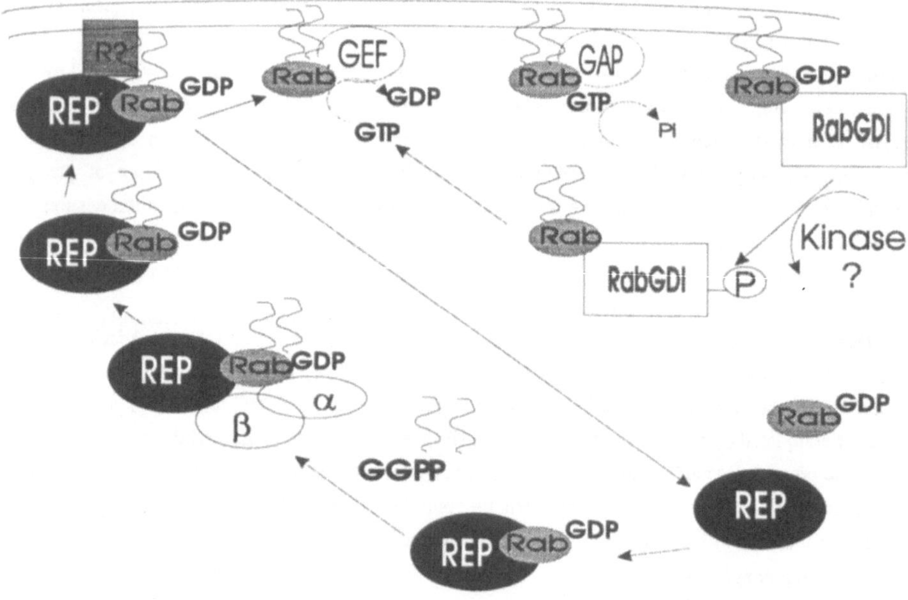

Fig.1 Proposed cycle of REP and Rab proteins.
GEF - Guanine nucleotide exchange factor; GAP - GTPase activating protein; α - the alpha subunit of GGTase II; β - the beta subunit of GGTase II; GGPP - geranylgeranyl pyrophosphate.

Prenylated Rabs can be extracted from membranes by the GDP dissociation inhibitor (GDI), a protein structurally and functionally related to REP (Waldherr et al., 1993b; Pfeffer et al., 1995). In contrast to REPs, the RabGDIs recognise only the prenylated form of a Rab protein, which then remains in the cytosol in a GDP-bound state. Although the sequence of events that ends with a specific membrane delivery of the Rab protein is not yet clarified, it appears that the first round of Rab membrane association is driven by the REP activity (Alexandrov et al., 1994; Wilson et al., 1996) while the RabGDI is likely to guide the further rounds of Rab membrane delivery (Soldati et al., 1994; Ullrich et al., 1994; Dirac-Svejstrup et al., 1997).

The identification of REP as an essential component of the GGTase II (Seabra et al., 1992) followed the mapping of the locus (CHM) responsible for a human X-linked retinal dystrophy called Choroideremia (Cremers et al., 1990; van den Hurk et al., 1997) that indeed turned out to encode the REP-1 protein (Andres et al., 1993). So far, two REPs have been found in mammals: REP-1 encoded by the CHM gene and REP-2 encoded by the CHML gene (Cremers et al., 1994). The molecular mechanism by which this disease develops is not known but the low affinity of REP-2 towards Rab27 (a small G protein abundant in eye tissues) suggests that the defect in prenylation of Rab27 in patients carrying mutations in REP-1 might result in the development of choroideremia (Seabra et al., 1995).

The only REP in the yeast *Saccharomyces cerevisiae* is encoded by the essential MRS6/MSI4 gene (Ragnini et al., 1994; Fujimura et al., 1994; Jiang et al., 1994). This gene was independently isolated as a high copy number suppressor of the respiratory deficient mutant *mrs2-1/wo* (Waldherr et al., 1993a; Ragnini et al., 1994) and as a suppressor of heat shock sensitivity due to the Ras-GAP mutation *ira1-1* (Fujimura et al., 1994). Later on it was observed that overexpression of MRS6 could also suppress the thermo-sensitive secretion mutant *ypt1*$^{N121I/A161V}$ as well as a mutant defective in the regulatory subunit of cAMP-dependent protein kinase Bcy1 (*bcy1*) that acts downstream of Ras (Fujimura et al., 1994). This wide spectrum of effects (e.g. biogenesis of mitochondria, Ras-mediated signalling, secretion) suggests that the Mrs6p function is located at a cross-road among different cellular pathways.

In order to unravel the different aspects of REP activity we have used the yeast *Saccharomyces cerevisiae*. With combined genetic and biochemical approaches we were able to provide some new information about REP localisation, domain structure and the intracellular regulation of REP activity.

The yeast REP is localised in both membrane and cytosolic fractions

The ability to deliver prenylated Rab proteins to their target membranes is one of basic aspects of REP function but the exact mechanism of this delivery process is still unclear. Studies on mammalian REPs have shown that these proteins have higher affinity for the GDP-bound than for the GTP-bound Rab proteins. Since there is an excess of GTP over GDP in the cells, it has been postulated that the binding of REP to Rab/Ypt proteins could occur immediately after Rab/Ypt protein synthesis. In this case, the REP could function as a bonafide chaperone by keeping the nascent Rab/Ypt protein in the GDP-conformation (Seabra, 1996).

In order to identify the functional localisation of the yeast REP, we investigated the intracellular distribution of the Mrs6 protein by biochemical fractionation and immunoblot analyses of fractionated yeast protein extracts (Miaczynska et al., 1997). We observed that while most of this protein is present in the cytosol, a significant fraction was found to be associated with intracellular membranes. Consistent with the hypothesis of an early interaction of REP with Ypt proteins, an enrichment was observed in the microsomal fraction containing ER-Golgi membranes, while no membrane-bound Mrs6p was recovered in plasma membranes, vacuoles, mitochondria or micosomal fractions associated with mitochondria.

Analysis of the membrane-bound form of Mrs6p was consistent with its tight peripheral association to membranes. The presence of proteins in the target membrane was not essential since their removal by proteinase K pretreatment only slightly affected REP membrane binding *in vitro*. Although these results suggested an overall hydrophobicity of the Mrs6 protein, neither the recombinant nor the endogenous Mrs6p fractionated with the detergent phase upon partitioning in Triton X-114.

Interestingly, we observed a dependence of the Mrs6p-membrane association on the presence of divalent cations. The importance of divalent cations such as Ca^{++} in vesicular transport has been extensively reported (Südhof, 1995; Baker et al., 1990). This observation raised the possibility that regulation of the concentration of divalent cations could be part of a mechanism that defines the membrane or cytosolic state of the REP protein.

By studying the membrane-cytosol distribution of different Mrs6 mutants and Mrs6/GDI chimeras, we observed that mutations affecting Ypt binding ability and changing the C-terminus of Mrs6p result in an association of the mutant proteins with the membrane fraction and the complete disappearance from the cytosol (Miaczynska et al., 1997; U. Bialek, unpublished results). This suggests that REP has a more soluble conformation upon binding of Ypt.

Based on these results, we expect that in the absence of intracellular compartmentalisation or of a regulated conformational switching, the Mrs6p would tend to interact unspecifically with intracellular membranes. Since we were unable to detect the Mrs6p in some of the membranes that represent the Ypt protein localisation targets (such as the plasma membrane for Sec4p or vacuoles for Ypt7p), we can exclude that the membrane-bound fraction of Mrs6p observed in our experiments could be attributed to the Ypt delivery activity postulated for REP (Wilson et al., 1996).

REP domain structure and Ypt binding sites

One important aspect of the REP function is how it contacts its substrate. While the possible regions in the Rab proteins required for REP binding were mapped (Stenmark et al., 1994; Beranger et al., 1994), nothing was known about the corresponding regions in members of the REP/GDI family of proteins.

One of our early findings was that the REP proteins belong to the same family of proteins as the RabGDIs (Waldherr et al., 1993b). RabGDI was first identified as a factor able to interact with the Rab/Ypt proteins keeping them soluble in a GDP-bound conformation (Sasaki et al., 1990; Araki et al., 1990). Later it became evident that the major role of GDIs is to re-cycle membrane-bound Rab/Ypt proteins to make them available for another round of membrane binding. It has been also proposed that the membrane-cytosol cycling of Rabs is required for defining the timing of vesicular docking and fusion events (Rybin et al., 1996).

We showed that three major blocks of similarities could be identified in the REP-GDI family of proteins (Fig. 2). These blocks, called <u>s</u>tructurally <u>c</u>onserved <u>r</u>egions (SCR) 1, 2 and 3, are located in the first two-thirds of these proteins. By the combined use of genetic and biochemical assays, we were able to show that the SCRs together with the carboxy-terminal region of the yeast REP were required for Ypt binding (Bauer et al., 1996). Co-immunoprecipitation of Ypt1p with Mrs6p was abolished by mutation of

Gly53 in the SCR1a as well as by deletion of the last 173 amino acids but not by a deletion of the last 33 amino acids. Using a similar approach we were recently able to define the C-terminal essential part of Mrs6p to the region between amino acids 500 and 573 (S. Lorenzetti, unpublished results). Mutations of His74 in SCR1b, Asp132 in SCR2 or Gln289 in SCR3b have less drastic effects than the mutation in SCR1a since REP-Ypt binding was only reduced (Bauer et al., 1996).

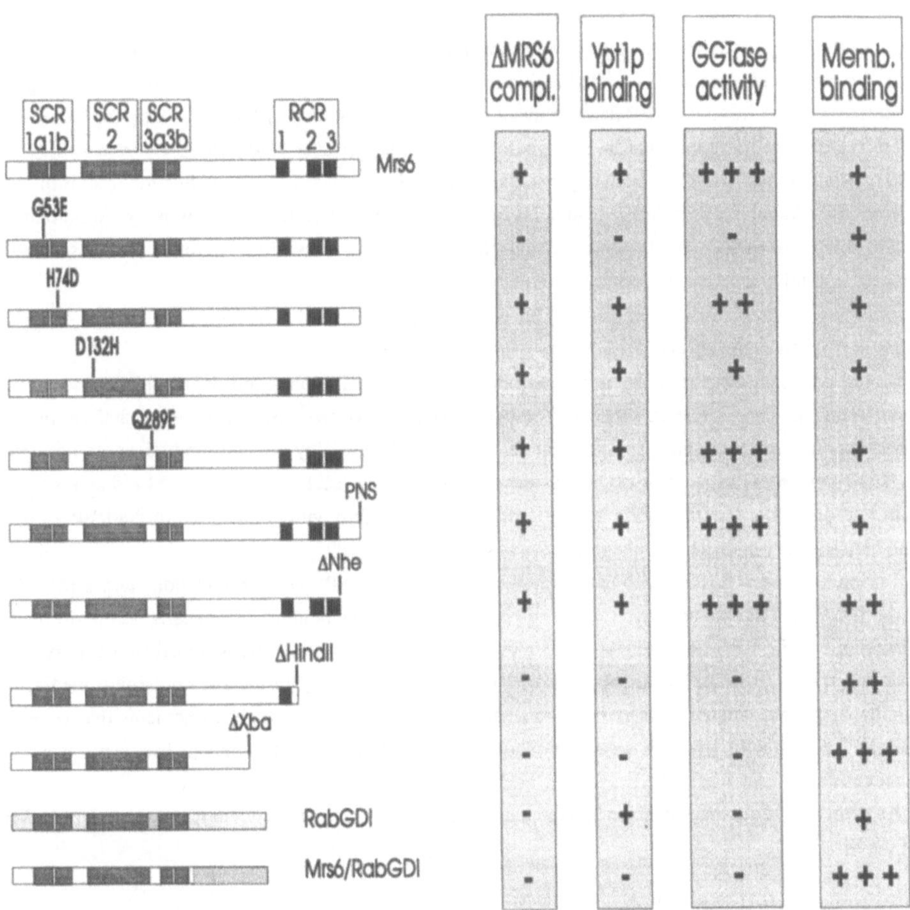

Fig. 2 Phenotypic characterisation of Mrs6 wild type and mutant proteins.
Left panel: schematic drawing of MRS6, Rab GDI proteins and their mutant forms. Right panel: summary of results showing the ability of Mrs6p constructs to: complement mrs6 null allele (first column), bind Ypt1p (second column), prenylate Ypt1p (third column) and bind intracellular membranes (fourth column). (-) no, (+) low, (++) middle and (+++) maximal activity or binding.

Similar results were obtained by mutagenesis analyses of the SCRs in the bovine RabGDI (Schalk et al., 1996). The crystallographic structure of the RabGDI obtained by Schalk and colleagues showed that the

SCRs are located on the protein surface that is thought to be highly conserved among the members of this family. The SCR1a, the SCR3b and the C-terminus of the RabGDI form the core of the Rab/Ypt-binding domain (with the C-terminus folding back towards the N-terminus). The SCR2 and SCR3a are instead thought to be less important for Rab binding and possibly have a function in interacting with other effectors (Schalk et al., 1996). The structural and functional similarities underlined above have, however, to take in consideration the fact that the C-terminus of REP-RabGDI proteins is the most variable region differing in length and amino acid composition.

Among REPs, some additional structurally conserved blocks (termed REP-conserved regions; RCR 1, 2 and 3) were identified in the C-terminal region of these proteins (Fig. 2; Bauer et al., 1996). Although the importance of the RCRs for REP function await more detailed studies, REP proteins carrying deletions in these regions are non-functional in both yeast and mammals (Bauer et al., 1996; van den Hurk et al., 1997). Random and site-directed mutagenesis studies are in progress to better define the role of the REP C-terminal domain. Preliminary results suggest that the C-terminus is extremely important for maintaining the Mrs6p tertiary structure and alterations in some of the C-terminal residues result in protein instability and/or a thermo-sensitive phenotype (see below).

Phenotypic characteristics of REP thermo-sensitive mutants

So far, only mutations resulting in a C-terminal truncated REP-1 protein or its complete absence have been found in choroideremia patients. These mutant proteins arise from genomic deletions or from nonsense, frameshift or splice-site mutations in the CHM locus. Why missense mutations do not appear in these patients is as yet unclear. It has been suggested that such mutations could either result in a different disease or that the REP-1 mutant protein could interfere with the function of the REP-2 protein and consequently have a lethal phenotype (van den Hurk et al., 1997).

Taking advantage of the yeast genetic system, we created REP missense mutations and screened for those that give a thermo-sensitive (ts) phenotype. Hydroxylamine random mutagenesis was performed on plasmids bearing the MRS6 gene and the mutated DNA pool was then transformed into a yeast strain carrying a Gal1/10-MRS6 gene and screened for thermo-sensitivity in glucose-containing media. In a similar approach, random mutagenesis was performed on internal MRS6 gene fragments that were then re-inserted in a wild type Mrs6 context. The transformants carrying the mutant fragments were then screened for thermo-sensitivity.

As shown in Table 1 we obtained 30 ts mutants. Of these at least 3 had mutations in the C-terminal part of the Mrs6p.

Table 1. Thermo-sensitive mrs6 mutants selected in the screening.

Copy number	Number of mutants selected	Localisation of the mutations
one genomic copy	2	C-terminal
high copy (YEp)	3	n.d.
low copy (YCp)	22	n.d.
low copy (YCp)	3	C-terminal

n.d. stands for sequences not determined

In agreement with the idea that the C-terminal part of the Mrs6p is required for GGTase II activity, we observed that the mutants carrying amino acid changes in this part of protein, have a barely detectable GGTase II activity even at the permissive temperature. Preliminary results show that one of these mutants, mrs6-2ts, cultured at the non-permissive temperature stops growing in an early budding stage with a G2/M DNA content. As the microtubular network appears altered at this stage in the mrs6-2ts strain (Bialek et al., manuscript in preparation) we are tempted to speculate that a failure in proceeding through the bud formation check point (Lew and Reed, 1995) prevents these cells from terminating the cell cycle.

Concluding remarks

The work done so far to understand Rab/Ypt protein prenylation has clarified the importance of the Rab escort protein for this process. REP is essential in both higher and lower eukaryotes for vesicular transport, being required for the substrate recognition by GGTase II and for keeping prenylated proteins in a soluble state before membrane-specific targeting. The basic characteristics of the REP domain structure have been unravelled by showing that the N-terminal structurally conserved regions and additional C-terminal blocks are required for Rab/Ypt binding and REP activity. The ability of REP to associate with microsomal membranes, possibly in a cation-dependent manner, suggests that this protein itself might have a regulated membrane-cytosolic cycle independent of the Rab/Ypt cycle. In this case, it is clear that such regulation would affect all intracellular trafficking and could be used as a hierarchical control of the secretory ability of the cell which is required during metabolic switches (such as carbon source changes or nutritional stress) or during sporulation/germination processes.

The hypothesis that REP function and prenylation could be required for processes other than vesicular transport was first put forward due to the ability of REP, when expressed in high copy, to bypass mutations in the Ras pathway (Fujimura et al., 1994) or in mitochondrial biogenesis (Ragnini et al., 1994). How processes not strictly related to intracellular trafficking could be affected by the differential expression of REP whose function depends on the formation of a multiprotein complex is still unclear. Does Mrs6p overexpression directly affect GGTase activity and if so how? Does the REP protein work only as a "Rab/Ypt-adapter" for GGTase II prenylation or has it *per se* a regulatory role in the formation and dissociation of the GGTase II multiprotein complex?

The morphological characteristics of the mrs6-2ts arrested cells suggest a failure in passing through the budding check-point (Bialek et al., manuscript in preparation). The existence of a linkage between morphogenesis and intracellular trafficking at the level of Rab/Ypt protein prenylation is indeed not surprising since during bud formation most of the newly synthesised products are directed towards the growing point. However, which of the geranylgeranylated proteins play a role in this process is as yet obscure. A candidate could be one of the late-acting Ypt proteins, Sec4, which has been implicated in the activation of a multiprotein complex called Exocyst (Bowser et al., 1992; TerBush et al., 1996). The Exocyst complex has been proposed to be involved in the rearrangement of cortical actin to provide access to the site of fusion during exocytosis (TerBush et al., 1996). The discovery that yeast casein kinases also possess a consensus for GGTase II prenylation (Vancura et al., 1994) opens the possibility that one of the members of this family could act as a messenger for connecting cell cycle and prenylation activity of the cells.

Acknowledgements

We thank Dr. Cathal Wilson for a critical reading of the manuscript. This work was supported by a grant from the Austrian Research Foundation. U. Bialek is supported by the Austrian Research Foundation within Vienna Biocenter Ph.D. Programm. M. Miaczynska was supported by Bertha von Suttner Ph.D. fellowship.

Literature

Alexandrov K, Horiuchi H, Steele-Mortimer O, Seabra MC and Zerial M (1994) Rab escort protein-1 is a multifunctional protein that accompanies newly prenylated rab proteins to their target membranes. EMBO J. 13: 5262-5273

Andres DA, Seabra MC, Brown MS, Armstrong SA, Smeland TE, Cremers FPM and Goldstein JL (1993) cDNA cloning of component A of Rab geranylgeranyl transferase and demonstration of its role as Rab escort protein. Cell 73: 1091-1099

Araki S, Kikuchi A, Hata Y, Isomura M and Takai Y (1990) Regulation of reversible binding of smg p25A, a ras p21-like GTP-binding protein, to synaptic plasma membranes and vesicles by its specific regulatory protein, GDP dissociation inhibitor. J. Biol. Chem. 265: 13007-13015

Armstrong SA, Hannah VC, Goldstein JL and Brown MS (1995) CAAX geranylgeranyl transferase transfers farnesyl as efficiently as geranylgeranyl to RhoB. J. Biol. Chem. 270: 7864-7868

Baker D, Wuestehube L, Schekman R, Botstein D and Segev N (1990) GTP-binding Ypt1 protein and Ca^{2+} function independently in a cell-free protein transport reaction. Proc. Natl. Acad. Sci. USA 87: 355-359

Bauer BE, Lorenzetti S, Miaczynska M, Minh Bui D, Schweyen RJ and Ragnini A (1996) Amino- and carboxy-terminal domains of the yeast Rab escort protein are both required for binding of Ypt small G proteins. Mol. Biol. Cell 7: 1521-1533

Benito-Moreno RM, Miaczynska M, Bauer BE, Schweyen RJ and Ragnini A (1994) Mrs6p, the yeast homologue of the mammalian choroideraemia protein: immunological evidence for its function as the Ypt1p Rab escort protein. Curr. Genet. 27: 23-25

Beranger F, Cadwallader K, Porfiri E, Powers S, Evans T, de Gunzburg J and Hancock JF (1994) Determination of structural requirements for the interaction of Rab6 with RabGDI and Rab geranylgeranyltransferase. J. Biol. Chem. 269: 13637-13643

Bhatnagar RS and Gordon JI (1997) Understanding covalent modifications of proteins by lipids: where cell biology and biophysics mingle. Trends in Cell Biol. 7: 14-20

Bowser R, Müller H, Govindan B and Novick P (1992) Sec8p and Sec15p are components of a plasma membrane-associated 19.5S particle that may function downstream of Sec4p to control exocytosis. J. Cell Biol. 118: 1041-1056

Casey PJ and Seabra MC (1996) Protein prenyltransferases. J. Biol. Chem. 271: 5289-5292

Cremers FPM, van de Pol DJR, van Kerkhoff LPM, Wieringa B and Ropers H-H (1990) Cloning of a gene that is rearranged in patients with choroideraemia. Nature 347: 674-677

Cremers FPM, Armstrong SA, Seabra MC, Brown MS and Goldstein JL (1994) REP-2, a rab escort protein encoded by the choroideremia-like gene. J. Biol. Chem. 269: 2111-2117

Del Villar K, Mitsuzawa H, Yang W, Sattler I and Tamanoi F (1997) Amino acid substitutions that convert the protein substrate specificity of farnesyltransferase to that of geranylgeranyltransferase type I. J. Biol. Chem. 272: 680-687

Dirac-Svejstrup AB, Sumizawa T and Pfeffer S (1997) Identification of a GDI displacement factor that releases endosomal Rab GTPases from Rab-GDI. EMBO J. 16: 465-472

Fujimura K, Kazuma T, Nakano A and Toh-e A (1994). The *Saccharomyces cerevisiae* MSI4 gene encodes the yeast counterpart of component A of the Rab geranylgeranyl transferase. J. Biol. Chem. 269: 9205-9212

Goldstein JL and Brown MS (1990) Regulation of the mevalonate pathway. Nature 343: 425-430

James GL, Goldstein JL and Brown MS (1995) Polylysine and CVIM sequences of K-RasB dictate specificity of prenylation and confer resistance to benzodiazeptine peptidomimetic in vitro. J. Biol. Chem. 270: 6221-6

Jiang Y and Ferro-Novick S (1994) Identification of yeast component A: reconstitution of the geranylgeranyl transferase that modifies Ypt1p and Sec4p. Proc. Natl. Acad. Sci. USA 91: 4377-4381

Lew DJ and Reed SI (1995) Cell cycle checkpoint monitors cell morphogenesis in budding yeast. J. Cell Biol. 129: 739-749

Miaczynska M, Lorenzetti S, Bialek U, Benito-Moreno RM, Schweyen RJ and Ragnini A (1997) The yeast Rab escort protein binds intracellular membranes *in vivo* and *in vitro*. J. Biol. Chem. 272: 16972-16977

Peter M, Chavrier P, Nigg EA and Zerial M (1992) Isoprenylation of rab proteins on structurally distinct cysteine motifs. J. Cell Sci. 102: 857-865

Pfeffer SR, Dirac-Svejstrup AB and Soldati T (1995) Rab GDP dissociation inhibitor: putting Rab GTPases in the right place. J. Biol. Chem. 270: 17057-17059

Ragnini A, Teply R, Waldherr M, Voskova A and Schweyen RJ (1994) The yeast protein Mrs6p, a homologue of the rabGDI and the human choroideraemia proteins, affects cytoplasmic and mitochondrial functions. Curr. Genet. 26: 308-314

Rybin V, Ullrich O, Rubino M, Alexandrov K, Simon I, Seabra MC, Goody R and Zerial M (1996) GTPase activity of Rab5 acts as a timer for endocytotic membrane fusion. Nature 383: 266-269

Sasaki T, Kikuchi A, Araki S, Hata Y, Isomura M, Kuroda S and Takai Y (1990) Purification and characterization from bovine brain cytosol of a protein that inhibits the dissociation of GDP from and the subsequent binding of GTP to *smg* p25A, a *ras* p21-like GTP-binding protein. J. Biol. Chem. 265: 2333-2337

Schalk I, Zeng K, Wu S-K, Stura EA, Matteson J, Huang M, Tandon A., Wilson IA and Balch WE (1996) Structure and mutational analysis of Rab GDP-dissociation inhibitor. Nature 381 42-48

Seabra MC, Brown MS, Slaughter CA, Sudhof TC and Goldstein JL (1992) Purification of component A of Rab geranylgeranyl transferase: possible identity with the choroideremia gene product. Cell 70: 1049-1057

Seabra MC, Ho YK and Anant JS (1995) Deficient geranylgeranylation of Ram/Rab27 in choroideremia. J. Biol. Chem. 270: 24420-24427

Seabra MC (1996) Nucleotide dependence of Rab geranylgeranylation. J. Biol. Chem. 271: 14398-14404

Shahinian S and Silvius JR 1995 Doubly-lipid-modified protein sequence motifs exhibit long-lived anchorage to lipid bilayer membranes. Biochemistry 34: 3813-3822

Shen F and Seabra MC (1996) Mechanism of digeranylgeranylation of Rab proteins. J. Biol. Chem. 271: 3692-3698

Soldati T, Shapiro AD, Dirac-Svejstrup AB and Pfeffer SR (1994) Membrane targeting of the small GTPase Rab9 is accompanied by nucleotide exchange. Nature 369: 76-78

Stenmark H, Valencia A, Martinez O, Ullrich O, Goud B and Zerial M (1994) Distinct structural elements of rab5 define its functional specificity. EMBO J. 13: 575-583

Südhof TC (1995) The synaptic vesicle cycle: a cascade of protein-protein interactions. Nature, 375: 645-653

TerBush DR, Maurice T, Roth D and Novick P (1996) The Exocyst is a multiprotein complex required for exocytosis in *Saccharomyces cerevisiae*. EMBO J. 15: 6483-6494

Ullrich O, Horiuchi H, Bucci C and Zerial M (1994) Membrane association of Rab5 mediated by GDP-dissociation inhibitor and accompanied by GDP/GTP exchange. Nature 368: 157-160

Vancura A, Sessler A, Leichus B and Kuret J (1994) A prenylation motif is required for plasma membrane localization and biochemical function of casein kinase I in budding yeast. J. Biol. Chem. 269: 19271-19278

van den Hurk JAJM, Schwartz M, van Bokhoven H, van de Pol TJR, Bogerd L, Pinckers AJLG, Bleeker-Wagemakers EM, Pawlowitzki IH, Rüther K, Ropers H-H and Cremers FPM (1997) Molecular basis of choroideremia (CHM): Mutations involving the Rab escort protein-1 (REP-1) gene. Human Mutation 9: 110-117

Waldherr M, Ragnini A, Jank B, Teply R, Wiesenberger G and Schweyen RJ (1993a) A multitude of suppressors of group II intron-splicing defects in yeast. Curr. Genet. 24: 301-306

Waldherr M, Ragnini A, Schweyen RJ and Boguski MS (1993b) MRS6 - yeast homologue of the choroideraemia gene. Nature Genet. 3: 193-194

Wilson AL, Erdman RA and Maltese WA (1996) Association of Rab1B with GDP-dissociation inhibitor (GDI) is required for recycling but not initial membrane targeting of the Rab protein. J. Biol. Chem. 271: 10932-10940

Zhang FL and Casey PJ (1996) Protein prenylation: molecular mechanisms and functional consequences. Annu. Rev. Biochem. 65: 241-269

Development of a Positive Screen for the Identification of Suppressive Mutations in Secretion Defective Strains of *Pseudomonas aeruginosa*

Romé Voulhoux, Denis Duché, Vincent Géli, Andrée Lazdunski and Alain Filloux

Laboratoire d'Ingéniérie des Systèmes Macromoléculaires, CNRS, 31 Chemin Joseph Aiguier, BP 71, 13402 Marseille cedex 20, France

Tel.: +33 0491164127. Fax: +33 0491712124. E-mail: filloux@ibsm.cnrs-mrs.fr

Introduction

Pseudomonas aeruginosa is a gram-negative bacterium, opportunistic pathogen, which secretes numerous proteins in the extracellular medium (Lazdunski *et al.*, 1990). Over the last decade, three main types of secretion pathway have been described. The envelope of gram-negative bacteria is composed of two membranes separated by the periplasm. Type I and III pathways do not involve a transient periplasmic stage. In these cases, the exoproteins contain a C- or N-terminal uncleavable secretion signal, and are directly routed from the cytoplasm to the exterior of the cell by mean of a proteinaceous complex which constitutes the secretion machinery. The machineries, rather simple in Type I with three components, can be highly complex in Type III with up to twenty components. Those pathways, exemplified with the *Escherichia coli* hemolysin for Type I (Holland *et al.*, 1990), and with the Yops of *Yersiniae* for Type III (Cornelis and Wolf-Watz, 1997), are used in *P.aeruginosa* by alkaline protease (Duong *et al.*, 1992) and exoenzyme S (Yahr *et al.*, 1996), respectively. However, all the other enzymes secreted by *P. aeruginosa*, including elastase, lipase, alkaline phosphatase, phospholipases or exotoxinA, are first translocated through the cytoplasmic membrane in a signal sequence-dependent manner using the Sec machinery (Filloux *et al.*, 1987; Douglas *et al.*, 1987). The periplasmic intermediate is further processed to the external medium *via* a machinery composed of twelve components, namely the Xcp machinery (Tommassen *et al.*, 1992). This constitutes the Type II pathway, also called main terminal branch of the General Secretory Pathway (GSP) (Pugsley, 1993). Curiously, the Xcp proteins are mainly localized in, or associated with, the cytoplasmic membrane (Filloux *et al.*, 1990; Bally *et al.*, 1991; 1992; Bleves *et al.*, 1996), with the exception of XcpQ (GspD family)

NATO ASI Series, Vol. H 106
Lipid and Protein Traffic
Pathways and Molecular Mechanisms
Edited by Jos A. F. Op den Kamp

localized in the outer membrane (Akrim *et al.*, 1992; Hardie *et al.*, 1996). Even though similarities exist between the Gsp components from various bacteria, they are not always exchangeable (de Groot *et al.*, 1991). Moreover heterologous secretion is ever hardly observed, even when two related enzymes such as cellulases of *Erwinia chrysanthemi* and *Erwinia carotovora* are swapped (Py *et al.*, 1991). The identification of the (specific) interactions which occur during the assembly of the secretion machinery on one hand, and the recognition of the exoprotein on the other hand, will be essential to improve our knowledge on the mechanistic of the process. We would like to discuss in this report the development of a genetic tool based on the lethality conferred by the expression of an hybrid exoprotein, containing the C-terminal pore forming domain of the colicin A, when secretion is hampered. This will be an unvaluable tool to quickly identify suppressive mutations and thus key interactions of the secretion event.

Description of the strategy: rationale of the selection process

Colicins are plasmid encoded toxins produced by *E. coli*, which kill bacteria from the same species which are themselves unable to produce colicins and the corresponding immunity protein. Colicin secretion is a special case, it involves a small lysis protein which permeabilizes the outer membrane (Lazdunski, 1988). Colicin A (ColA) produces its lethal effect by the insertion of its C-terminal domain (pore-forming domain: PfcolA) in the cytoplasmic membrane (Baty *et al.*, 1990). Insertion of PfcolA generates a depolarisation of the cytoplasmic membrane *via* ion efflux (Bourdineaud *et al.*, 1990), and results in cell death. The uptake of ColA depends on two outer membrane proteins, OmpF and BtuB (Cavard and Lazdunski, 1981), for reception, and the Tol/PAL complex for translocation (Bénédetti and Géli, 1996) (Fig. 1A). Beside the C-terminal pore forming domain, the N-terminal and central domains of ColA are involved to achieve the sequential uptake process (Bénédetti *et al.*, 1991).

It has been reported that the pore forming domain of ColA could exert its activity on the producing strain when it was fused with mitochondrial sorting signals or a signal peptide (sp-PfcolA) (Espesset *et al.*, 1994; 1996). sp-PfcolA is targeted to the periplasm, where it can insert from the external side of the cytoplasmic membrane and kill the cell (Fig. 1B). The cytotoxic activity is thus completely independent of the reception (OmpF/BtuB) and translocation (Tol/PAL) steps, and only those strains which also produce the immunity protein (Cai) will survive (Géli *et al.*, 1989).

The intragenic information which is contained within the exoproteins, for their specific recognition by the secretion machinery, is currently under investigation. The idea is that it probably involves structural motifs, rather than linear stretches of amino-acid residues. In some reports, mention is made that full size proteins such as pullulanase

(PulA) from *Klebsiella oxytoca* (Sauvonnet and Pugsley, 1996), or exotoxin A (ETA) from *P. aeruginosa* (Lu and Lory, 1996), fused to the whole ß-lactamase (Bla) are efficiently transported by their cognate secretion machinery. In the case of ETA, it has also been shown that an hybrid protein containing the first 120 residues (ETA_{120}) fused to Bla is properly secreted (Lu and Lory, 1996).

With respect to those data, our strategy will consist in constructing hybrid proteins containing an exoprotein of *P. aeruginosa* at the N-terminus and PfcolA at the C-terminus. The N-terminal part of the hybrids will allow export of the chimeras to the periplasm. At this stage, one could hypothesize that if the hybrid is properly and quickly secreted *via* the Xcp machinery, no insertion of the PfcolA part will take place and the cells will survive (Fig. 1C). However, if a mutant strain producing the hybrid is hampered for secretion, then the PfcolA part of the chimera, resident in the periplasm, will insert in the inner membrane and will kill the cell (Fig. 1D). Only the mutants that will have corrected their secretion defect will be able to grow and will deserve further investigations.

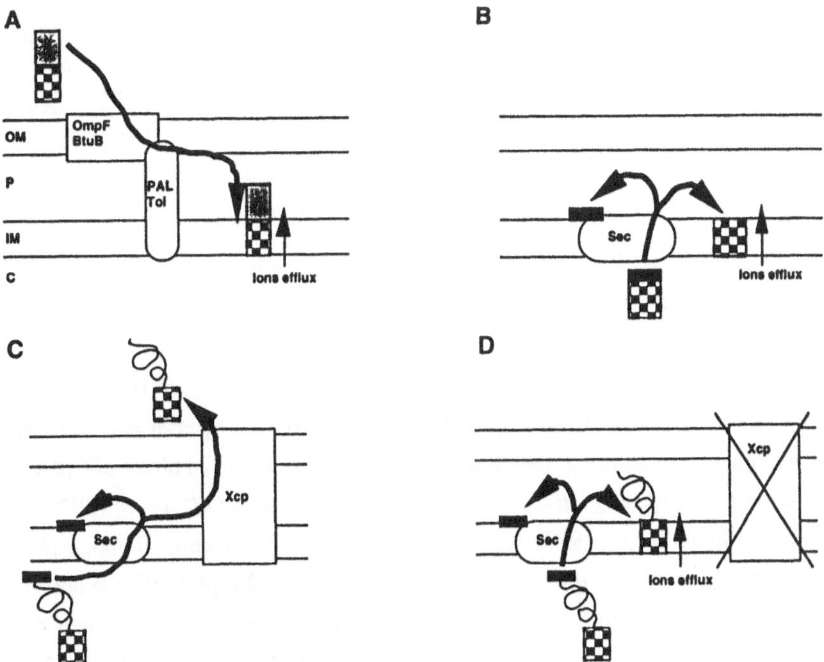

Fig. 1: Schematic representation of the mode of action of ColA and hybrids containing PfcolA in various backgrounds of *E. coli* (A & B) and *P. aeruginosa* (C & D). **A.** Previously to its insertion in the cytoplasmic membrane, ColA is taken up *via* the Omp/Tol system. **B.** When PfcolA is fused to a signal peptide, the producing strain is killed when the immunity protein is not coexpressed. **C.** Hypothetical Xcp-dependent secretion of PfcolA when fused to a *P. aeruginosa* exoprotein and, **D.** membrane insertion when the secretion machinery is defective. ▣: ColA, ▣:PfColA,: ■ signal peptide,: ➰ *P. aeruginosa* exoprotein.

Is PfcolA able to kill *P. aeruginosa* ?

The externally added colicin A cannot be taken up in *P. aeruginosa* because the OmpF/BtuB/Tol/PAL system is different from that of *E. coli* (Fig. 2A).

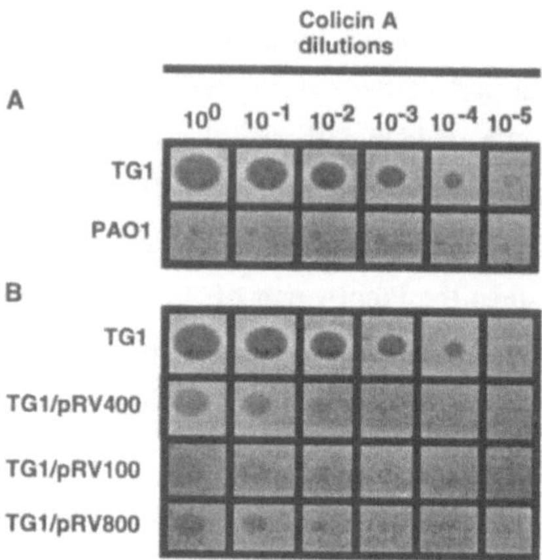

Fig 2: Sensitivity to externally added colicin A. A solution of purified colicin A (1 mg/ml) was diluted 1:10 serially from 10^0 to 10^{-5} in water and samples (1 µl) of each dilution were spotted on a lawn of strain and zones of growth inhibition were observed. A *P. aeruginosa* PAO1 strain is resistant to Colicin A. B *E. coli* TG1 strain is protected against Colicin A by immunity protein, Cai. The level of protection is depending on the copy number of the vector containing the *cai* gene. pRV400 is a low copy plasmid. pRV100 and pRV800 are medium copy plasmids.

In order to test the killer effect of ColA on *Pseudomonas*, we thus recloned the previously constructed *sp-pfcolA* hybrid gene (Espesset *et al.*, 1996), on the broad host range plasmid pMMB67 (Fürste *et al.*, 1986), yielding pRV300. The hybrid protein consists of the signal peptide of pectate lyase B of *Erwinia carotovora* (Lei *et al.*, 1987), fused to PfcolA. In parallel, we also recloned, from pImTc (Espesset *et al.*, 1994), the gene encoding the immunity protein (*cai*), directed against the lethal effect of ColA, on the compatible broad host range vectors, pBBRM1CS-3 (Kovach *et al.*, 1995) (pRV800) and pLAFR (Friedman et al., 1982) (pRV400), medium and low copy number plasmids, respectively. The results on the effect of sp-PfcolA encoded by pRV300 are described in Table 1 and Figure 3. Briefly, only few *E. coli* colonies were obtained when pRV300 was used for transformation, as compared to the control vector pMMB67. Moreover, those clones were detected only when cells were cotransformed with pImTc, even though no selection (Tc^R) for this plasmid was used upon transformation. This demonstrates that pRV300 had a similar effect as pCT1 (Espesset *et al.*, 1996) from which *sp-pfcolA* was recloned. We also clearly noticed that the balance between ColA and Cai is critical, and the higher amount of Cai the better level of protection. Figure 2B shows that Cai encoded from plasmids with medium copy number (pRV100, pRV800) confers higher resitance to ColA than Cai encoded from low copy number plasmid (pRV400).

We further demonstrated that PfColA is equally able to exert its effect on *P. aeruginosa* (Table 1 and Fig. 3), since pRV300 could not be introduced in PAO1 as compared to the control vector. However we did not found yet conditions where the effect of ColA, in *P. aeruginosa*, could be efficiently abolished by introducing one of our constructs carrying the *cai* gene (pRV100, pRV400 or pRV800).

Strains / Plasmids	TG1		PAO1	
	Ap	Cotransformation of plmTc	Cb	Selective Medium
pMMB67 (Ap) plmTc (Tc)	>1000	0%	>1000	Number of transformants
pRV300 (Ap) plmTc (Tc)	3	100%	0	

Table1: sp-PfColA is cytotoxic to *E. coli* and *P. aeruginosa*. Plasmids pMMB67 and pRV300 were introduced into *E. coli* (TG1) by transformation (mixed DNA preparation also containing plmTc), and into *P. aeruginosa* (PAO1) by conjugation (*E. coli* donator cells also contained plmTc). Cells were plated on NA, Ap (50 μg/ml), IPTG (1 mM) for *E. coli*, and PIA, glycerol, Cb (500 μg/ml), IPTG (2 mM) for *P. aeruginosa*. After overnight incubation at 37°C (Figure 3), clones were counted and replated on tetracycline medium to check for the cointroduction of plmTc. plmTc cannot replicate in *P. aeruginosa*. All three clones, obtained with pRV300 in *E. coli*, were Tc resistant.

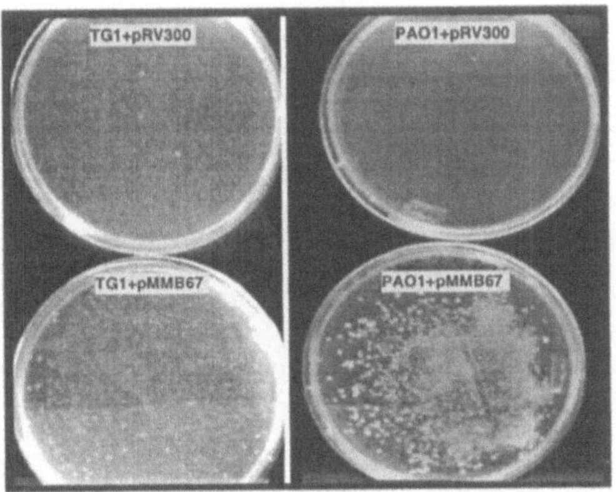

Figure 3: Plates resulting from the TG1 transformation and the PAO1 conjugation, as described in Table 1.

Construction and activity of three new PfcolA hybrids

We have shown that PfcolA is deleterious to *P. aeruginosa* and the subsequent step in the strategy was to produce active hybrids containing PfcolA and susceptible to be secreted by *Pseudomonas*.

Construction of the gene fusions and expression in E. coli

ETA and elastase (LasB), both secreted by *P. aeruginosa*, were chosen as carriers for secretion of the PfcolA passenger to the extracellular medium. The construction of three gene fusions, *eta120-pfcolA* (pRV500), *eta-pfcolA* (pRV600) and *lasB-pfcolA* (pRV700), on plasmid pMMB67, is shown in figure 4A. The corresponding products were identified by western-blotting using a polyclonal antiserum directed against PfcolA (Fig. 4B).

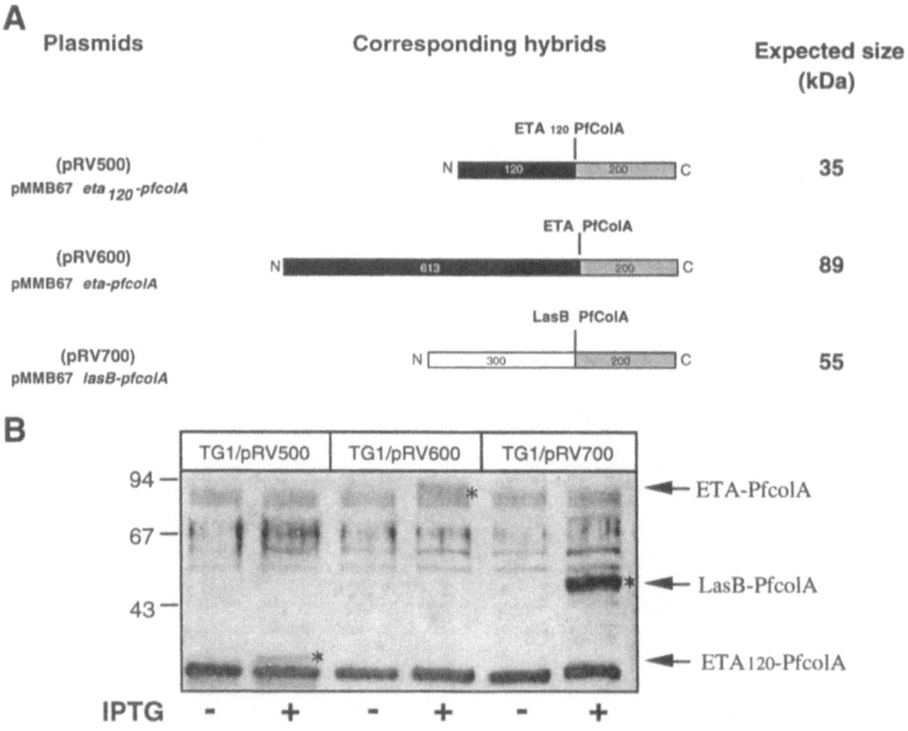

Figure 4: Construction and expression of hybrids containing PfcolA. **A.** Gene fusions were constructed by PCR on the part encoding the domains of interest, followed with a two steps cloning at *EcoRI/HindIII* of pMMB67 under control of the *tac* promoter. **B.** Identification of the hybrid proteins was done by analyzing whole cells extracts of TG1 containing the *cai* gene (pImTc) and the various constructs described in A. Western blots were revealed using an antibody directed against PfcolA and the ECL chemiluminescence system (Pierce). Hybrid proteins are indicated with an asterisk.

Pore-forming activity of hybrid proteins

We investigated whether the hybrid proteins could still be active in term of pore formation, since PfColA will remain attached to a large N-terminal domain consisting of the exoprotein carrier. As previously described with pRV300, we noticed that the efficiency in recovering clones when cells were transformed with either pRV500, pRV600 or pRV700 was lower as compared to the vector control (Table 2). In the case of pRV500 (*eta120-pfcolA*) no clones were ever obtained. This could be related to a higher insertion efficiency of PfcolA, since this particular chimera carries the shortest N-terminal extension as compared to ETA-PfcolA and LasB-PfColA. In addition, and similarly to pRV300 previously described, clones with pRV600 and pRV700 were only obtained when cotransformed with pImTc containing the *cai* gene.

Plasmids	pMMB67 (Ap) pImTc (Tc)	pRV500 (Ap) pImTc (Tc)	pRV600 (Ap) pImTc (Tc)	pRV700 (Ap) pImTc (Tc)
Selective Medium	Ap	Ap	Ap	Ap
Number of transformants	>1000	0	100	50
Cotransformation of pImTc	0%	/	100%	100%

Table 2: PfcolA containing hybrids are cytotoxic in *E. coli*. TG1 cells were transformed with DNA preparation containing pImTc (*cai*) and one of the following plasmids, pMMB67 (control), pRV500 (*eta120-pfcolA*), pRV600 (*eta-colA*) or pRV700 (*lasB-colA*). Colonies were selected on NA, Ap (50 µg/ml), IPTG (1 mM) and further checked on plates containing Tc (15 µg/ml) to observe pImTc cotransformation.

Modulation of the expression level of the hybrid proteins

In addition to the variable killing efficiency of the different hybrids, we also noticed that their level of expression had a marked effect on lethality. As an example, when the *lasB-pfcolA* gene fusion was cloned on the low copy number plasmid pLAFR (pRV200), TG1 clones were obtained even if the *cai* gene was not cointroduced. It should be noticed that the gene fusion was cloned behind the *lac* promoter and those clones appeared only when the *plac* inducer IPTG was omitted from the plates. However, we were able to demonstrate that the gene fusion did not contain any mutations abolishing activity or expresssion of the hybrid protein. Indeed, when cells carrying pRV200 were grown in liquid medium, a drastic arrest in the growth was observed upon IPTG addition

(Fig. 5). This arrest was correlated with higher SDS sensitivity of the cells, which was previously used as a test to evidence Col A activity (Cavard and Lazdunski, 1979).

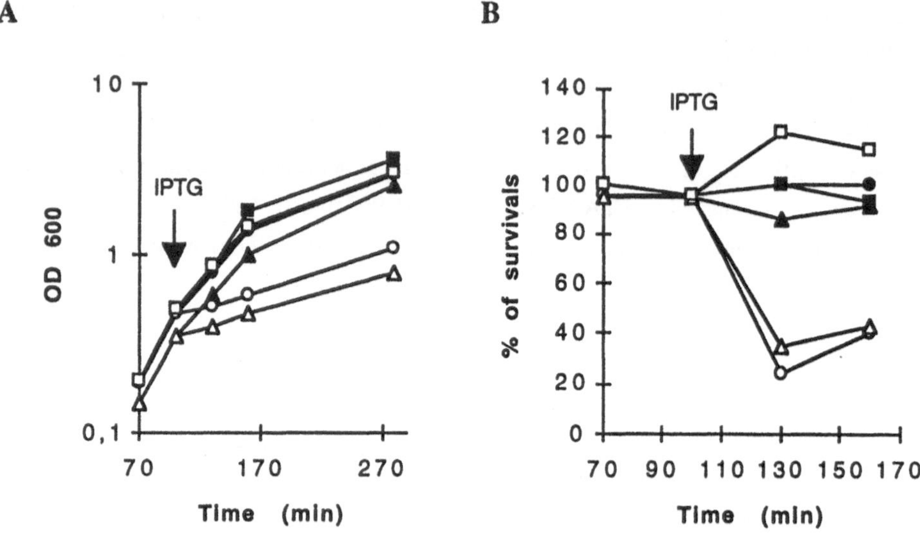

Fig. 5: Effect of sp-PfcolA (circles) and LasB-ColA (triangles) on *E. coli* TG1 cells (squares for vector control) grown in LB medium. Expression of the protein of interest was induced upon addition of IPTG where indicated. Open symbols: + IPTG, filled symbols: - IPTG. A. Cell growth curves. B. % of survivals after SDS treatment (0,5 mg/ml, 10 min) and OD_{600} measurement.

Conclusions and perspectives

We demonstrated in this study that Colicin A is an efficient killing toxin for *P. aeruginosa*. Three hybrids were constructed based on our knowledge on intragenic information for protein secretion, and ability of PfcolA to reinsert from the external face of the inner membrane once exported to the periplasm. The hybrids were shown to be properly expressed and to keep the pore-forming activity of ColA, which is essential for the rest of the study. The secretion efficiency of those hybrids in *Pseudomonas* wild type strains and various *xcp* mutants should now be tested. The expectation is that, if the wild-type resisted the effect of PfcolA due to efficient secretion of the hybrid to the extracellular medium, *xcp* mutants should not survive except if a secondary mutation corrected the defect in the Xcp machinery. The level of production and the secretion kinetics will be important factors for the successful development of this genetic screen. We already presented some evidences showing that we have the possibility to modulate the expression level of the hybrids. It should be possible to obtain levels of hybrids production which will be at the lower limit to cause a lethal effect. The kinetics of secretion, which will determine the time during which the hybrid will reside in the periplasm, are also highly important. Theoretically, exotoxin A, which is hardly detectable in the periplasm (Lory *et al.*, 1983) under normal conditions, should be a better candidate as compared to elastase. However, if the kinetics are anyhow too slow, chances to obtain hypersecretor mutants do exist.

Acknowledgments

We would like to thank Danièle Cavard for her help in the establishment of conditions which evidenced ColA sensitivity, and Claude Lazdunski and Sophie Bleves for careful reading of the manuscript. This work was partly supported by the french association for cystic fibrosis (AFLM) and Rome Voulhoux is supported by the Minister of Research and Technology.

REFERENCES

Akrim M, Bally M, Ball G, Tommassen J, Teerink H, Filloux A and Lazdunski A (1993) Xcp-mediated protein secretion in *Pseudomonas aeruginosa*: identification of two additional genes and evidence for regulation of *xcp* gene expression. Mol. Microbiol. 10: 431-443.

Bally M, Ball G, Badère A and Lazdunski A (1991) Protein secretion in *Pseudomonas aeruginosa*: the *xcpA* gene encodes an integral inner membrane protein homologous to *Klebsiella pneumoniae* secretion function protein PulO. J. Bacteriol. 173: 479-486.

Bally M, Filloux A, Akrim M, Ball G, Lazdunski A and Tommassen J (1992) Protein secretion in *Pseudomonas aeruginosa*: characterization of seven *xcp* genes and processing of secretory apparatus components by prepilin peptidase. Mol. Microbiol. 16: 1121-1131.

Baty D, Lakey J, Pattus F and Lazdunski C (1990) A 136-amino-acid-residue COOH-terminal fragment of colicin A is endowed with ionophoric activity. Eur. J. Biochem. 189: 409-413.

Bénédetti H, Frenette M, Baty D, Knibielher M, Pattus F and Lazdunski C (1991) Individual domains of colicins confer specificity in colicin uptake, in pore-properties and in immunity requirement. J. Mol. Biol. 217: 429-439.

Bénédetti H and Géli V (1996) Colicin transport, channel formation and inhibititon. In "Transport processes in eucaryotic and procaryotic organisms". Handbook of Biological Physics. Eds Konings, Kaback and Loklema. Vol. 2, pp. 665-691.

Bleves S, Lazdunski A and Filloux A (1996) Membrane topology of three Xcp proteins involved in exoprotein transport by *Pseudomonas aeruginosa*. J. Bacteriol. 178: 4297-4300.

Bourdineaud JP, Boulanger P, Lazdunski C and Letellier L (1990) In vivo properties of colicin A: channel activity is voltage dependent but translocation may be voltage independent. Proc. Natl. Acad. Sci. USA. 87: 1037-1041.

Cavard D and Lazdunski C (1979) Purification and molecular properties of new colicin. Eur. J. Biochem. 96: 517-524.

Cavard D and Lazdunski C (1981) Involvement of BtuB and OmpF proteins in binding and uptake of colicin A. FEMS Microbiol. Lett. 12: 311-316.
Cornelis GR and Wolf-Watz H (1997) The *Yersinia* Yop virulon: a bacterial system for subverting eucaryotic cells. Mol. Microbiol. 23: 861-867.
de Groot A, Filloux A and Tommassen J (1991) Conservation of *xcp* genes, involved in two-step protein secretion process, in different *Pseudomonas* species and other gram-negative bacteria. Mol. Gen. Genet. 222: 278-284.
Douglas CM, Guidi-Rontani C and Collier RJ (1987) Exotoxin A of *Pseudomonas aeruginosa*: active cloned toxin is secreted into the periplasmic space of *Escherichia coli*. J. Bacteriol. 169: 4962-4966.
Duong F, Lazdunski A, Cami B and Murgier M (1992) Sequence of a cluster of genes controlling synthesis and secretion of alkaline protease in *Pseudomonas aeruginosa*: relationships to other secretory pathways. Gene 121: 47-54.
Espesset D, Corda Y, Cunningham K, Bénédetti H, Lloubès R, Lazdunski C and Géli V (1994) The colicin A pore-forming domain fused to mitochondrial intermembrane space sorting signals can be functionally inserted into the *Escherichia coli* plasma membrane by a mechanism that bypasses the Tol proteins. Mol. Microbiol. 13: 1121-1131.
Espesset D, Duché D, Baty D and Géli V (1996) Colicin A is inhibited by its cognate immunity protein through direct interaction in *Escherichia coli* inner membrane. EMBO J. 15: 2356-2364.
Filloux A, Murgier M, Wretlind B and Lazdunski A (1987) Characterization of two *Pseudomonas aeruginosa* mutants with defective secretion of extracellular proteins and comparison with other mutants. FEMS Microbiol. Lett. 40, 159-163.
Filloux A, Bally M, Ball G, Akrim M, Tommassen J and Lazdunski A (1990) Protein secretion in Gram-negative bacteria: transport across the outer membrane involves common mechanisms in different bacteria. EMBO J. 9: 4323-4329.
Friedman AM, Long SR, Brown SE, Binkema WJ and Ausubel FM (1982) Construction of a broad host range cosmid cloning vector and its use in the genetic analysis of *Rhizobium* mutants. Gene 18: 289-296.
Fürste JP, Pansegrau W, Frank R, Blöcker H Scholz P, Bagdasarian M and Lanka E (1986) Molecular cloning of the plasmid RP4 primase region in a multi-host-range *tacP* expression vector. Gene 48: 119-131.
Géli V, Baty D, Pattus F and Lazdunski C (1989) Topology and function of the integral membrane protein conferring immunity to colicin A. Mol. Microbiol. 3: 679-687.
Hardie KR, Lory S and Pugsley AP (1996) Insertion of an outer membrane protein in *Escherichia coli* requires a chaperone-like protein. EMBO J. 15: 978-988.
Holland IB, Kenny B and Blight M (1990) Haemolysin secretion from *Escherichia coli*. Biochimie 72: 131-141.
Kovach ME, Elzer PH, Hill DS, Robertson GT, Farris MA, Roop II RM and Peterson KM (1995) Four new derivatives of the broad-host-range cloning vector pBBR1MCS, carrying different antibiotic-resistance cassettes. Gene 166: 175-176.
Lazdunski A, Guzzo J, Filloux A, Bally M and Murgier M (1990) Secretion of extracellular proteins by *Pseudomonas aeruginosa*. Biochimie 72: 147-156.
Lazdunski CJ (1988) What can we learn from colicins about the dynamics of insertion and transfer of proteins into and across membranes. In: Membrane Biogenesis (Op den Kamp, J. A. F., Ed), NATO ASI Series, Vol. H16, pp: 375-393. Springer Verlag, Berlin Heidelberg.
Lei SP, Lin HC, Wang SS, Callaway J and Wilcox G (1987) Characterization of the *Erwinia carotovora pelB* gene and its product pectate lyase J. Bacteriol. 169: 4379-4383.
Lory S, Tai PC and Davis BD (1983) Mechanism of protein excretion by gram-negative bacteria: *Pseudomonas aeruginosa* exotoxin A. J. Bacteriol. 156: 695-792.
Lu HM and Lory S (1996) A specific targeting domain in mature exotoxinA is required for its extracellular secretion from *Pseudomonas aeruginosa*. EMBO J. 15: 429-436.

Pugsley AP (1993) The complete general secretion pathway in Gram-negative bacteria. Microbiol. Rev. 57: 50-108.

Py B, Salmond GPC, Chippaux M and Barras F (1991) Secretion of cellulases in *Erwinia chrysanthemi* and *E. carotovora* is species-specific. FEMS Microbiol. Lett. 79: 315-322.

Sauvonnet N and Pugsley AP (1996) Identification of two regions of *Klebsiella oxytoca* pullulanase that together are capable of promoting ß-lactamase secretion by the general secretory pathway. Mol. Microbiol. 22, 1-7.

Tommassen J, Filloux A, Bally M, Murgier M and Lazdunski A (1992) Protein secretion in *Pseudomonas aeruginosa*. FEMS Microbiol. Rev. 103: 73-90.

Yahr TL, Goranson J and Frank DW (1996) Exoenzyme S of *Pseudomonas aeruginosa* is secreted by a type III pathway. Mol. Microbiol. 22: 991-1003.

Mitotic Division of the Golgi Apparatus

Graham Warren
Imperial Cancer Research Fund
PO Box 123
44 Lincoln's Inn Fields
London WC2A 3PX

Introduction

Of all the organelles in the animal cell, the Golgi apparatus undergoes the most dramatic transformation during cell division (Warren, 1993; Warren and Wickner, 1996). During interphase the Golgi stacks are linked laterally by tubules and networks connecting equivalent cisternae in the adjacent stacks (Rambourg and Clermont, 1990). This yields a ribbon-like structure which bifurcates and rejoins, generating a compact reticulum (Louvard, et al., 1982; Lucocq and Warren, 1987) most often found in the peri-centriolar region, adjacent to the cell nucleus (Ho, et al., 1990).

At the onset of mitosis this organisation is lost. The links connecting the Golgi stacks are broken and the discrete stacks assume a more peri-nuclear distribution (Burke, et al., 1982). Each stack then undergoes fragmentation, converting the cisternae into collections of tubules and vesicles, some of which are shed into the surrounding cytoplasm (Lucocq, et al., 1989; Lucocq, et al., 1987). These fragments are more or less completely dispersed throughout the mitotic cell cytoplasm by metaphase/anaphase and this is followed, during telophase, by reassembly of a Golgi apparatus in each of the two daughter cells (Lucocq, et al., 1989; Souter, et al., 1993). Though still unproven, it is thought that this fragmentation process helps ensure partitioning of the Golgi apparatus between the two daughter cells. In other words, it is thought to ensure division of the Golgi apparatus into two nearly equal parts (Birky, 1983; Warren, 1993).

Using cell-free assays and microscopy, we have shown that fragmentation occurs by two different pathways (Fig. 1)(Warren, et al., 1995). The COPI-dependent pathway interrupts intra-Golgi transport by COPI vesicles (Misteli and Warren, 1994). These vesicles normally bud from the dilated cisternal rims and fuse with their target cisterna in either the anterograde or retrograde direction depending on the proteins they are carrying. Anterograde vesicles carry cargo for the cell surface, lysosomes and secretory granules; retrograde vesicles carry

NATO ASI Series, Vol. H 106
Lipid and Protein Traffic
Pathways and Molecular Mechanisms
Edited by Jos A. F. Op den Kamp

Fig. 1. Model for disassembly and reassembly of the mammalian Golgi apparatus during mitosis

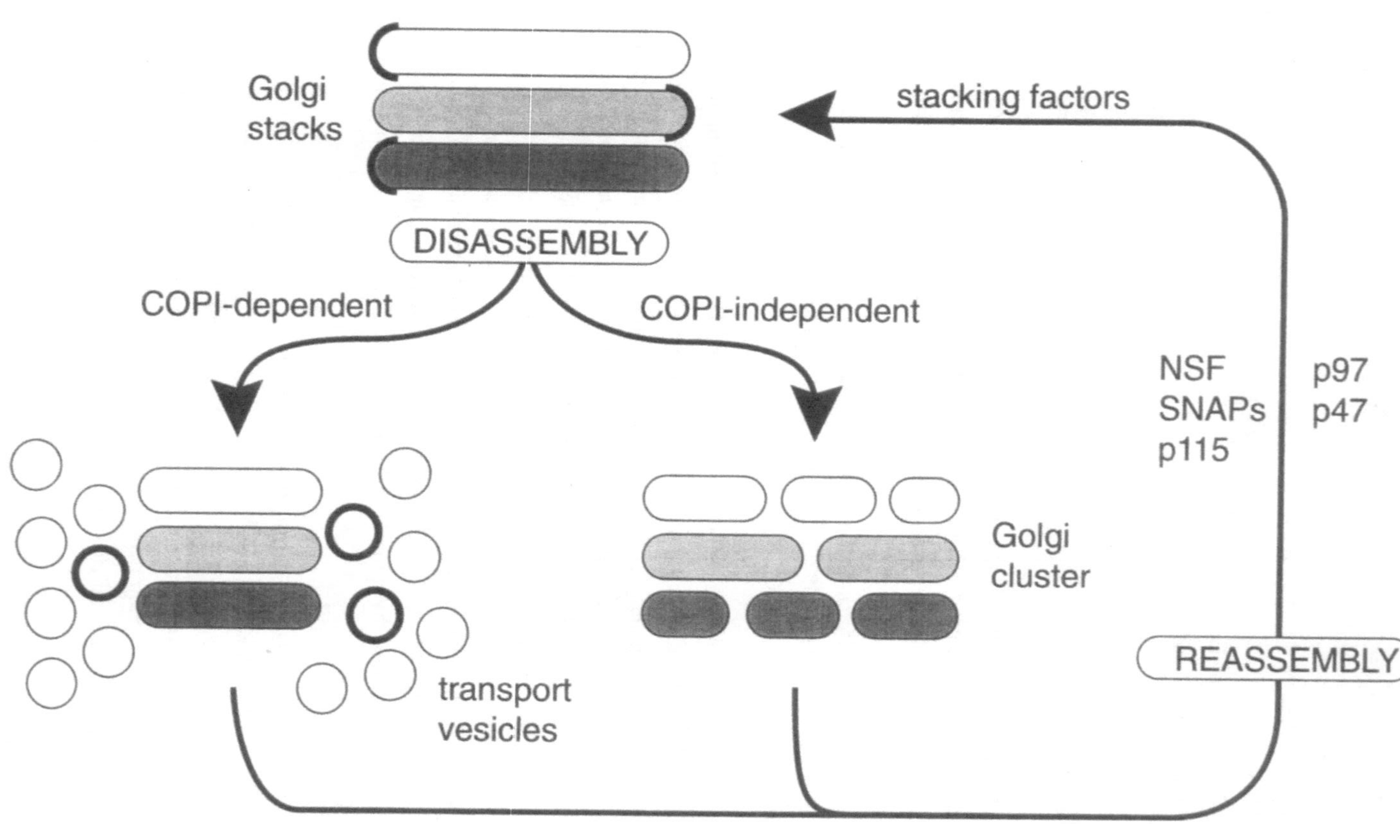

salvaged and recycling proteins (Pelham, 1994; Rothman and Wieland, 1996). During mitosis, budding continues but fusion is inhibited thereby converting the cisternal rims into transport vesicles (Warren, 1985). Quantitation shows that up to two-thirds of Golgi membrane can be consumed by this pathway (Misteli and Warren, 1995).

The second or COPI-independent pathway accounts for the remaining third of membrane but is much less well-characterised (Misteli and Warren, 1995). The cisternal cores, thought to contain the resident Golgi enzymes (Sönnichsen, et al., 1996), become increasingly fenestrated and tubulated converting stacked cisternae into mitotic Golgi clusters (Misteli and Warren, 1995). This may also result from an inhibition of membrane fusion but this would be an inhibition of homotypic fusion, not the heterotypic fusion exemplified by the fusion of transport vesicles with their target membrane. Further work is still needed to confirm this hypothesis (Rothman and Warren, 1994).

Mitotic Golgi fragments can be reassembled in vitro into large, stacked, cisternal structures (Rabouille, et al., 1995b). Re-growth of cisternae requires two fusion ATPases (Rabouille, et al., 1995a), both members of the AAA family. The first is NSF (NEM-sensitive factor) which appears to reconstruct the cisternal rims by restoring the docking and fusion of transport vesicles with their target membrane. To do this it needs several soluble, accessory proteins including SNAPs (Soluble NSF Attachment Proteins) and the p115 vesicle docking protein. The second is the NSF-like ATPase, p97, first characterised because it was a large and abundant cytosolic protein (Peters, et al., 1990). It is barrel-shaped, the six subunits comprising the staves of the barrel (Peters, et al., 1992). It is thought to reconstruct the cisternal cores, a homotypic fusion event. This is consistent with the known function of the yeast homologue (CDC48p) which is involved in the homotypic fusion of outer nuclear envelopes during yeast karyogamy (Latterich, et al., 1995). In addition to the two fusion ATPases, factors are also needed to stack the cisternal membranes. These are still being characterised.

The partitioning mechanism: Stochastic or semi-ordered?

Stereological analysis of HeLa cells undergoing mitosis originally showed that up to 10,000 vesicles could be formed by fragmentation of the single copy Golgi apparatus (Lucocq, et al., 1989). This gave rise to the idea that partitioning was a stochastic process, the laws of probability determining the accuracy of the division event (Birky, 1983). Such a large number of fragments, randomly distributed, would ensure a theoretical accuracy of partitioning to within a few percent of 50:50. However, recent experiments have suggested that the process is more ordered than we had previously thought.

The experiments that have altered our view arose from efforts to visualise the Golgi apparatus in living cells. This has been made possible by the Green Fluorescent Protein (GFP) which we have attached to the retention signal of a resident Golgi enzyme and stably expressed in HeLa cells (Shima, et al., 1997). The tag was shown to localise to the Golgi apparatus using immuno-EM permitting studies of the Golgi apparatus during the cell cycle in single cells. GFP also permitted quantitation of both the number and amounts of Golgi membrane in interphase and mitotic cells.

The first unexpected result was the finding that the Golgi cluster is the end product of the fragmentation process, not free vesicles. Earlier we had thought that the two fragmentation pathways would continue until the entire Golgi was converted into tubules and vesicles which would then become dispersed throughout the mitotic cell cytoplasm. The constancy of cluster number in metaphase cells (130±2 (n=5)) revealed by our GFP studies strongly suggests that the cluster is an end-product, not an intermediate. The simplest explanation is to suggest that the free vesicles observed in mitotic cells are transport vesicles produced by the COPI-dependent pathway whereas the clusters represent fragmented core cisternae, the product of the second fragmentation pathway.

The second unexpected result was that the clusters were ordered, not scrambled, structures. Using a variety of different Golgi markers we could show that cis, medial and trans markers maintained their relative distribution despite the conversion of stacks into clusters. This strongly suggests that the cisternae fragment in situ, and the resulting fragments do not change their relative location.

The third unexpected result was that partitioning was more ordered than expected for a stochastic process. During prophase, fragmentation was an orderly process and the dispersed clusters were surprisingly immobile until telophase signalled congregation in each daughter cell. Furthermore, quantitation showed that the amount of Golgi membrane in each daughter cell was more equal than would be predicted on the basis of stochastic partitioning of 130 randomly dispersed units. Our present guess is that microtubules are somehow involved and experiments are in progress to test this. This observation is not without precedent. Wilson and colleagues in the early part of this century showed that mitochondrial partitioning was more accurate than would have been predicted by a stochastic mechanism (Birky, 1983; Wilson, 1916).

Mitotic regulation of vesicle tethering

We had earlier focused our attention on the vesicle docking protein p115 and had shown that it bound much less avidly to Golgi membranes under mitotic conditions (Levine, et al., 1996). p115 appears to be responsible for the initial tethering of transport vesicles to membranes. This would permit a two-dimensional search of the membrane for the t-SNARE that identifies it. If it is the cognate SNARE for the v-SNARE on the vesicle then a SNARE pair would form triggering downstream fusion events (Sollner, et al., 1993a; Sollner, et al., 1993b). If the correct t-SNARE is not on the membrane then the vesicle could detach and move to other, nearby membranes. Under mitotic conditions, such tethering would not occur so none of the downstream events could occur. This would explain the mitotic consumption of cisternal rims by the continued budding of COPI vesicles.

These earlier experiments also showed that the target for mitotic modification was not the p115 itself but the membrane receptor for this protein. This receptor has now been identified by affinity chromatography as GM130, a protein we had earlier cloned and sequenced as a component of the Golgi matrix (Nakamura, et al., 1995). The sequence of GM130 predicts a long, rod-like protein with extensive regions of coiled-coil in the central domain. The C-terminal half was shown to bind tightly to Golgi membranes and the N-terminal 75 amino acids to p115. A peptide synthesised to this region mimicked the binding to p115. Mitotic phosphorylation of this peptide or GM130 prevented binding to p115 in a reversible manner (Nakamura, et al., 1997). We are now trying to determine precisely which of the two $p34^{cdc2}$ kinase phosphorylation sites in this N-terminal region of GM130 are responsible for the observed inhibition of binding. Using a cell-free assay that mimics reassembly of Golgi cisternae, we have been able to show that mitotic regulation of p115/GM130 binding can explain the COPI-dependent fragmentation pathway (Nakamura, et al., 1997).

A co-factor for the p97 ATPase

In marked contrast to NSF, the p97 fusion ATPase initially appeared to require no soluble accessory proteins for function. We noticed, however, that the molecular weight of pure p97 was about 150kDa smaller than cytosolic p97 suggesting the presence of a tightly-bound co-factor. We have now identified this co-factor as a protein of 47kDa (Kondo, et al., 1997). A trimer of this protein sits on top of the p97 barrel partially occluding the central cavity. p47 is essential for the p97-mediated re-growth of Golgi cisternae. It is also widely distributed in rat tissues. There is even a homologue in the budding yeast, *S. cerevisiae* suggesting that p47 may be needed for all fusion reactions carried out by p97. The precise function of p47 is still unknown but our preliminary results suggest that it is involved in delivering the p97 to membrane receptors.

Summary

Figure 1 summarises our present working model for the disassembly and reassembly of Golgi stacks. Many of the components have now been identified at the molecular level using the combined approaches of microscopy and biochemistry. There is every reason to believe that other proteins such as those involved in stacking Golgi cisternae will soon yield to the same approaches.

Acknowledgements

I would like to thank all those members of my laboratory who, over the years, have contributed to the work described in this review.

References

Birky, C.W. 1983. The partitioning of cytoplasmic organelles at cell division. *Int. Rev. Cytol.* 15:49-89.

Burke, B., G. Griffiths, H. Reggio, D. Louvard and G. Warren. 1982. A monoclonal antibody against a 135-K Golgi membrane protein. *EMBO. J.* 1:1621-1628.

Ho, W.C., B. Storrie, R. Pepperkok, W. Ansorge, P. Karecla and T.E. Kreis. 1990. Movement of interphase Golgi apparatus in fused mammalian cells and its relationship to cytoskeletal elements and rearrangements of nuclei. *Eur. J. Cell Biol.* 52:315-327.

Kondo, H., C. Rabouille, R. Newman, T.P. Levine, D. Pappin, P. Freemont and G. Warren. 1997. p47 is a co-factor for p97-mediated membrane fusion. *Nature*. in press

Latterich, M., K.U. Frohlich and R. Schekman. 1995. Membrane fusion and the cell cycle: Cdc48p participates in the fusion of ER membranes. *Cell.* 82:885-893.

Levine, T.P., C. Rabouille, R.H. Kieckbusch and G. Warren. 1996. Binding of the vesicle docking protein p115 to Golgi membranes is inhibited under mitotic conditions. *J. Biol. Chem.* 271:17304-17311.

Louvard, D., H. Reggio and G. Warren. 1982. Antibodies to the Golgi complex and the rough endoplasmic reticulum. *J Cell Biol.* 92:92-107.

Lucocq, J.M., E.G. Berger and G. Warren. 1989. Mitotic Golgi fragments in HeLa cells and their role in the reassembly pathway. *J. Cell Biol.* 109:463-474.

Lucocq, J.M., J.G. Pryde, E.G. Berger and G. Warren. 1987. A mitotic form of the Golgi apparatus in HeLa cells. *J. Cell Biol.* 104:865-874.

Lucocq, J.M. and G. Warren. 1987. Fragmentation and partitioning of the Golgi apparatus during mitosis in HeLa cells. *EMBO J.* 6:3239-3246.

Misteli, T. and G. Warren. 1994. COP-coated vesicles are involved in the mitotic fragmentation of Golgi stacks in a cell-free system. *J. Cell Biol.* 125:269-282.

Misteli, T. and G. Warren. 1995. A role for tubular networks and a COP I-independent pathway in the mitotic fragmentation of Golgi stacks in a cell-free system. *J. Cell Biol.* 130:1027-1039.

Nakamura, N., M. Lowe, T.P. Levine, C. Rabouille and G. Warren. 1997. The vesicle docking protein p115 binds GM130, a cis-Golgi matrix protein, in a mitotically regulated manner. *Cell.* 89:445-455.

Nakamura, N., C. Rabouille, R. Watson, T. Nilsson, N. Hui, P. Slusarewicz, T.E. Kreis and G. Warren. 1995. Characterization of a cis-Golgi matrix protein, GM130. *J. Cell Biol.* 131:1715-1726.

Pelham, H.R.B. 1994. About turn for the COPs. *Cell.* 79:1125-1127.

Peters, J.-M., J.R. Harris, A. Lustig, S. Muller, A. Engel, S. Volker and W.W. Franke. 1992. Ubiquitous soluble Mg2+-ATPase complex: A structural study. *J Mol Biol.* 223:557-571.

Peters, J.M., M.J. Walsh and W.W. Franke. 1990. An abundant and ubiquitous homo-oligomeric ring-shaped ATPase particle related to the putative vesicle fusion proteins, Sec18p and NSF. *Embo J.* 9:1757-1767.

Rabouille, C., T.P. Levine, J.M. Peters and G. Warren. 1995a. An NSF-like ATPase, p97, and NSF mediate cisternal regrowth from mitotic Golgi fragments. *Cell.* 82:905-914.

Rabouille, C., T. Misteli, R. Watson and G. Warren. 1995b. Reassembly of Golgi stacks from mitotic Golgi fragments in a cell-free system. *J. Cell Biol.* 129:605-618.

Rambourg, A. and Y. Clermont. 1990. Three-dimensional electron microscopy: structure of the Golgi apparatus. *Eur J Cell Biol.* 51:189-200.

Rothman, J.E. and G. Warren. 1994. Implications of the SNARE hypothesis for intracellular membrane topology and dynamics. *Curr. Biol.* 4:220-233.

Rothman, J.E. and F.T. Wieland. 1996. Protein sorting by transport vesicles. *Science.* 272:227-34.

Shima, D.T., H. Haldar, R. Pepperkok, R. Watson and G. Warren. 1997. Partitioning of the Golgi apparatus during mitosis in living HeLa cells. *J. Cell Biol.* in press

Sollner, T., M.K. Bennett, S.W. Whiteheart, R.H. Scheller and J.E. Rothman. 1993a. A protein assembly-disassembly pathway in-vitro that may correspond to sequential steps of synaptic vesicle docking, activation, and fusion. *Cell.* 75:409-418.

Sollner, T., S.W. Whitehart, M. Brunner, H. Erdjumentbromage, S. Geromanos, P. Tempst and J.E. Rothman. 1993b. SNAP receptors implicated in vesicle targeting and fusion. *Nature.* 362:318-324.

Sönnichsen, B., R. Watson, H. Clausen, T. Misteli and G. Warren. 1996. Sorting by COP I-coated vesicles under interphase and mitotic conditions. *J Cell Biol.* 134:1411-1425.

Souter, E., M. Pypaert and G. Warren. 1993. The Golgi stack reassembles during telophase before arrival of proteins transported from the endoplasmic reticulum. *J Cell Biol.* 122:533-40.

Warren, G. 1985. Membrane traffic and organelle division. *Trends Biochem. Sci.* 10:439-443.

Warren, G. 1993. Membrane partitioning during cell division. *Annu. Rev. Biochem.* 62:323-348.

Warren, G., T. Levine and T. Misteli. 1995. Mitotic disassembly of the mammalian Golgi-apparatus. *Trends Cell Biol.* 5:413-416.

Warren, G. and W. Wickner. 1996. Organelle Inheritance. *Cell.* 84:395-400.

Wilson, E.B. 1916. The distribution of the chondriosomes to the spermatozoa in scorpions. *Proc. Natl. Acad. Sci. USA.* 2:321-324.

Vacuole Inheritance: A New Window on Interorganelle Traffic

Zuoyu Xu, Christian Ungermann, and William Wickner
Department of Biochemistry
Dartmouth Medical School
7200 Vail Building
Hanover, N.H. 03755-3844

For a low copy number organelle to be maintained during repeated cell divisions, its division must be regulated by the cell cycle and its segregation between the daughter cells be spatially controlled (Warren and Wickner, 1996). Cytological, genetic, and biochemical studies of vacuole inheritance in S. cerevisiae have uncovered many of the proteins of this trafficking reaction and begun to reveal how it is regulated.

The yeast vacuole (lysosome) is a large, low copy number organelle (Weisman et al., 1987). Early in S. phase, just after bud emergence, the vacuole projects a membranous tubule or stream of vesicles, the "inheritance structure," into the bud (Weisman and Wickner, 1988). There these vesicles fuse, establishing the new, daughter cell vacuole. Genetic studies have revealed over a dozen VAC genes which, directly or indirectly, support vacuole inheritance (Weisman et al., 1990). In vitro studies have revealed the stages and catalysts of vacuole fusion, the final step of the inheritance reaction.

When purified vacuoles are incubated with cytosol and ATP, they form tubular projections which connect the vacuoles and lead to fusion (Conradt et al., 1992). Though the formation of these projections can be quantified, this assay is tedious and has been replaced by assay of the vacuole fusion per se (Haas et al., 1994). With this assay, the reaction has been divided into successive stages and the proteins defined which participate in each stage.

Virtually all of the proteins which participate in this reaction are peripheral or integral proteins of the vacuole membrane. This membrane has two integral "SNARE:" proteins (Nichols et al., 1997), homologs of neural syntaxin and synaptobrevin. The syntaxin homolog, the "t-SNARE" Vam3p, is initially found in a complex with the synaptobrevin homolog, or "v-SNARE", Nyv1p. Sec17p, the yeast homolog of α-SNAP, is also part of this complex (C.U., in preparation).

NATO ASI Series, Vol. H 106
Lipid and Protein Traffic
Pathways and Molecular Mechanisms
Edited by Jos A. F. Op den Kamp

Working model of Vacuole Priming, Docking and Fusion

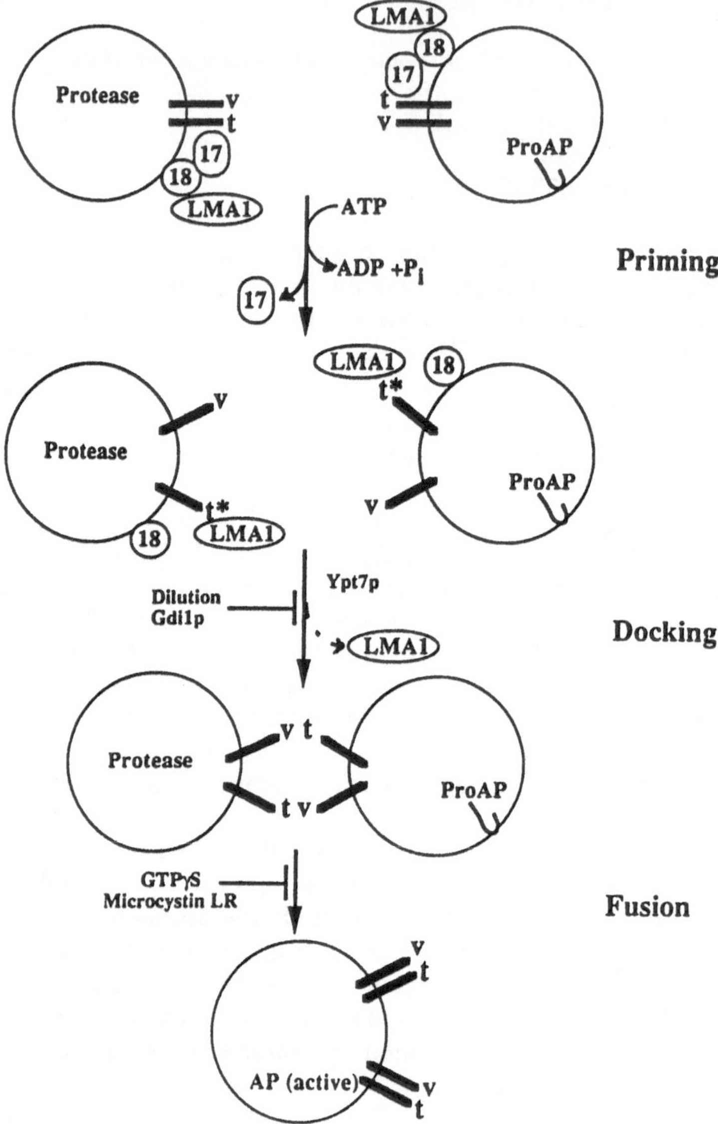

Figure Legend

Our current working model of in vitro vacuole fusion. Fusion is monitored by assaying the alkaline phosphatase activity which results from vacuole fusion allowing the proteases in 1 vacuole population to gain access to the [catalytically-inactive] pro-phosphatase (pro-AP) in the other population.

Also on the vacuole are the ras-like G protein Ypt7p (Haas et al., 1995), the NSF homolog Sec18p (Haas and Wickner, 1996), and a small heterodimeric protein which we term "low Mr activity 1", or LMA1 (Xu and Wickner, 1995; Xu et al., 1997). The LMA1 is bound via Sec18p, which serves as its receptor (Z.X., in preparation).

The first stage of the *in vitro* vacuole inheritance reaction is "priming", in which the two vacuole SNAREs are prepared for docking. Sec18p, upon binding and hydrolyzing ATP, initiates each phase of priming: 1. The v-SNARE and the t-SNARE are separated (C.U., in preparation). 2. The t-SNARE is activated for docking. 3. Sec17p is released. 4. LMA1 is transferred from Sec18p to the activated tSNARE, stabilizing it (Xu et al., 1997; Z.X., in preparation). After this priming stage, neither ATP, Sec18p, nor Sec17p participate in the further steps of the reaction which lead to fusion.

Primed vacuoles are competent to "dock", defined as the formation of a stable association of vacuoles which can go on to fuse (Mayer and Wickner, 1997). Docking is assayed either as the acquisition of dilution-resistance or morphologically as the formation of large clusters, or "rafts", of associated vacuoles (when fusion per se is blocked). Docking can only occur after priming, and requires the pairing of v-SNAREs and t-SNAREs from opposing vacuoles in the presence of functional Ypt7p. After docking, the reaction becomes insensitive to antibody to the t-SNARE or to Ypt7p; of course, this does not constitute proof that these molecules don't participate in the later, fusion reaction. Preliminary results suggest that the LMA1 is released from t-SNARE association upon docking, which may represent t- and v-SNARE pairing.

Though our knowledge of the priming and docking reactions is still primitive, our ignorance is almost complete with regard to the mechanism of fusion per se. Fusion is blocked by GTPγS or by mastoparans, suggesting regulation by a second G protein. It is also blocked by microcystin LR, a protein phosphatase inhibitor, suggesting that some crucial vacuolar protein must be kept in a dephosphorylated state for fusion to proceed (Conradt et al., 1994; Haas et al., 1994).

The in vitro studies of vacuole inheritance have already contributed to our understanding of organelle fusion reactions, as they've shown that NSF and SNAP can act to prepare SNAREs for docking rather than to catalyze fusion (Mayer et al., 1996). Even more striking, preliminary studies (V. Fanton and W.W., unpublished) suggest that the in vitro formation of "segregation structures," structures which mediate vacuole inheritance in vivo, may require the Sec18-dependent priming step. This result beckons us to further explore the molecular basis of segregation structure formation. Furthermore, the availability of assays of these early priming and docking stages suggests that the system is ripe for a frontal, enzymological approach in which detergents are used to solubilize the vacuole membrane and reconstituted proteoliposomes are assayed for recovery of functions such as Sec18 binding, Sec17 release, LMA1 binding and release, v- and t-SNARE pairing, etc. Ultimately, the proteins which support this reaction will have to be purified and their genes identified to allow studies of the spatial and cell cycle regulation of these processes, aspects which first drew us to study this problem.

References

Conradt, B., Haas, A., and Wickner, W. (1994). Determination of four biochemically distinct, sequential stages during vacuole inheritance *in vitro*. J. Cell Biol. 126: 99-110.

Conradt, B., Shaw, J., Vida, T., Emr, S.D., and Wickner, W. (1992). *In vitro* reactions of vacuole inheritance. J. Cell Biol. 119: 1469-1479.

Haas, A., Conradt, B., and Wickner, W. (1994). G-protein ligands inhibit *in vitro* reactions of vacuole inheritance. J. Cell Biol. 126: 87-97.

Haas, A., Scheglmann, D., Lazar, T., Gallwitz, D., and Wickner, W. (1995) The GTPase Ypt7p of *Saccharomyces cerevisiae* is required on both partner vacuoles for the homotypic fusion step of vacuole inheritance. EMBO J. 14: 5258-5270.

Haas, A. and Wickner, W. (1996). Homotypic vacuole fusion requires Sec17p (yeast a-SNAP) and Sec18p (yeast NSF). EMBO J.15: 3296-3305.

Mayer, A. and Wickner, W. (1997). Docking of yeast vacuoles is catalyzed by the Ras-like GTPase Ypt7p after symmetric priming by Sec18p (NSF). J. Cell Biol. 136: 307-317.

Mayer, A., Wickner, W., and Haas, A. (1996). Sec18p (NSF)-driven release of Sec17p (α-SNAP) can precede docking and fusion of yeast vacuoles. Cell 85: 83-94.

Nichols, B.J., Ungermann, C., Pelham, H.R.B., Wickner, W.T., and Haas, A. (1997) Homotypic vacuolar fusion mediated by t- and v-SNAREs. Nature 387: 199-202.

Warren, G. and Wickner, W. (1996). Organelle inheritance. Cell 84: 395-400.

Weisman, L.S., Bacallao, R. and Wickner, W. (1987). Multiple methods of visualizing the yeast vacuole permit evaluation of its morphology and inheritance during the cell cycle. J. Cell Biol. 105: 1539-1548.

Weisman, L.S., Emr, S.D., and Wickner, W.T. (1990). Mutants of Saccharomyces cerevisiae which Block Intervacuole Vesicular Traffic and Vacuole Division and Segregation. Proc. Natl. Acad. Sci. USA, 87: 1076-1080.

Weisman, L.S. and Wickner, W. (1988). Intervacuole exchange in the yeast zygote defines a new pathway in organelle communication. Science, 241: 589-591.

Xu, Z., Mayer, A., Muller, E., and Wickner, W. (1997). A heterodimer of thioredoxin and I2B Cooperates with Sec18p (NSF) to promote yeast vacuole inheritance. J. Cell Biol. 136: 299-306.

Xu, Z. and Wickner, W. (1995). Thioredoxin is required for vacuole inheritance in S. cerevisiae. J. Cell Biol. 132: 787-794.

Sea Urchin Nuclear Envelope Assembly In Vitro

Philippe Collas

Department of Biochemistry
Norwegian College of Veterinary Medicine,
PO Box 8146 Dep., 0033 Oslo, Norway

Abstract. We have investigated the formation of the male pronuclear envelope in a cell-free system consisting of permeabilized sea urchin sperm nuclei incubated in fertilized sea urchin egg cytosolic extract together with purified cytoplasmic membrane vesicles. In fertilized egg cytosol, the sperm nuclear lamina disassembles as a result of lamin B phosphorylation mediated by egg protein kinase C. The conical sperm nucleus decondenses into a spherical pronucleus in a cytosol- and ATP-dependent manner. Assembly of the pronuclear envelope is a lamin-independent process involving ATP-dependent binding of vesicles to chromatin and GTP-dependent fusion of vesicles to each other and to specialized polar lipophilic structures of the sperm nucleus. Three egg cytoplasmic membrane vesicle fractions with distinct biochemical, chromatin-binding and fusion properties, are required for pronuclear envelope assembly. Targeting of the bulk of nuclear envelope vesicles to chromatin is mediated by a lamin B receptor (LBR)-like integral membrane protein. During the last step of pronuclear formation involving nuclear swelling, lamin B is imported into the nucleus from a soluble pool, and physically associates with LBR. Our results indicate that male pronuclear envelope assembly in vitro involves a highly ordered series of reactions.

Keywords. Nuclear envelope, chromatin, vesicle targeting, vesicle fusion, lamina

1 Introduction

The nuclear envelope (NE) consists of a double membrane fenestrated by nuclear pores. The nuclear membrane can be divided into three distinct domains, each containing integral and peripheral proteins. The outer nuclear membrane is continuous with the endoplasmic reticulum (ER), bears ribosomes and contains markers common to those of the ER. The inner membrane is associated on its nucleocytoplasmic side with the nuclear lamina, a network of intermediate filaments called lamins. The third membrane domain connects the outer and the inner membranes and is associated with nuclear pore complexes. The NE is

NATO ASI Series, Vol. H 106
Lipid and Protein Traffic
Pathways and Molecular Mechanisms
Edited by Jos A. F. Op den Kamp

essential for various nuclear functions including chromatin organization, DNA replication and protein import.

Cell-free systems derived from eggs or embryos constitute powerful tools to investigate NE formation. Extracts from *Xenopus* eggs, *Drosophila* embryos, surf clam eggs and sea urchin eggs have been developed, that support sperm chromatin decondensation, NE formation and nuclear lamina assembly. An advantage of in vitro systems is the potential to independently evaluate cytosolic and membrane vesicle contributions to nuclear assembly. Both cytosolic and membrane fractions can be chemically or enzymatically manipulated. Such manipulations have led to the identification of distinct steps and factors required for the formation of the NE in vitro in *Xenopus* (Lohka and Masui, 1984; Wilson and Newport, 1988), *Drosophila* (Ulitzur and Gruenbaum, 1989), surf clam (Longo et al., 1994) and sea urchin (Cameron and Poccia, 1994; Collas and Poccia, 1995a,b; 1996; Collas et al., 1995; 1996; 1997).

The transformation of the sea urchin sperm nucleus into a pronucleus at fertilization provides an opportunity to investigate the assembly of the NE. Sea urchin eggs are fertilized in G1 phase of the first cell cycle after the oocyte has completed both meiotic divisions. At fertilization, the sperm NE rapidly vesiculates inside the egg cytoplasm, and a new NE reforms around the male pronucleus as the sperm chromatin decondenses (Longo and Anderson, 1968). The male pronuclear envelope originates from endoplasmic reticulum (ER)-derived vesicles and to a limited extent from de novo membrane synthesis. Remnants of the sperm NE localized at the base and the tip of the sperm nucleus, which resist disassembly at fertilization, are incorporated into the pronuclear envelope (Longo and Anderson, 1968). The NE apparently forms by fusion of vesicles along the chromatin surface.

To investigate the mechanisms of NE assembly in the sea urchin, we have developed a cell-free system. The system consists of demembranated sperm nuclei, a fertilized egg cytosolic extract, membrane vesicles, and an energy-regenerating system. Sea urchin sperm nuclei are demembranated by solubilization with 0.1% of a non-ionic detergent such as Triton X-100. Demembranated nuclei contain two detergent-resistant polar lipophilic structures (LSs; Collas and Poccia, 1995a), reminiscent of the NE remnants identified by Longo and Anderson (1968). As described in this paper, these LSs are required for vesicle binding to chromatin in vitro. The cytosolic extract consists of a 150,000-*g* supernatant of fertilized and homogenized sea urchin eggs (Collas et al., 1996). Membranes contained in the 150,000-*g* pellet are recovered by centrifugation through sucrose, and either used as a whole or fractionated into distinct populations.

Using this sea urchin cell-free system, we have investigated several aspects of male pronuclear envelope assembly. They include solubilization of the sperm nuclear lamina, decondensation of the sperm chromatin, targeting of NE precursor vesicles to the chromatin, binding of these vesicles to chromatin, fusion of the vesicles to form a sealed envelope, assembly of the new nuclear lamina, and swelling of the pronucleus associated with growth of the NE. These aspects are reviewed in the present communication.

2 Disassembly of the Sperm Nuclear Lamina

Sea urchin sperm nuclei have been shown to contain a lamina (Collas et al., 1995). This lamina consists of a major 65 kDa B-type lamin, and several minor uncharacterized lamin epitope-containing molecules of 49, 54, 72 and 84 kDa identified by immunofluorescence and immunoblotting using five anti-lamin antibodies (Collas et al., 1995). Triton X-100-permeabilized sperm nuclei retain their lamina, including the 65 kDa lamin B.

2.1 Phosphorylation and Solubilization of Sperm Lamin B in Fertilized Egg Cytosol

In vivo, the first step of sea urchin male pronuclear formation inside the egg cytoplasm is the removal of the sperm nuclear envelope, a process completed within minutes of fertilization (Longo and Anderson, 1968). The first step of male pronuclear formation in vitro is the disassembly of the sperm nuclear lamina. Immunofluorescence and immunoblotting data using an anti-sea urchin lamin B antibody show that sperm lamina disassembly occurs within 10 min of incubation of nuclei in egg cytosol (Collas et al., 1995; 1997). Concomitant with disassembly from nuclei, lamin B appears progressively in the cytosol in a phosphorylated 68 kDa form. Solubilization of sperm lamin B is preceded by its phosphorylation, as demonstrated by rapid incorporation of ^{32}P into sperm lamin B, immediately followed by the release of phosphorylated lamin into the cytosol. Lamin B phosphorylation and solubilization in fertilized egg cytosol require Ca^{2+}.

2.2 Role of PKC in Sperm Lamina Solubilization

Several lamin kinases have been identified, which include cdc2, the Ca^{2+}-dependent protein kinase PKC, and the cAMP-dependent kinase PKA. We recently attempted to determine the kinase mediating sperm lamin B phosphorylation in interphase egg cytosolic extract, using protein kinase inhibitors (Collas et al., 1997). The general kinase inhibitors 6-dimethylaminopurine and staurosporine inhibit lamin B phosphorylation and solubilization. These events are not affected by cyclin-dependent kinase (cdk) (including cdc2)-specific inhibitors such as olomoucine and roscovitine, or the PKA inhibitor PKI. The most effective inhibitor of lamin B phosphorylation and solubilization tested was chelerythrine, a specific inhibitor of PKC, suggesting a role of PKC in sperm lamin B solubilization in vitro. Inhibition of lamin solubilization by kinase inhibitors systematically prevents sperm chromatin decondensation, suggesting that both events may be related.

Further evidence for the involvement of PKC in interphase sperm lamin B phosphorylation and solubilization in vitro was provided by the inhibition of these processes by a PKC pseudosubstrate inhibitor peptide (PKC [19-31]), a competing PKC substrate peptide ((Ser^{25}) PKC [19-36]), or immunodepletion of egg soluble PKC from the extract. $p13^{suc1}$-agarose beads, which specifically bind cdc2 kinase, or autocamtide 3, a CaM kinase II-specific substrate peptide, do not inhibit the egg lamin kinase.

Sperm lamin B can be phosphorylated and solubilized in vitro by purified mammalian PKC. Two-dimensional phosphopeptide mapping of lamin B phosphorylated by mammalian PKC and by the egg cytosolic kinase reveal virtually identical phosphopeptides, indicating that the cytosolic interphase lamin kinase accounting for the lamin phosphopeptides detected is PKC. Sperm lamina disassembly promoted by purified PKC does not seem to be sufficient to induce chromatin decondensation.

3 Decondensation of the Sperm Chromatin

The morphology of sperm chromatin decondensation in vitro mimics decondensation in vivo, although the process is notably slower in vitro. Sperm chromatin decondensation in vivo has been documented at the electron and light microscopy level (Longo and Anderson, 1968; Cothren and Poccia, 1993).

Two phases of chromatin decondensation have been identified in vitro: a membrane-independent decondensation phase and a membrane-dependent swelling phase (Cameron and Poccia, 1994; Collas and Poccia, 1995b). The first phase occurs in membrane-free cytosol, and converts the conical ~1.5 x 4 μm sperm nucleus through an ovoid intermediate to a sphere of ~4 μm. This transformation requires cytosol and ATP hydrolysis, provided by the addition of an ATP-regenerating system. Decondensation is promoted at alkaline pHs and in activated egg cytosol (Cameron and Poccia, 1994). The second phase involves swelling of the decondensed chromatin. If membrane vesicles are present in the reaction, nuclei will bind membranes and if GTP is provided, they will form an NE lacking nuclear lamins. Nuclear swelling requires additional input of ATP, soluble lamins, Ca^{2+} and cytosolic factors sensitive to heat and NEM (Collas et al., 1995; 1996). The final nuclear diameter is limited by suboptimal amounts of vesicles present, the amount of ATP in the extract, or depletion of lamins from the extract. Thus chromatin decondensation seems to be a property of the chromatin and cytosol, whereas nuclear selling requires a complete NE with a lamina. It may be driven by growth of the lamina or import of nuclear proteins after a functional NE has formed.

4 Formation of the Pronuclear Envelope

4.1 Role of LSs in Membrane Vesicle Binding to Chromatin

Isolated sperm nuclei permeabilized with 0.1% Triton X-100 lose their NE except for remnants, which we named lipophilic structures or LSs (Collas and Poccia, 1995a) at the tip and the base of the nucleus that correspond to those NE portions retained in vivo (Longo and Anderson, 1968). It is at those tips that membrane vesicles are initially targeted to the chromatin in vitro (Fig. 4.1). Binding then progresses towards the equator of the nucleus. Upon addition of GTP, the vesicles fuse with one another and with the LSs. Whether polar vesicle binding also occurs in vivo is unclear because difficult to determine by electron microscopy.

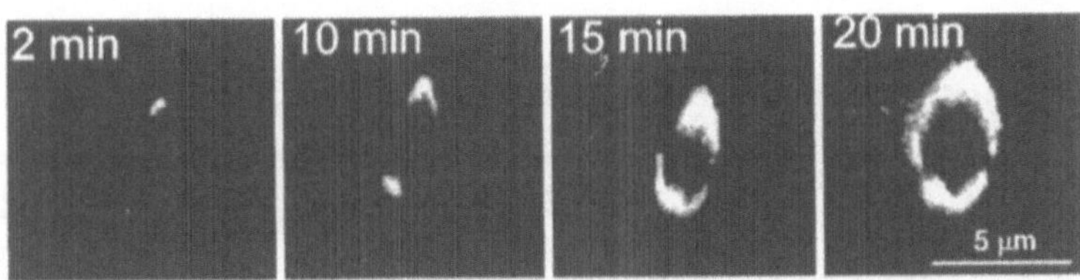

Fig. 4.1. Polarized membrane vesicle binding around decondensing sea urchin sperm chromatin in vitro. Membrane vesicles, labeled with the lipophilic dye $DiOC_6$, bind initially to the nuclear poles and progressively around the nuclear periphery. Taken from Collas and Poccia (1995a) with permission.

The LSs are not mere membrane remnants that serve as targets for vesicle binding, but are structures required for NE assembly in vitro (Collas and Poccia, 1995a). Extraction of 0.1% Triton X-100-permeabilized nuclei with 1% Triton X-100 removes LSs. The solubility characteristics of the solubilized material are not typical of normal membranes, and may originate from unusual lipid composition or association with proteins. Removal of LSs abolishes membrane binding to chromatin. When LSs are added back to stripped nuclei, they re-bind specifically to their sites of origin, at one or both nuclear poles. Unipolar LS reconstitution allows vesicle binding to one pole only, and bipolar reconstitution allows vesicle binding to both poles and formation of a complete NE (Collas and Pocciam 1995a).

4.2 Distinct Egg Vesicle Populations Contribute to the NE

In addition to the sperm LSs, three populations of egg membrane vesicles (MV1, MV2α and MV2β) have been separated by buoyant density, that contribute to the male pronuclear envelope in vitro (Collas and Poccia, 1996). Binding of each of these fractions to sperm chromatin requires LSs. The fractions display distinct biochemical, binding and fusion properties as summarized in Table 4.2. MV1 binds exclusively in the polar LS regions and is required for fusion of the vesicles to each other and to the LSs. MV2β, by far the most abundant fraction in the egg cytoplasm, binds over the entire chromatin surface. MV2β is enriched in an ER marker enzyme, and contains marker proteins such as lamin B receptor (LBR) and lamin B. MV2α binds only in the LS regions, is enriched in a Golgi marker and is required together with MV1 for fusion of MV2β. All three fractions are required to form a complete NE excluding high molecular weight dextrans. These data suggest the contribution of multiple vesicle populations from both the sperm and the egg to form an NE.

Table 4.2 Properties of membrane vesicle fractions

Membrane fraction	Surface binding	Protease-sensitive	NEM-sensitive	Binding protein[b]	Fusigenic	Marker proteins
MV1	Tips[a]	Yes	Yes	PMP	Yes[c]	None
MV2α	Tips[a]	Yes	No	PMP	No	Golgi[d]
MV2β	All	Yes	No	IMP	No	ER[e], LBR, Lamin B

[a] Binding in the polar LS regions. [b] PMP, peripheral membrane protein; IMP, integral membrane protein. [c] Requires LSs to fuse. [d] α-D-mannosidase. [e] α-D-glucosidase.

4.3 Targeting of the Bulk of NE Precursors to Chromatin

Although all membrane vesicle fractions require LSs to bind sperm chromatin in vitro, targeting to chromatin of the bulk of NE precursor vesicles (MV2β) is mediated by an integral membrane protein (IMP). This IMP displays biochemical characteristics of the mammalian lamin B receptor (LBR; Worman et al., 1988). LBR is an IMP whose NH_2-terminal domain binds lamin B, DNA and heterochromatin protein HP1. LBR is essential for membrane targeting to the nucleus after mitosis (Chaudhary and Courvalin, 1993).

We have recently identified a 56 kDa protein (p56) in sea urchin sperm and eggs, that cross-reacts with an antibody against the NH_2-terminal domain of human LBR (Collas et al., 1996). In sperm, p56 is localized exclusively in the tips (LS regions) from where it is removed by egg cytosolic extract. In eggs, p56 is present in a subset of the major vesicle fraction MV2β which contributes to the bulk of the NE. p56 is an IMP, as it resists extraction from vesicles with high salt and alkali but is solubilized by a non-ionic detergent. Immunodepletion of p56-containing vesicles using anti-LBR antibodies abolishes vesicle binding to chromatin, indicating that p56-containing vesicles are required for NE assembly. Anti-LBR antibodies included in the binding reaction also prevent vesicle binding, suggesting that MV2β vesicles have a chromatin-binding capacity mediated by p56. p56-containing vesicles do not contain lamin B, which is localized in a distinct subset of MV2β vesicles. Lamin B-containing vesicles are not required for vesicle binding to chromatin or fusion, as these processes are not affected by immunodepletion of lamin B-containing vesicles. Likewise, immunodepletion of egg soluble lamin B does not impair vesicle binding and fusion.

p56 and lamin B are differentially targeted to chromatin in vitro. Whereas p56-containing vesicles bind to chromatin in the early stages of pronuclear envelope formation, lamin B is incorporated into the nucleus at a later stage of NE assembly (see Section 5). Only then do p56 and lamin B physically associate in sea urchin the pronuclear envelope (Collas et al., 1996). These observations suggest that p56 is a sea urchin LBR homologue that targets membrane vesicles to chromatin and later tethers the membrane to the nuclear lamina.

4.4 Membrane Vesicle Fusion

The formation of a sealed NE requires fusion of the chromatin-bound vesicles with one another. These vesicles also fuse with the sperm polar LSs which become incorporated into the pronuclear envelope.

Fusion of vesicles to form the NE necessitates co-operation between the different membrane fractions participating in NE assembly (Collas and Poccia, 1996). The sperm LSs are required for fusion of vesicles around the chromatin but are not fusigenic by themselves. LSs fuse with MV1 but not MV2α or MV2β alone. MV1 fuses with LSs and is required for fusion of LSs with MV2α . MV2α acts as an intermediate for fusion of MV1 (fused with LSs) with MV2β. The nature of the molecules involved in the fusion process in these membrane fractions and in LSs has yet to be determined.

Fusion of chromatin-bound vesicles in vitro is promoted by GTP hydrolysis. Because binding of fluorescently-labeled membrane vesicles to chromatin produces a fluorescent rim around the nucleus, a single fluorescent dye is not sufficient to assess vesicle fusion. Thus, we have developed a fluorescent membrane fusion assay, based on fluorescence energy transfer (Collas and Poccia, 1995a; 1996). Two sets of membrane vesicles are labeled with two different fluorescent lipophilic dyes, one green ($DiOC_6$) the other red ($DiIC_{18}$) and bound to chromatin. Fusion of the green- and red-labeled vesicles is promoted with 100 μM GTP and visualized by the mixing of the green and red dyes within the fused membranes. Before fusion, fluorescence patterns of the green- and red-labeled vesicles around the chromatin are different. Upon GTP addition, fluorescence patterns in the green and red channels appear identical, due to the redistribution of the dye-associated membrane phospholipids within the fused membranes. Furthermore, emission in the green channel appears yellow-orange instead of green, as a result of both green and red emission detected in the green channel. We have interpreted this color change to result from fluorescence energy transfer from the green dye ($DiOC_6$) to the red dye ($DiIC_{18}$), a process requiring the close proximity of the two fluorophores within the same membrane (Collas and Poccia, 1997). Fluorescence can also be transferred from a labeled vesicle to fused unlabeled vesicles.

The NE excludes non-karyophilic high molecular weight components such as dextrans from the nucleus. The formation of a continuous NE as a result of vesicle fusion has also been monitored by the exclusion of a 150-kDa FITC-labeled dextran from nuclei (Cameron and Poccia, 1994). Fluorescence and dextran exclusion fusion assays have been verified by electron microscopy (Collas and Poccia, 1995a; 1997).

5 Assembly of the Nuclear Lamina and Nuclear Swelling

In the sea urchin, formation of the male pronuclear envelope in vitro is a lamin-independent process. Targeting of vesicles to chromatin, binding and fusion occur in lamin B-depleted cytosol and in the absence of lamin B-containing vesicles. However, the nuclei are small, devoid of a nuclear lamina and the NE is probably devoid of pores since these nuclei are incapable of macromolecule import (Collas and Aleström, 1996). The nuclear lamina assembles during a later stage of pronuclear development, and is associated with swelling of the nucleus to an ~8 μm pronucleus upon additional input of ATP (Collas et al., 1995).

Nuclear swelling requires soluble lamin B, and membrane vesicles which do not necessarily contain lamin B. Immunoblotting studies have shown that lamin B is stored in egg cytoplasm in soluble and membrane-associated pools. We recently determined the relative contributions of each lamin pool to the nuclear lamina by incubating small lamina-less nuclei in intact cytosol or cytosol immunodepleted of soluble lamin B, each with intact vesicles or vesicles depleted of lamin B-containing vesicles (Collas et al., 1996). In the absence of soluble lamin B, no lamina forms and nuclei do not swell even if lamin B-containing vesicles are present. In contrast, in intact cytosol a lamina assembles and nuclear swelling occurs even if lamin B-containing vesicles are absent. However, lamina assembly and nuclear swelling do not occur when vesicles are omitted from the reaction. These experiments indicate that nuclear lamin B incorporation into the nucleus as well as nuclear swelling require soluble lamin B and additional membranes not necessarily containing lamin B.

The mode of incorporation of soluble lamin B into the nucleus was investigated by blocking nuclear pore function with wheat germ agglutinin (WGA) (Collas et al., 1996). With WGA, nuclear incorporation of lamin B and nuclear swelling are prevented, even though vesicle fusion with nuclear membranes occurs. Inhibition of lamin import can be reversed by the addition of the WGA ligand *N,N',N''*-triacetylchitotriose. Thus nuclear lamina assembly and nuclear swelling depend on active import of soluble lamin B via nuclear pore complexes.

What remains to be elucidated is the timing of formation of nuclear pores with respect to NE and lamina assembly, and the origin of the numerous nuclear pore components. Small nuclei formed in vitro are incapable of import, suggesting that they lack functional pores, whereas swollen nuclei contain functional nuclear pores (Collas and Poccia, 1997). Since in the sea urchin in vitro, the lamina assembles by nuclear import of soluble lamin, the first functional nuclear pores must form prior to the initiation of lamina formation. Several studies in mitotic somatic cells in vivo (Chaudhary and Courvalin, 1993; Buendia and Courvalin, 1997) and in *Xenopus* cell-free extracts (Goldberg et al., 1997) suggest that nuclear pore assembly occurs during the late stages of NE reconstitution. Our results support these findings. Successive stages of nuclear pore assembly have been recently documented by electron microscopy in *Xenopus* (Goldberg et al., 1997). The origin of the numerous pore components, however, remains to be determined. Which components originate from soluble or membrane-associated sources, and whether membrane-associated pore components are segregated into distinct vesicles prior to NE and pore assembly remains to be investigated.

References

Buendia, B., and Courvalin, J.-C. (1997) Domain-specific disassembly and reassembly of nuclear membranes during mitosis. Exp. Cell Res. 230: 133-144.

Cameron, L.A., and Poccia, D.L. (1994) In vitro development of the sea urchin male pronucleus. Dev. Biol. 162: 568-578.

Chaudhary, N., and Courvalin, J.-C. (1993) Stepwise reassembly of the nuclear envelope at the end of mitosis. J. Cell Biol. 112: 295-306.

Collas, P., and Poccia, D.L. (1995a) Lipophilic organizing structures of sperm nuclei targets membrane vesicle binding and are incorporated into the nuclear envelope. Dev. Biol. 169: 123-135.

Collas, P., and Poçcia, D.L. (1995b) Formation of the sea urchin male pronucleus in vitro: membrane-independent chromatin decondensation and nuclear envelope-dependent nuclear swelling. Mol. Reprod. Devel. 42: 106-113.

Collas, P., Pinto-Correia. C., and Poccia, D.L. (1995) Lamin dynamics during sea urchin male pronuclear formation in vitro. Exp. Cell Res. 219: 687-698.

Collas, P., and Aleström, P. (1996) Nuclear localization signal of SV40 T antigen directs import of plasmid DNA into sea urchin male pronuclei in vitro. Mol. Reprod. Devel. 45: 431-438.

Collas, P., and Poccia, D.L. (1996) Distinct egg membrane vesicles differing in binding and fusion properties contribute to sea urchin male pronuclear envelopes in vitro. J. Cell Sci. 109: 1275-1283.

Collas, P., Courvalin, J.-C., and Poccia, D.L. (1996) Targeting of membranes to sea urchin sperm chromatin is mediated by a lamin B receptor-like integral membrane protein. J. Cell Biol. 135: 1715-1725.

Collas, P., and Poccia, D.L. (1997) Methods for studying in vitro assembly of male pronuclei using oocyte extracts from marine invertebrates: sea urchins and surf clams. Meth. Cell Biol.: in press.

Collas, P., Thompson, L., Fields, A., Poccia, D.L., and Courvalin, J.-C. (1997) Protein kinase C-mediated interphase lamin B phosphorylation and solubilization. J. Biol. Chem.: in press.

Cothren, C.C, and Poccia, D.L. (1993) Two steps required for male pronucleus formation in the sea urchin egg. Exp. Cell Res. 205: 126-133.

Goldberg, M.W., Wiese, C., Allen, T.D., and Wilson, K.L. (1997) Dimples, pores, star-rings, and thin rings on growing nuclear envelopes: evidence for structural intermediates in nuclear pore complex assembly. J. Cell Sci. 110: 409-420.

Lohka, M.J., and Masui, Y. (1984) Roles of cytosol and cytoplasmic particles in nuclear envelope assembly and sperm pronuclear formation in cell-free preparations from amphibian eggs. J. Cell Biol. 98: 1222-1230.

Longo, F.J., and Anderson, E. (1968) The fine structure of pronuclear development and fusion in the sea urchin, *Arbacia punctulata*. J. Cell Biol. 39: 335-368

Longo. F.J., Matthews, L., and Palazzo, R.E. (1994) Sperm nuclear transformations in cytoplasmic extracts from surf clam (*Spisula solidissima*). Dev. Biol. 162: 245-258.

Ulitzur, N., and Gruenbaum, Y. (1989) Nuclear envelope assembly around sperm chromatin in cell-free preparations from *Drosophila* embryos. FEBS Lett. 1: 113-116.

Wilson, K.L., and Newport, J.W. (1988) A trypsin-sensitive receptor on membrane vesicles is required for nuclear envelope formation in vitro. J. Cell Biol. 107: 57-68.

Worman, H.J., Yuan. J., Blobel, G., and Georgatos, S.D. (1988) A lamin B receptor in the nuclear envelope. Proc. Natl. Acad. Sci. USA 85: 8531-8534.

Cell Vesicle Trafficking and Bacterial Protein Toxins

Cesare MONTECUCCO

1Centro CNR Biomembrane and Dipartimento di Scienze Biomediche, Universita` di Padova, Via G, Colombo 3, 35121 Padova, Italy

Summary

Eukaryotic cells are characterized by an intense trafficking of proteins, peptides and chemicals in and out of the cell as well as by trafficking within the various cell organelles. A group of bacterial protein toxins interfere with vesicular trafficking inside cells. Clostridial neurotoxins affect mainly the highly regulated fusion of neurotransmitter and hormones containing vesicles with the plasma membrane. They cleave the three SNARE proteins: VAMP, SNAP-25 and syntaxin, and this selective proteolysis results in a blockade of exocytosis. The Helicobacter pylori cytotoxin is implicated in the pathogenesis of gastroduodenal ulcers. It causes a progressive and extensive vacuolation of cells followed by necrosis, consequent to a cytotoxin induced alteration of membrane trafficking at the level of late endosomes. Vacuoles origine from this compartment in a rab7 dependent process and swell because they are acidic and accumulate membrane permeant amines.

Key words: tetanus / botulism / neurotoxins / proteases / exocytosis / endosomes / membrane fusion

Introduction

During the course of evolution a limited number of bacteria "have explored" all animal tissues as possible environments able to support their growth, proliferation and spreading. Some of these bacteria have adopted aggressive strategies which include the modification of the host physiology to their advantage, i.e. to increase their proliferation and/or spreading (1). Such objectives are pursued with a variety of different means. A very popular one is that of producing protein toxins. Hundreds of different toxins have been identified so far. They display a variety of activities: 1) they may alter cells of the host in such a way as to improve bacterial multiplication and/or diffusion or 2) they act against inflammatory or lymphoid cells or 3) they may foolish the immune systems (i.e. superantigens) (2,3).

There are several reasons to study the mechanism of action of bacterial protein toxins: a) toxin are the products of a long term co-evolution of the toxin producing bacteria with the animal host. Hence, each toxin activity has been "shaped" around key physiological aspects of cells. Then, by studying the mechanism of cell intoxication of these toxins we may as well learn about fundamentaf features of cells. b) the molecular understanding of cell intoxication leads to a deeper comprehension of the pathogenesis of the disease in which toxins are implicated. c) in the case of toxins playing a major role as virulence factors, the study of their mode of action provides information for the design of preventive or therapeutic vaccine against the disease.

Our studies have centered around some toxins belonging to the first group and in particular on protein toxins with intracellular targets. All of them have "chosen" to affect

NATO ASI Series, Vol. H 106
Lipid and Protein Traffic
Pathways and Molecular Mechanisms
Edited by Jos A. F. Op den Kamp

fundamental aspects of cell physiology (2,3). Recently we have concentrated on toxins that interfere with different aspects of vesicle traffic inside cells.

Tetanus and Botulinum Neurotoxins and exocytosis

These toxins are produced by toxigenic strains of bacteria of the genus *Clostridium* and are released upon bacterial lysis. They are the most poisonous sustances known: their mouse LD50 is comprised between 0.1 and 1 ng/Kg body weight. Tetanus toxin (TeNT) and botulinum neurotoxins (abbreviated BoNT, seven different types indicated with letters from A to G) are the sole cause of the neuroparalytic syndromes of tetanus and botulism, respectively (4-6). These toxins spread in the organism from the site of entrance and intoxicate nerve terminals, which become incapable of releasing neurotransmitters. BoNTs affect specifically the release of acetylcholine, particularly at neuromusular junctions. TeNT acts on central synapses, particularly on the inhibitory synapses of the Renshaw cells of the spinal cord.

These neurotoxins bind with high specificity to the presynaptic membrane. Notwithstanding a long search begun in the previous century, the receptors of these neurotoxins are not known. Available experimental results can be rationalized by implicating in the neurospecific binding both a protein and polysialogangliosides, which are particularly enriched at presynaptic terminals. Recently, Nishiki et al. (7) have reported evidence that the amino-terminal portion of synaptotagmin, a transmembrane protein of small synaptic vesicles, together with gangliosides, binds BoNT/B with high affinity. The receptors of the other neurotoxins is not yet known, but it can be anticipated that they have to be essential proteins that cannot be altered by a nerve cell without affecting its vitality. In fact, identified toxin or virus receptors are proteins acting as growth factor or cytokine receptors, transporters, ion channels, i.e. proteins playing essential roles in cell physiology.

Toxin binding at unmyelinated nerve terminals is followed by endocytosis inside vesicles whose nature has been only recently determined for the TeNT at CNS neurons (8), but remains not characterized at motoneurons terminals (9) (fig. 1). TeNT binds to an acceptor(s) molecole present on the lumen of small synaptic vesicle during the short time that a vesicle remains fused to the plasma membrane to discharge its content of neurotransmitter. The vesicle is then internalized in a process requiring the dynamic interaction of amphiphysin, dynamin and clathrin (10). Similar data are not yet available for the motoneuron peripheral terminals. But, it is clear that neurotoxin internalization is linked to the electrical activity of the synapse because animals kept under exercise intoxicate more rapidly (reviewed in 11). After internalization at the neuromuscular junction of peripheral motor neurons, BoNT and TeNT follow a different destiny. The former enters the cytosol of these nerve terminals, whereas the latter is transported inside vesicles along the axons up to the spinal cord (see fig. 1) and migrates trans-synaptically into inhibitory interneurons (12). It is then internalized inside small synaptic vesicles of the spinal cord inhibitory interneurons. Hence, TeNT and BonTs follow different pathways of vesicular trafficking inside peripheral motor neurons that leads them to act at different synapses. However, their mode of cell entry appears to be essentially similar.

The third step of neuron intoxication is the movement of the catalytic light chain of the toxin (chain L, 50 kDa) from the vesicle lumen, across the membrane of the vesicle, into the cytosol. There is considerable evidence that the heavy chain (chain H, 100 kDa) is mainly responsible for such a process, which is driven by the acidification of the vesicular

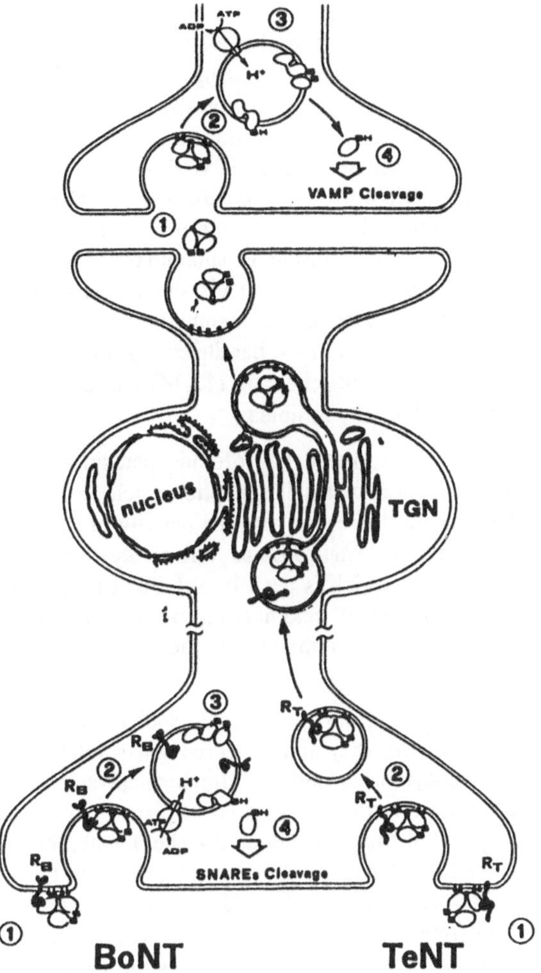

Figure 1. Tetanus and botulinum neurotoxins cell trafficking and activity. 1) BoNTs bind to the presynaptic membrane at the neuromuscular junction and are consequently internalized inside vesicles; 2) vesicle lumen is acidified by a vacuolar ATPase proton pump. 3) at acid pH, BoNTs change conformation, penetrate the lipid bilayer of the vesicle and translocate the L chain into the cytosol, 4). L displays a metalloproteolytic activity specific for three protein components of the neuroexocytosis apparatus, termed SNARE proteins. BoNT/B, /D, /F and /G attack VAMP, a protein of synaptic vesicles; BoNT/A and /E cut SNAP-25 and BoNT/C cleaves SNAP-25 and syntaxin. At variance from BoNT, TeNT binds to the presynaptic membrane and is internalized inside a vesicle, which transports TeNT all along the perypheral motor neuron axon and releases it into the synapse with Renshaw cells of the spinal cord. Here it is taken up by these inhibitory interneurons inside synaptic vesicles, enters the cytosol and cleaves VAMP.

lumen brought about by the action of a vacuolar-type ATPase proton pump residing on the vesicle membrane (8). It is also well documented that these neurotoxins exhibit a low pH driven conformational change from a water-soluble "neutral" conformation to a hydrophobic "acid" conformation (6). The latter one partitions into the lipid bilayer and somehow assists the translocation of the L chain. Several studies have documented the ability H to form ion channels (references in 3 and 6). While most researchers agree on the possibility that this channel is related to the process of translocation of the L polypeptide chain into the cytosol, there are different views on the mode of translocation: a) via a pore inside H subunit(s), b) via a cleft formed by H subunit(s) at the lipid-protein boundary and c) via disruption of the membrane vesicle (reviewed in 6). On the basis of the results of membrane photolabelling experiments, we proposed model b) and this view is now paralleled by the finding that the protein translocating pores of the endoplasmic reticulum (13,14) and of the plasma membrane of prokaryotes (15) are also open laterally to lipids.

Hence, to enter into cells, these neurotoxins parasitize a specialized form of endocytosis, i.e. the internalization of small synaptic vesicle, a process that takes place continuosly at nerve terminal and increases with the activity of the synapse. By using this way to enter into cells clostridial neurotoxins effectively "choose" the more active neurons. Once inside the vesicle, which is used by the toxin as a cellular form of Trojan horse, they exploit the physiological acidification of the vesicular lumen to change conformation and move across the membrane inside the cytosol, where, the catalytic chain displays its enzymic activity.

The L chains of TeNT and BoNTs are highly specific metalloproteinases. Among the thousands of neuronal proteins, they appear to recognize solely three proteins that are key players of the neuroexocytosis squadron. Biochemical and genetic evidence have shown that the release of neurotransmitter that takes place at the active zones of nerve terminals is mediated by: a) proteins residing on the membrane of neurotransmitter containing vesicles (termed v-SNAREs), b) proteins residing on both vesicular and presynaptic membranes (termed t-SNARES) and c) a set of soluble proteins that are recruited from the cytosol (reviewed in 16-19). Such multi-protein binding gives rise to a 20S complex that is energized by ATP hydrolysis, docks the vesicle to specialized "active zones" of the presynaptic membrane. Vesicle fusion with the presinaptic membrane, and release of neurotransmitter, takes place within hundreds of microseconds from the rise of local calcium concentration following the opening of active zones calcium channels governed by membrane potential.

The eight clostridial neurotoxins are very specific for VAMP, a 13 kDa vesicle protein and for two mainly presynaptic membrane proteins: SNAP-25 and syntaxin. These proteins are the only known neurotoxin proteolytic substrates and are cleaved at different peptide bonds by the different neurotoxins. TeNT and BoNT/B, /D, /F and /G recognize and cleave VAMP thus releasing a large portion of the VAMP molecule in the cytosol (6,20,21). The residual membrane bound VAMP fragment is presumably unable to carry out the function(s) of VAMP and this leads to inhibition of exocytosis. The role(s) of VAMP is not yet known, but its fundamental importance is apparent from the fact that VAMP is not specific of the small synaptic vesicles of nerve terminals, but it is present on a variety of vesicles in almost every cell which has been looked at (19,22 and references cited therein). VAMP is proposed to be responsible for addressing the vesicle to the target membrane (16,18) but this suggestion is not supported by the morphology of TeNT intoxicated synapses which show an apparently normal docking of small synaptic vesicles to active zones of the presynaptic membrane (23,24).

BoNT/C cleaves syntaxin near the membrane surface and thus releases in the cytosol most of the molecule, again causing a long lasting inhibition of neurotransmitter release (25,26). At variance, BoNT/A, /C and /E remove only a short COOH-terminal peptide from SNAP-25 (6,27-31). This indicates that the carboxyl-terminus of SNAP-25 plays an essential role in neuroexocytosis, one that is not yet understood. The fact that an excitatory toxin such as the spider alpha-latratoxin can overcome the blockade of neuroexocytosis caused by BoNT/A (32), but not that caused by the VAMP specific toxins suggests that this carboxylterminal region of SNAP-25 is not part of the fusion pore itself.

A stricking feature of this novel group of zinc-endopeptidases is their specificity. Among the thousand of proteins present in the cytosol, no other substrate has been found in addition to the three SNAREs. Clearly, the segments around the cleavage sites cannot be the sole responsible for this recognition because they differ among each other (6). On the other hand, these the L chains of these neurotoxins are very similar both in terms of their amino acid sequence and of their predicted secondary structure. This suggested that they recognize the tertiary structure of the SNAREs and that their three substrates have a common toxin recognition motif (33). Indeed, the SNARE proteins share multiple copies of a motif characterized by the presence of three negative charges spaced with hydrophobic residues (33). The three negatively charged residues appear to be particularly important (33-36). Recent evidence indicate that the main determinant of the specific recognition of the three SNARE by the neurotoxins is a double interaction with: a) a segment that includes the peptide bond to be cleaved and b) another segment closely similar in VAMP, SNAP-25 and syntaxin, which accounts for antibody cross-reactivity and cross-inhibition of the different neurotoxin types. The relative contribution of segments a) and b) to the specificity and strenght of neurotoxin binding remains to be determined. It also remains to identify other segments that modulate individual interactions of each toxin. In the case of TeNT and BoNT/B it is clear that the cluster of positively charged residues present in VAMP after the cleavage site is important for maximal rate of cleavage (36). Peptide bond hydrolysis within region a) leaves the toxin bound to its substrate only via its interaction with b) and other regions. This is espected to cause a large decrease in binding affinity, which in turn should lead to a rapid release of the hydrolysed substrate. The physiological role of the SNARE motif is not yet clear. Recent studies of the laboratory of Kelly (37) indicate clearly that it is involved in the regulation of the insertion of VAMP in vesicles. Further studies are necessary to determine the role in SNAP-25 and syntaxin.

The vacuolating cytotoxin of Helicobacter pylori

Gastroenterology has been recently revolutionized by the demonstration that gastroduodenal ulcers have an infectious aethiology. Clear evidence associate prolonged infection of the stomach mucosa with toxigenic strains *Helicobacter pylori* with the development of atrophic gastritis, gastroduodenal ulcers and stomach adenocarcinoma in humans (38-40). A major virulence factor produced by *H. pylori* is a cytotoxin, termed vacA, that cause vacuolar degeneration of cells in vitro (41-43). Oral administration of purified vacA to mice is sufficient to induce degeneration of the gastric mucosa and recruitment of inflammatory cells, two key events in the process that eventually leads to gastric ulcers. *H. pylori* strains with an altered vacA gene are non cytotoxic. These and other evidence implicate vacA in the pathogenesis of human gastroduodenal ulcers.

VacA is released as a 95 kDa protein and can be cleaved by bacterial proteinases into two fragments with no change of biological activity (43). The carboxyl terminal 58 kDa fragment is able at low pH to increase the permeability of liposomes to monovalent cations (44), as seen above for the clostridial neurotoxins. VacA oligomerises into heptamers and hexamers, whose structure has been studied by electron microscopy (45). VacA shows the remarkable property of being activated by short exposures to very acidic pH values and to be able to resist to pepsin at pH 2 for prolonged periods of time: conditions mimicking the intragastric environment (46).

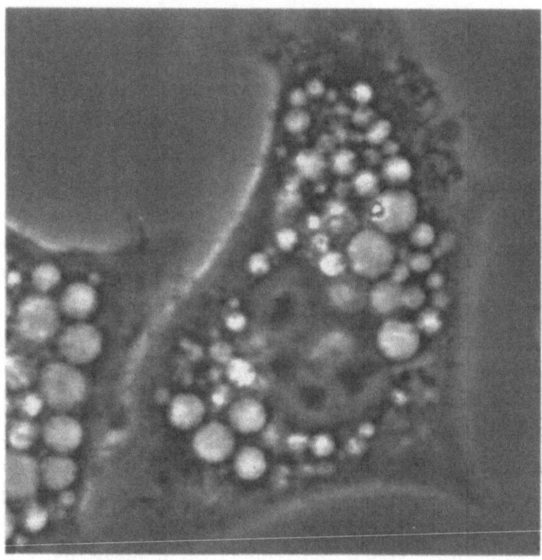

Fig. 2. HeLa cells incubated with 100 nM vacA for 4 hours.

VacA causes the formation of large vacuoles in cultured cells (fig. 2) as well as in vivo (41,42,46-48). The vacuolar lumen is acidic (83) as a result of the activity od a vacuolar-type ATPase proton pump present on the vacuolar membrane (84,85). Vacuoles first appear in the perinuclear area and then grow in size up to several micrometers of diameter. The prcess(es) leading to the formation of such gigantic membranous compartments are not known. Massive fusion of smaller compartments promoted directly or indirectly by the toxin may be involved. Alternatively, vacA could inhibit directly or indirectly the fission/maturation of a cell vesicular/tubular/cysternal compartment. H. pylori-induced vacuoles contain fluid phase markers and their membrane is highly enriched in rab7 (86), a small GTP-binding protein previously shown to be associated with late endosomal compartments (87). Rab proteins are known to regulate extent and specificity of intracellular membrane traffic in eukariotic cells (88,89) (fig. 3). These proteins are anchored to specific intracellular compartments by geranyl-geranylation at their C-terminal and cycle between GDP-bound and GTP-bound forms (90). Similarly to what found for ras protein, point mutations in highly conserved regions can stabilize rab proteins in either the GDP or the GTP-bound conformation and cause them to act as dominant interfering mutants. Several inactive rab mutants have been generated that show a low affinity binding

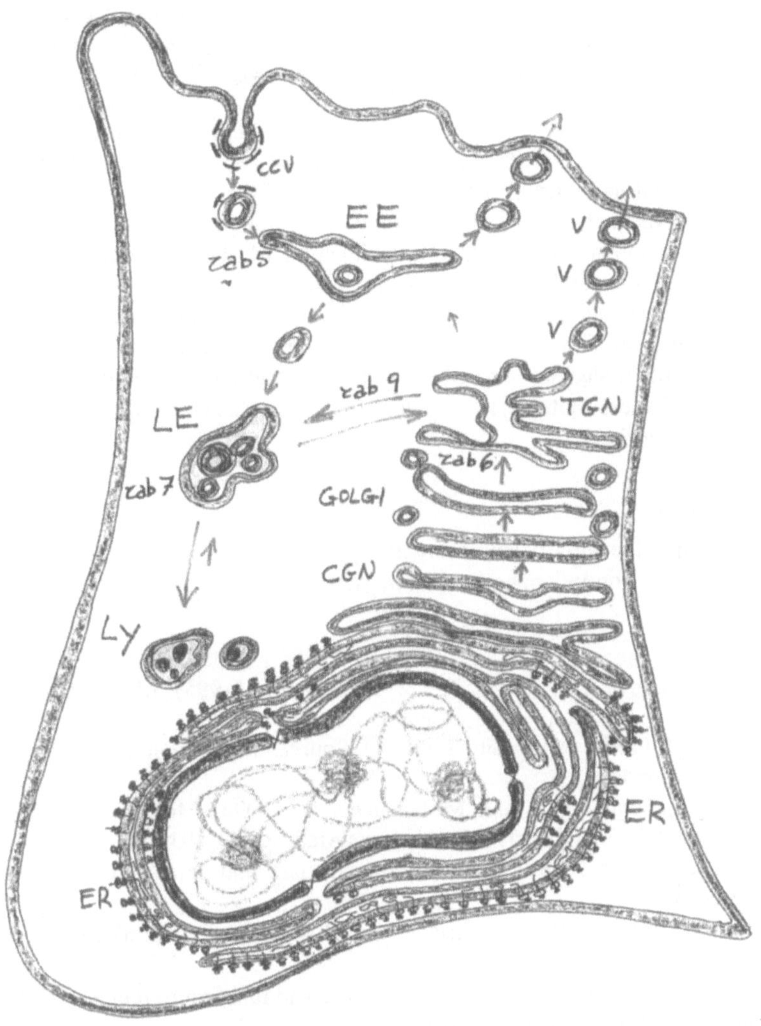

Fig. 3. Schematic picture of membrane trafficking. CCV, clathrin coated vesicles, EE, early endosomes, LE, late endosomes, TGN, trans Golgi network, ER, endoplasmic reticulum.

a low affinity binding of both GDP and GTP or a high preference for GDP. Conversely, permanently activated rab mutants can be obtained by stabilization of the GTP-bound form via replacement of an active site Gln residue with Ile, which blocks intrinsic GTPase activity, or by substitutions which compromises the interaction with GAP, a catalyst of GTP hydrolysis (see ref. 91, for a complete coverage of this subject). The accumulation of rab7 on vacuole membrane could be an epiphenomenon or it could involve directly this small GTPase protein. To discriminate between these possibilities, HeLa cells were transfected with rab7 mutants that are fixed in a permanently active or inactive form. Cells overexpressing active rab7 show a modest increase in cell vacuolization upon exposure to vacA. On the contrary, cells transfected with inactive rab7 mutants do not vacuolate. In other words, inactive rab7 mutants display a dominant negative phenotype with respect to vacuolization (52). Comparison of the effect of overexpression of rab5 and rab9 mutants indicate that vacA affects membrane trafficking at the late endosomal stage, possibly at the level of lysosome formation (fig. 3). Recently, we found that in vacuolated cells: a) the transferrin recycling is unaffected; b) EGF degradation in late endosomes/lysosomes is impaired and c) catepsin D targeting to lysosomes is diminished and is escreted in the medium. These results support the conclusion that vacA leads to a structural and functional alteration of the late endosomal/pre-lysosomal compartment. Identification of the target(s) of vacA should contribute to the understanding of the complex events underlying vesicular trafficking in eukariotic cells and should provide the basis for the use of vacA as a novel tool for further studies, as it has been the case for the clostridial neurotoxins. At the same time, an understanding at the molecular level of the acitivty of vacA will provide the basis for the rational design of vacA mutants to be evaulated as candidate components of an "anti-ulcer" theraputic vaccine.

Acknowledgements
Work carried out in the authors laboratory id supported by Telethon-Italia grant n. 763, by MURST 40% Project on Inflammation and by NIRECO.

References

1.Mims, C., Dimmock, N., Nash, A. & Stephen, J. (1995) Mims'Pathogenesis of Infectiuous Disease. Academic Press, London.
2.Alouf, J.E. and Freer, J.H. Editors (1991) Sourcebook of bacterial protein toxins. Academic Press, London.
3.Menestrina, G., Schiavo, G. and Montecucco, C. (1994) Molecular Mechanisms of Action of Bacterial Protein Toxins. Mol. Aspects Med. 15, 79-193.
4.Simpson, L.L. Editor (1989) Botulinum neurotoxin and tetanus toxin. San Diego: Academic Press
5.Montecucco, C. Editor (1995) Clostridial Neurotoxins. Curr. Top. Microbiol. Immunol. vol 195, Springer, Heidelberg
6.Montecucco, C. and Schiavo, G. (1995) Structure and function of tetanus and botulinum neurotoxins. Quart. Rev. Biophys. 28, 423-472
7.Nishiki, T., Kamata, Y., Nemoto, Y., Omori, A., Ito, T., Takahashi, M. and Kozaki, S. (1994) Identification of protein receptor for Clostridium botulinum type B neurotoxin in rat brain synaptosomes. J. Biol. Chem. 269, 10498-10503.
8. Matteoli, M., Verderio, C., Rossetto, Iezzi, N., Coco, S., Schiavo, G. & Montecucco, C. (1996) Synaptic vesicle endocytosis mediates the entry of tetanus neurotoxin into hippocampal neurons. Proc. Natl. Acad. Sci. USA 93, 13310-13315.
9.Dolly, J.O., Black, J., Williams, R.S. & Melling, J. (1984) Acceptors for botulinum neurotoxin reside on motor nerve terminals and mediate its internalization. Nature 307, 457-460.
10.Shupliakov, O., Low, P., Grabs, D., Gad, H., Chen, David, C., Takei, K., De Camilli, P. & Brodin, L. (1997) Synaptic vesicle endocytosis impaired by disruption of dynamin-SH3 domain interactions. Science 276, 259-263.
11.Wellhoner, H.H. (1992) Tetanus and botulinum neurotoxins. In Handbook of Experimental Pharmacology (eds. H. Herken & F. Hucho), vol 102, pp. 357-417. Berlin: Springer-Verlag.

12.Schwab, M.E., Suda, K. and Thoenen, H. (1979) Selective retrograde trans-synaptic transfer of a protein, tetanus toxin, subsequent to its retrograde axonal transport. J. Cell. Biol., 82, 798-810.

13.Martoglio, B., Hofmann, M. W., Brunner, J. & Dobberstein, B. (1995). The protein-conducting channel in the membrane of the endoplasmic reticulum is open laterally toward the lipid bilayer. Cell 81, 207-214.

14.Martoglio, B. and Dobberstein, B. (1996). Snapshots of membrane translocating proteins. Trends Cell Biol. 6, 142-147.

15.Driessen, A. J. M., (1996) Prokaryotic protein translocation. In D. A. Phoenix (ed.), Protein Targeting, Portland Press, London, Chapter 5.

16.Ferro-Novick, S. & Jahn, R. (1994) Vesicle fusion from yeast to man. Nature 370, 191-193.

17.Sudhof, T.C. (1995) The synaptic vesicle cycle: a cascade of protein-protein interactions. Nature 375, 645-653.

18.Rothman, J.E. & Wieland, F.T. (1996) Protein sorting by transport vesicles. Science 222, 227-234.

19.Bock, J.B. & Scheller, R.H. (1997) A fusion of new ideas. Nature 387, 133-135.

20.Schiavo, G., Poulain, B., Rossetto, O., Benfenati, F., Tauc, L. & Montecucco, C. (1992) Tetanus toxin is a zinc protein and its inhibition of neurotrasmitter release and protease activity depend on zinc. EMBO J. 11, 3577-3583.

21.Schiavo, G., Benfenati, F., Poulain, B., Rossetto, O., Polverino de Laureto, P., DasGupta, B.R. & Montecucco, C. (1992) Tetanus and botulinum-B neurotoxins block neurotransmitter release by a proteolytic cleavage of synaptobrevin. Nature 359, 832-835.

22.Rossetto, O., Gorza, L., Schiavo, G., Schiavo, N., Scheller, R.H. & Montecucco, C. (1995) VAMP/Synaptobrevin isoforms 1 and 2 are widely and differentially expressed in nonneuronal tissues. J. Cell Biol. 132, 167-179.

23.Mellanby, J., Beaumont, M.A. & Thompson, P.A. (1988) The effect of lantanum on nerve terminals in goldfish muscle after paralysis with tetanus toxin. Neuroscience 25, 1095-1106.

24.Hunt, J.M., Bommert, K., Charlton, M.P., Kistner, A., Habermann, E., Augustine, G.J. & Betz, H. (1994) A post-docking role for synaptobrevin in synaptic vesicle fusion. Neuron 12, 1269-1279.

25.Blasi, J., Chapman, E.R., Yamasaki, S., Binz, T., Niemann, H., Jahn, R. (1993b) Botulinum neurotoxin C blocks neurotransmitter release by means of cleaving HPC-1/syntaxin. EMBO J. 12, 4821-4828.

26.Schiavo, G., Shone, C.C., Bennett, M.K., Scheller, R.H. & Montecucco, C. (1995) Botulinum neurotoxin type C cleaves a single Lys-Ala bond within the carboxyl-terminal region of syntaxins. J. Biol. Chem. 270, 10566-10570.

27.Blasi, J., Chapman, E.R., Link, E., Binz, T., Yamasaki, S., DeCamilli, P., S□dhof, T.C., Niemann, H. & Jahn, R. (1993a) Botulinum neurotoxin A selectively cleaves the synaptic protein SNAP-25. Nature 365, 160-163.

28.Schiavo, G., Rossetto, O., Catsicas, S., Polverino de Laureto, P., DasGupta, B.R., Benfenati, F. & Montecucco, C. (1993) Identification of the nerve-terminal targets of botulinum neurotoxins serotypes A, D and E. J. Biol. Chem. 268, 23784-23787.

29.Osen-Sand, A., Staple, J. K., Naldi, E., Schiavo, G., Rossetto, O., Petitpierre, S., Malgaroli, A., Montecucco, C. & Catsicas, S. (1996) Common and distinct fusion-proteins in axonal growth and transmitter release. J. Comp. Neurol. 367, 222-234.

30. Foran, P., Lawrence, G. W., Shone, C. C., Foster, K. A. & Dolly, J. O. (1996) Botulinum neurotoxin C1 cleaves both syntaxin and SNAP25 in intact and permeabilized

chromaffin cells: correlation with its blockade of catecholamine release. Biochemistry 35, 2630-2636.

31.Williamson, L. C., Halpern, J. L., Montecucco, C., Brown, J. E. and Neale, E. A.(1996) Clostridial neurotoxins and substrate proteolysis in intact neurons: botulinum neurotoxin C acts on SNAP-25. J. Biol. Chem. 271, 7694-7699.

32. Gansel, M., Penner, R. & Dreyer, F. (1987) Distinct sites of action of clostridial neurotoxins revealed by double poisoning of mouse motor-nerve terminals. Pflugers Arch. 409, 533-539.

33. Rossetto, O., Schiavo, G. Montecucco, C., Poulain, B., Deloye, F., Lozzi, L. & Shone, C.C. (1994) SNARE motif and neurotoxin recognition. Nature 372, 415-416.

34. Witcome, M., Rossetto, O., Montecucco, C. & Shone, C.C. (1996) Substrate residues distal to the cleavage site of botulinum type B neurotoxin play a role in determining the specificity of its endopeptidase activity. FEBS Lett. 386, 133-136.

35. Pellizzari, R., Rossetto, O., Lozzi, L., Giovedi, S., Johnson, E., Shone, C.C. & Montecucco, C. (1996) Structural determinants of the specificity for VAMP/synaptobrevin of tetanus and botulinum type B and G neurotoxins. J. Biol. Chem., 271, 20353-20358.

36. Cornille, F., Loic, M., Lenoir, C., Cussac, D., Roques, B. & Fournie-Zaluski, M.C. (1997). Cooperative expsite-dependent cleavage of synaptobrevin by tetanus tocin light chain. J. Biol. Chem. 272, 3459-3464.

37. Grote, E., Hao, J.C., Bennett, M.K. & Kelly, R.B. (1995) A targeting signal in VAMP regulating transport to synaptic vesicle. Cell 81, 581-589.

38.Marshall, B.J. & Warren, J.R. (1983) Unidentified curved bacilli in the stomach of patients with gastric and peptic ulceration. Lancet i, 1311-1315.

39. Marshall, B.J., Armstrong, J.A., McGechie, D.B. & Glancy, R.J. (1985) Attempt to fulfil Koch's postulates for pyloric campylobacter. Med. J. Austr. 142, 436-439.

40. Blaser, M.J. (1996) The bacteria behind ulcers. Sci. Am. 274, 104-107.

41.Leunk, R.D., Johnson, ,P.T., David, B.C., Kraft, W.G. & Morgan, D.R. (1988) Cytotoxic activity in broth-culture filtrates of Campylobacter pylori. J. Med. Microbiol. 26, 93-99.

42.Cover, T.L. & Blaser, M.J. (1992) Purification and characterization of the vacuolating toxin from Helicobacter pylori. J. Biol. Chem. 287, 10570-10575.

43. Telford, J.L., Ghiara, P., Dell'Orco, M., Comanducci, M., Burroni, D., Bugnoli, M., Tecce, M.F., Censini, S., Covacci, A., Xiang, Z., Papini, E., Montecucco, C. & Rappuoli, R. (1994) Gene structure of the Helicobacter pylori cytotoxin and evidence of its key role in gastric disease. J. Exp. Med. 179, 1653-1658

44. Moll, G., Papini, E., Colonna, R., Burroni, D., Telford, J., Rappuoli, R. & Montecucco, C. (1996) Lipid interaction of the 37 kDa and 58 kDa fragments of the Helicobacter pylori cytotoxin. Eur. J. Biochem. 234, 947-952

45.Lupetti, P., Heuser, J.E., Manetti, R, Massari, P., Lanzavecchia, S., Bellon, P.L., Dallai, R., Rappuoli, R. & Telford, J.L. (1996) Oligomeric and subunit structure of the Helicobacter pylori vacuolating cytotoxin. J. Cell Biol. 133, 801-807.

46.de Bernard, M., Papini, E., De Filippis, E., Gottardi, E., Telford, J., Manetti, M., Fontana, A., Rappuoli, R. & Montecucco, C. (1995) Low pH activates the Vacuolating Toxin of Helicobacter pylori which becomes acid and pepsin resistant. J. Biol. Chem. 270, 23937-23940

47.Papini E., Bugnoli M., de Bernard M., Figura N., Rappuoli R. and Montecucco C. (1993) Bafilomycin A1 inhibits Helicobacter pylori induced vacuolization of HeLa cells. Mol. Microbiol. 7, 323-327.

48.Papini, E., De Bernard, M., Milia, E., Bugnoli, M., Zerial, M., Rappuoli, R. & Montecucco, C. (1994) Cellular vacuoles induced by Helicobacter pylori originate from late endosomal compartments. Proc. Natl. Acad. Sci. U.S.A. 91, 9720-9724.

49.Nuoffer, C. & Balch, W.E. (1994) GTPases: multifunctional molecular switches regulating vesicular traffic. Annu. Rev. Biochem. 63, 949-990.

50.Simons, K. & Zerial, M. (1993) Rab proteins and the road maps for intracellular transport. Neuron 11, 789-799.

51.Zerial, M. & Huber, L. (1996) Guidebook to the Small GTPases. Oxford University Press, Oxford

52. Papini, E., Satin, B., Bucci, C., de Bernard, M., Telford, J.L., Manetti, R., Rappuoli, R., Zerial, M. & Montecucco, C. (1997) The small GTP binding protein rab7 is essential for cellular vacuolation induced by *Helicobacter pylori* cytotoxin. EMBO J. 16, 15-24.

Pore-Forming Antimicrobial Peptides and Polypeptides: Mechanism of Insertion into Membranes and Mode of Action

Lucienne Letellier
Laboratoire des Biomembranes, URA CNRS 1116, Université Paris Sud,
Bât 430, 91405 Orsay Cedex, France

1 Introduction

1.1 Antibacterial peptides

Antibacterial peptides are part of the immune response of almost all organisms ranging from mammalian cells to insects, plants and microorganisms. The role of these peptides in host defence was almost unknown ten years ago but their fundamental importance in microbial defence is now recognized. The interest for these peptides has also considerably increased during the last few years because of their potential use as antibiotics.

Antibacterial peptides, although of diverses origins and differing in their primary sequence and structure share many common features. Most of these peptides are cationic, amphiphilic and contain up to 40 amino acids. Almost all the peptides are active against bacteria, some of them are active against fungi and viruses but they cause little or no harm to eucaryotic cells (Hoffman and Hetru, 1992). Most of them kill Gram-positive and Gram-negative bacteria by permeabilizing their cytoplasmic membrane. Peptides with only D-amino acids have equal activity as the natural ones suggesting that their activity is not dependent on recognition of any bacterial receptor.

The first antimicrobial peptides were characterized in the 1980's. They were found in mammalian cells by Lehrer's group and called *defensins*. (Ganz and Lehrer, 1994). *Xenopus* are widely used as laboratory animals from which oocytes are removed surgically. Yet these frogs can thrive in water quite dense in bacteria and the wounds from their operation never become infected.This observation led Zasloff to the finding of the *magainins* . During the same period Boman and associates succeeded in isolating an antibacterial peptide from the moth Hyalophora cecropia, which they named *cecropin* (reviewed in Boman, 1991).

NATO ASI Series, Vol. H 106
Lipid and Protein Traffic
Pathways and Molecular Mechanisms
Edited by Jos A. F. Op den Kamp

Insect antimicrobial peptides are inducible and synthesized upon injury. They are found in some cases in the blood and in the fat body, at a concentration ranging from 1 to 10 μM. Their synthesis appears very shortly after the bacterial challenge.

Mammalian peptides are constitutively expressed from specialized cells either within the blood (phagocytes), the skin or the intestine, the genital tract or the lung eptihelium, all of which are susceptible to encounter bacteria. In frog skin they are released from specialized cells and diffuse through the skin. Peptides originating from mammalian and from frog skin are constitutively synthesized independently of an immunological challenge.

Bacteria synthesize antibacterial peptides which are currently referred to as *bacteriocins*. Bacteriocins are active only on Gram-positive bacteria. They share common structural and functional features with the the other antibacterial peptides. Since the 80's they have attracted the interest of numerous research groups because of the potential practical application to food preservation or the prevention and treatment of bacterial infections. Their origin, biosynthesis and biological activity have been described in recent reviews (Jack et al, 1995; Sahl et al., 1995).

1.2 Colicins

The first characterized antimicrobial polypeptides produced by bacteria were of the colicin type. In contrast to the bacteriocins, colicins are produced by and are active against Gram-negative bacteria.They are characterized by their relatively high molecular mass, their narrow range of activity, their complex mode of action including requirements for receptors and complex machineries for translocation. Colicins are not only interesting *per se* but they allow us to address many fundamental questions relating to the mechanisms of translocation and insertion of proteins in membranes. They are easily manipulated both genetically and biochemically and allow *in vitro* and *in vivo* studies.

2 defensins: peptides with intramolecular disulfide bridges and β-sheet structure

Unlike α-helical amphiphatic peptides whose interactions with membrane model systems have been extensively studied, only a few studies have been devoted to the mode of insertion into membranes and activity of β–sheet organized peptides .

2.1 Structure of defensins

Defensins are small cationic peptides that contain 29 to 42 amino acid residues including six invariant cysteine residues that form three intramolecular disulfide bridges. They belong to three different groups, the mammalian α-defensins, the β-defensins and the insect defensins. Their antimicrobial spectrum includes Gram-positive and Gram-negative bacteria, envelope viruses and fungi. Insect defensins are predominantly active against Gram-positive organisms. They differ from each other in the spacing and connectivity of the cysteines and by their 3D structure (White et al., 1995).

The solution structure of three α defensins and one insect defensin have been determined by NMR. A common feature of these defensins is their high content of β structure and the presence of a β hairpin. Defensin A is found in the blood of bacteria-challenged larvae of the fleshfly *Phormia terranovae*. (Hoffmann and Hetru,1992). It consists of three distinct domains: (i) a N-terminal loop formed by residues 1 to 13 which has a certain degree of flexibility; (ii) a central amphipathic α-helix consisting of residues 14 to 24; (iii) a C-terminal antiparallel β–sheet (residues 27 to 40) with a turn (residues 31-34). The α helix is stabilized by two disulfide bridges to one of the strands of the β sheet and the N-terminal loop is connected to the other β strand by the third disulfide bridge (Cornet et al., 1995). The structural organization of the insect defensin differs markedly from that of the mammalian defensin which lack the α helix.

2.2 Liposome assays for studying the binding and pore-forming activity of human defensins

Binding of the human defensin HNP-2 to large unilamellar vesicles of variable composition was analyzed using equilibrium dialysis and reverse phase HPLC (Wimley et al., 1994). Binding was not observed with zwitterionic lipids and only occurred if acidic lipids were present suggesting that binding is initiated through electrostatic interaction with lipids. The requirement for acidic lipids for binding appears a common feature of almost all families of cationic peptides. From a fluorescence quenching assay it was concluded that the defensin HNP-2 induced leakage of the dye from vesicles in an all or none manner. Dextrans of variable MM were entrapped into the liposomes. Large solutes did not leak significantly. Together these results suggest that defensin forms multimeric pores but only if a minimum number of defensins are present for a pore to assemble. The upper limit of the pore diameter, estimated from the Stokes radius of the largest solute able to diffuse through the channel, was to the order of 2.5nm. A speculative model was proposed in which an annular pore would be formed by a hexamer of dimers.

2.3 *In vivo* and *in vitro* studies of the mode of action of defensins

The first barrier encountered by defensins which are active against Gram negative bacteria is the polyanionic lipopolysaccharide (LPS) located on the outer leaflet of the outer membrane. Diffusion through the outer membrane is very limited because of the charged LPS. In addition only those hydrophilic compounds with a MM of *circa* 800-1000 Da can diffuse through the porin pores. This raises the question of how cationic peptides cross the outer membrane. It has been proposed that the net positive charge of the polycationic peptide would favour the displacement of the divalent cations that cross-link the adjacent LPS molecule and consequently increase the permeability of the outer membrane (Viljanen et al., 1988). The aggregating capacity of the peptide is also likely to define its permeating properties.Indeed, if high cooperativity in binding occurs then the peptide should remain on the surface and aggregate. Alternatively if the peptide remains bound without aggregating then the monomer are likely to diffuse through to reach the inner membrane.

The leakage of the bacterial content induced by human defensins was an indication that the target of the peptide was the cytoplasmic membrane (Lehrer et al.,1989) The mechanism by which the insect defensin A permeabilizes the cytoplasmic membrane of the Gram-positive bacteria *Micrococcus luteus*, was recently documented (Cociancich et al., 1993). Defensin A caused a loss of almost all the cytoplasmic potassium in less than 2 min. Permeabilization was only observed above a minimum defensin to cell ratio of 10^5. The rate of efflux was at least one order of magnitude larger than that caused by carriers. K^+ efflux took place concomitantly with a decrease of the membrane potential from 197 mV (negative inside) to a threshold value of 110 mV. Furthermore, if the membrane potential was decreased to this threshold value with a protonophore before addition of defensin, then efflux of K^+ was totally prevented. Taken as a whole these data suggested that K^+ efflux is channel-mediated and that opening and closing of the channel is regulated by the membrane potential. Disulfide bonds played an esssential role in maintaining the active conformation *in vivo*, since the activity of defensin was lost upon treatment with dithiothreitol (DTT). Channel formation was also prevented by divalent cations. The *M. luteus* membrane contains a high percentage of acidic lipids. Divalent cations may mask the negative charges of the membrane and prevent the binding of defensins to lipids.

There are two possible models for channel formation: either pre-existing oligomers insert spontaneously in the membrane, or defensin monomers first bind to the membrane and then diffuse laterally to form oligomers. In the second case, if lateral diffusion of the monomers is the rate limiting step in oligomerization and hence in channel formation, the time needed for oligomerization to take place is expected to decrease with increasing probability of collision between monomers in the membrane and therefore with increasing number of added defensin. This proposal is supported by the observation that the lag time which preceeds K^+ efflux decreased with increasing number of defensins.

A direct proof of the channel activity of defensin A was given by using the patch-clamp technique (Sackmann and Neher, 1983). Defensin A was inserted into giant azolectin liposomes where it formed channels. There was a reasonable correlation between the value of the membrane potential at which K^+ efflux stopped *in vivo* and the voltage at which the channels became silent *in vitro*. Channels were not well-behaved, showing heterogeneity both in conductances and opening and closure durations (Cociancich et al., 1993). This behaviour has been also reported for the rabbit neutrophil defensin NP 1 which formed voltage-dependent and weakly anion selective channels in lipid bilayers (Kagan et al., 1990). Two explanations, not necessarily mutually exclusive, may account for this: the pores may be formed by multimers of variable numbers of defensin molecules, or each multimer may have different conductance levels.

Defensin A also inhibited several essential cellular functions. Respiratory activity was inhibited but well after depolarization and K^+ efflux had been observed.The peptide might directly affect components of the respiratory chain. Alternatively, opening of the defensin channels may lead to a release of co-factors essential for respiration. ATP depletion, but not leakage, was also rapidly observed. The channel-forming colicins and the bacteriocin nisin also caused an ATP depletion in *E. coli* cells and in *M. luteus* respectively which was associated with an efflux of inorganic phosphate through the channels. This phosphate efflux resulted in a shift of the ATP hydrolysis equilibrium (Guihard *et al.*, 1993; Abee et al, 1994). It is likely that the defensin-induced ATP depletion takes place by a similar mechanism. This ATP depletion is presumably pivotal for cell killing by toxins.

3 cecropins and magainins: peptides forming amphipathic α helices, but devoid of cysteines

3.1 Cecropins

Cecropins are positively-charged peptides of 31 to 39 residues. They were first identified as part of the immune response of certain silkworm species and were also found in pig intestines. They are synthesized as prepropeptides of 62-64 residues. The 3D structure as deduced from NMR studies consists of two α-helices joined by a hinge region containing a glycine-proline doublet. The N-terminal is perfectly amphipathic (Holak et al., 1988). Cecropins are among the most potent antibacterial peptides. They are very active against Gram-negative bacteria the cytoplasmic membrane of which they permeabilize (reviewed in Boman, 1991).

Cecropins form voltage-dependent ion channels in lipid bilayers with single channel conductance of up to 2.5 nS (Christensen et al., 1988). A comparative study of their properties and those of synthetic analogs has allowed the determination of the structural requirements for binding and pore formation. (i) Shorter amphipathic peptides consisting of only 29 and 11 residues did not form channels although they adsorbed to the bilayer. (ii) The channel-forming activity of the peptides with a flexible hinge formed by the helix breaking Gly-Pro sequence was voltage dependent. The peptides with a more rigid chain were characterized by a lack of voltage dependence for their conductance which suggests that they insert directly to form channels. (iii) A positive surface charge or cholesterol in the bilayer reduced the macrosocopic conductance of the cecropin channels. These results are consistent with the known insensitivity of eukaryotic cells towards cecropin. A model for channel formation was proposed. It involves (i) oligomerization of the peptide followed by electrostatic adsorption on the bilayer-water interface, (ii) insertion of the largely hydrophobic C-terminal region leaving the amphiphatic helix at the interface, (iii) channel formation upon applying a positive voltage (refered to the side of peptide addition). The authors further speculate about the size of the channel formed. They assume that if the peptides form a water -filled cylindrical pore of about 5 nm long, then the pore diameter estimated for the channels with the largest conductance, would be 4 nm. Such large pores require the association of many molecules.

Atomic scale computer models of the cecropin channels were developed by Durell et al (1992). Cecropin peptides would assemble in membranes to form two types (I and II) of channels . Cecropin molecules were first arranged as antiparallel dimers. Dimers then bound together to the membrane so that the NH_2-terminal helices were sunk into the lipid head group layer and the COOH helices were spanning the lipid hydrocarbon chain region. At least four dimers had to assemble to form a pore. In the type I model the pore was formed by the carboxy terminal helices. A conformational change, possibly triggered by voltage, would convert type a I channel into a type II channel the pore of which would be formed from the NH_2 terminal helices rather than from the COOH terminal ones.

3.2 Magainins

Magainins are 23 residues in length, positively charged and form an amphipathic α helix. They disrupt the permeability barrier of the *Escherichia coli* inner membrane (Westerhoof et al., 1989). Ion channel formation in planar lipid bilayers was demonstrated in the case of magainin 1 (Duclohier et al., 1989). Liposome permeability assays have not led to a clear model of the mode of action of the peptide. Grant and collaborators summarize their experiments by saying: magainin 2a can permeabilize membranes via bilayer destabilization or by forming channels. Which mechanism predominates will depend on the membrane system, composition and degree of polarization (Grant et al., 1992). Whereas it is reasonably accepted that magainins form a multimeric channel, the characteristics of the channel and the mechanism by which it forms remain less defined. The experiments of Juretic et al (1994) with cytochrome oxidase-containing liposomes have led to the conclusion that at room temperature most of the magainin is in the membrane-bound monomeric form and that the membrane permeability complex may be a pentamer or hexamer or at least a trimer, but that at low temperature some of the membrane-bound magainin might be a dimer. Recently the following model was proposed: (a) the peptide binds to the outer surface of the bilayer and lies parallel to the surface in a monomer-dimer equilibrium; (b) transient pores of multimers are formed; (c) upon closing the peptide translocates across the membrane and is again in the monomer-dimer equilibrium (Matsuzaki et al., 1995).

4 Antibacterial peptides of bacterial origin

Most of the bacteriocin peptides are small (MM in the range of a thousand Daltons), heat stable, cationic and amphiphilic molecules. Many bacteriocins have two or more cysteines connected by a disulfide bridge and are membrane active. They cause an efflux of ions and small solutes, hydrolysis and partial efflux of cytoplasmic ATP, inhibition of the respiratory activity and depolarization of the cytoplasmic membrane. This has been shown for Pep5, nisin, subtilin, epidermin, gallidermin, streptococcin and lactacin (for a recent review see Sahl et al., 1995). The divalent cations Mg^{2+} and Ca^{2+} have an inhibitory effect on the *in vivo* action of nisin Z and lactacin F against *Listeria monocytogenes* and *lactobacillus* strains. Gadolinium is even more efficient (in the μM range) in preventing or arresting the permeabilizing effect of these bacteriocins (Abee et al., 1994). The inhibitory effect of gadolinium is not specific to these bacteriocins since it was also observed with defensin A (Cociancich et al., 1993) and colicin A and N (Bonhivers et al., 1995). The permeabilizing effect of these peptides was confirmed from studies on membranes vesicles and liposomes (Gao et al., 1991; Garcera et al., 1993; Driessen et al., 1995). Planar lipid bilayer experiments indicate that these bacteriocins form pores. Most of these pores show features common to those of the antibacterial peptides described above: conductance fluctuations between different levels with a burstlike character were more frequently found than steplike current increments. This is examplified in the case of Pep5 (Kordel et al., 1988). Models for the formation of the pores have also been derived from *in vivo* and vitro data and from analysis of the 3D structure of some of the peptides (Sahl et al., 1995).

Although of different phylogenetic origins the antimicrobial peptides described here show common features: they are all cationic and amphiphilic. They permeabilize membranes probably by forming multi-conductance pores. The insertion/opening of these pores requires acidic lipids and in most cases an energized membrane or application of voltage. Liposome permeability assays using fluorescence probes have been extensively used to investigate their mode of action. The peptide-induced permeability of liposomes towards fluorescent probes has often been considered as a criterion for channel formation. These data have to be interpreted with caution since leakage of vesicle content is often best explained by membrane destabilization. The abundant literature has unfortunately not allowed the situation to be clarified. The questions with which we are still faced are the following: do the peptides interact with membranes as monomers or multimers? Do they lay or not parallel to the membrane surface? Is the membrane potential responsible for membrane insertion and/or change in the state of oligomerization? Is membrane destabilization or channel formation responsible for permeabilization?

5. Colicins: translocation across the E.coli envelope, insertion in membranes and channel activity

5.1. Introduction

Colicins are toxins produced by a wide variety of *Enterobacteriacea* . These soluble proteins of high molecular weight (40 to 70 kDa) are synthesized in large quantities, generally secreted in the culture medium together with their immunity protein and easily purified. Their genes have been cloned and their nucleotide sequence has been known since 1982 which makes their genetic and chemical manipulation possible (for a review see Lazdunski et al., 1988). The cytoplasmic membrane of *Escherichia coli* is the target of the pore-forming colicins A, B, N, E1, Ia, Ib and K.To reach this target colicins first bind to the outer membrane by parasitising receptors whose physiological function is the transport of metabolites. Then they are translocated through the outer membrane and the periplasm. This translocation requires the participation of bacterial proteins: Tol Q, TolR, TolA and TolB (for colicins A, E1 and K) or TonB (for colicins B, Ia, Ib). The polypeptide chain of colicins comprises three linearly organized functional domains that play a specific role in each of these steps: the central and the N-terminal domains are involved in receptor binding and translocation respectively. The C-terminal domain carries the pore-forming activity (Lazdunski et al., 1988).

The paradox with colicins is that they have characterisitics of both water-soluble and membrane proteins. They therefore provide a very interesting model for the study of protein translocation and insertion into membranes.

5.2 *In vitro* studies

Colicins insert spontaneously into planar lipid bilayers where they form ion channels with small conductances (15-20pS in 1M NaCl at neutral pH), and which show poor selectivity between anions and cations. Channels are almost always open at large negative voltage. The formation of the channels requires an acidic pH (reviewed in Pattus et al., 1990)· Both the results of Schein et al (1978) and of Slatin (1988) suggest that the colicin channels are formed by a monomer. The C-terminal peptide fragment of different colicins, which can be isolated by proteolytic cleavage, also carries channel-activity (Bullock et al., 1983). However there are some discrepancies between the electrophysiological characteristics of the channel formed by the fragment and by the whole molecule.

The 3-dimensional structure of the C-terminal domain of colicin A (204 aa) was determined by X-ray crystallography at 2.4 A° resolution. It consists of a ten-helix bundle roughly arranged in three layers. The central helices 8 and 9 are buried in the protein and form a very hydrophobic hairpin. They are surrounded by helices 1-7 and 10 which are strongly amphipathic and constitute the interface between the hydrophobic hairpin and the aqueous medium (Parker et al., 1992). Disulfide bond engineering of the pore-forming domain of colicin A based of the 3-D structure has provided important information on the conformation changes associated with the binding and unfolding of colicin A both *in vivo* (see below) and *in vitro*. Three double-cysteine mutants were constructed, whose disulfide bridges connected helix 1 with helix 9, helix 5 with helix 6 and helix 9 with helix 10, respectively. The disulfide bond which connected helix 1 with helix 9 prevented the potential-independent insertion of colicin A into membrane vesicles whereas the disulfide bridges which connected helix 5 with helix 6 or helix 9 with helix 10 prevented the potential-dependent channel opening in planar lipid bilayers (Duché *et al.*, 1994). These data imply that unfolding or conformational change of the 1-2 hairpin relative to the 8-9 hairpin is required for binding and insertion.

Conformational changes responsible for channel activity of colicins has been demonstrated for the C-terminal domain of colicin Ia (Slatin et al., 1994). Single cysteines were introduced by site directed mutagenesis into specific sites. The purified proteins were biotinylated and their channel activity studied in planar lipid bilayer. After insertion of the protein in the bilayer the voltage was varied to allow opening and closing of the channel. The water soluble streptavidin which binds biotin was added to the solution bathing the membrane. Streptavidin added to the *cis* side bound these amino acids when the channel was closed but not when it was open. Further binding of streptavidin to these amino acids prevented channel opening. Streptavidin did not bind these amino acid when put on the trans side and when the channel was close but bound to a stretch of 37 residues when the channel was open. Again binding of streptavidin when the channel was in a closed configuration prevented its opening. These results indicate that a large domain comprising 31 amino acids of the protein undergoes important conformational change and is even able to flip across the membrane upon opening and closing of the channel. This constitutes strong evidence that even large domains of membrane proteins can be reversibly inserted in membranes.

5.3 Insertion of colicins in whole cells and pore forming activity

Pore-forming colicins cause a leakage of cytoplasmic K^+. A K^+-selective electrode was used to analyze quantitatively this process. This technique allows continuous measurements without requiring any manipulation of the cell suspension such as centrifugation or filtration. Provided that this efflux can be dissociated from that induced by the K^+ transporters and that one knows the number of proteins susceptible to form channels, it is possible to determine if K^+ efflux is channel-mediated (Boulanger and Letellier, 1988; Bourdineaud et al., 1990). Furthermore this technique allows us to monitor "on line" the progression of the protein through the envelope and therefore to analyse the translocation steps.

5.3.1 Colicin A forms channels in the E.coli cytoplasmic membrane

Addition of colicin A to *E.coli* cells caused, after a time ag, a net efflux of K^+ whose rate (3 x 10^5 K^+ ions .s^{-1} per colicin) suggests that it is channel-mediated. Furthermore the linear dependance of the rate of efflux with the number of colicins is an indication that each colicin forms a single channel. The voltage gating observed *in vitro* also exists *in vivo* since colicin channel opening and closing depended on the value of the membrane potential applied (efflux of K^+ cessed when $\Delta\Psi$ was decreased below a value of -85 mV) (Bourdineaud et al., 1990).

5.3.2 translocation of colicin A through the envelope

The K^+ efflux caused by colicin A is preceded by a lag period. During this period, the colicin binds to its receptor and is translocated through the envelope. From the comparison of the lag time required to initiate K^+ efflux under energized and de-energized conditions it appears that one of the steps following binding occurs in de-energized cells. This suggests either that colicin A is translocated and inserted into the de-energized membrane and $\Delta\Psi$ is necessary for channel opening or that $\Delta\Psi$ is required for insertion and channel opening (Bourdineaud et al., 1990).

5.3.3. Role of negatively charged phospholipids in translocation and insertion of colicin A

In *Escherichia coli* cells the overall negative charge is mainly due to phosphatidyl glycerol and di-phosphatidyl glycerol. Strains have been constructed in which the PG content can be controlled and varied (Kusters et al.1991). By measuring the K^+ efflux induced by colicin A and the lag time preceding this efflux, it was possible to gain information on the effect of acidic lipids on the translocation of the toxin. Reducing the PG content from 13.5% down to 1.2% led to an increase of the time needed for translocation and insertion into the inner membrane from 30s to 65s. Such acidic lipid dependence was not observed wih colicin N. The pI of the pore-forming domain of colicin N is far higher (pI = 10.25) than that of colicin A (pI = 5.82). Therefore the amount of negatively charged lipids required for the insertion of the pore-forming domain of colicin N is likely to be lower than that of colicin A. Therefore in contrast to colicin A, it would insert in cells having a low level of acidic lipids.

5.3.4 Colicin A unfolds during its translocation and spans the whole cell envelope once forming its channel

At high colicin concentration, the measurement of the lag time before K^+ efflux is a good approximation of the time needed for translocation. The translocation time of colicin A denaturated in 8M urea prior to being added to cells, was half that of native colicin A. This suggests that the rate-limiting step in translocation may be the partial unfolding of the polypeptide chain. Unfolding of colicin was further analyzed using the three double-cysteine mutants which cross-linked and rendered neighbouring helices of the colicin A C-terminal domain non-dissociable (see § 5.2). The mutated colicins formed channels only after their disulfide bonds had been reduced by DTT. In solution, the reduction of the disulfide bonds by DTT was a slow process whose kinetics depended on the position of the disulfide bond ($t_{1/2}$ varying between 85 and 100 s). This $t_{1/2}$ was strongly decreased ($t_{1/2}$ = 8 to 9 s) upon predenaturation of the mutated colicins with urea. The $t_{1/2}$ of reduction of the mutants bound to *Escherichia coli* sensitive cells were similar to those of predenaturated colicins. This suggested that the interaction of the oxidized double-cysteine mutants with the *Escherichia coli* envelope triggered their unfolding. Interestingly the presence of the disulfide bonds slowed down the rate of translocation of the toxin. Furthermore, the rate of translocation increased with decreasing length of the loop connecting the two cysteine residues forming the disulfide bond. One important feature of these results is that a modification in the pore-forming domain has a pronounced effect on the translocation which suggests that the different domains of colicin A are interdependent.

Proteolysis experiments together with K^+ efflux experiments have shown that colicins are still accessible from the external medium when forming a pore in the inner membrane Bénédetti et al., 1992). Furthermore several data suggest that colicin A remains in close contact with its receptor and machinery of translocation once forming its pore (Guihard et al., 1994; Duché et al., 1995). The recent determination of the crystal structure of the pore-forming colicin Ia at 3 A° resolution (Wiener et al., 1997) give strong support to this proposal. The molecule is 210 A° long. The channel-forming and translocation domains are separated from the receptor binding domain by two 160 A° long helices that form a hairpin, with the receptor -binding domain at the bend. If one supposes that the periplasm has an average width of 150 A° then the two helices would be long enough to span the periplasm and to remain in contact with both the inner and outer membrane.

Acknowledgements

A major part of the work presented in this review is the result of long terms collaborations with the groups of C. Lazdunski (H. Bénédetti, R. Lloubès) D. Baty (D. Duché), F. Pattus (G. van der Goot) and J. Hoffmann (C. Hetru) and with A. Ghazi in the Laboratoire des Biomembranes. Their contribution was invaluable. Louise Evans is kindly acknowledged for reading of the manuscript.

References

Abee, T., Klaenhammer, T., and Letellier, L., (1994), Kinetics studies of the action of lactacin F., a bacteriocin produced by *Lactobacillus johnsonii* that forms poration complexes in the cytoplasmic membrane; Appl. Environm. Microbiol., 60: 1006-1013

Abee, T., Rombouts, F., Hugenholtz, J., Guihard, G. and Letellier, L. (1994), Mode of action of nisin Z against *Listeria monocytogenes* Scott A grown at high and low temperature; Appl. Environm. Microbiol, 60: 1962-1968

Bénédetti, H., R. Lloubès, C. Lazdunski, and Letellier, L.(1992) Colicin A unfolds during its translocation in *Escherichia coli* cells and spans the whole cell envelope when its pore has formed. EMBO J.,11: 441-447

Boman, H.G. (1991), Antibacterial peptides:key components needed in immunity Cell, 65: 205-207

Bonhivers, M., Guihard, G., Pattus, F. and Letellier, L. (1995), *In vivo* and *in vitro* studies of the inhibition of the channel activity of colicins by gadolinium, Eur. J. Biochem., 229: 155-163

Boulanger, P. and Letellier, L. (1988), Characterization of ion channels involved in the penetration of phage T4 DNA into *E.coli* cells. J. Biol. Chem., 263: 9767-9775

Bourdineaud, J.P., Boulanger, P., Lazdunski, C. and Letellier, L. (1990), *In vivo* properties of colicin A: channel activity is voltage-dependent but translocation may be voltage independent. Pro. Natl. Acad. Sci.US, 87: 1037-1041

Bullock, J., Cohen, F., Dankert, J. and Cramer, W., (1983) Comparison of the macroscopic and single channel conductance properties of colicin E1 and its COOH-terminal tryptic peptide, J. Biol. Chem., 258: 9908-9912

Christensen, B., Fink, J., Merrifield, R.B. and Mauzerall, D. (1988), Channel-forming properties of cecropins and related model compounds incorporated in planar lipid membranes, Proc. Natl. Acad. Sci., US, 85: 5072-5076

Cociancich, S., Ghazi, A., Hetru, C., Hoffmann, J., and Letellier, L. (1993), Insect defensin, an inducible antibacterial peptide, forms voltage-dependent channels in *Micrococcus luteus* J. Biol. Chem., 268: 19239-19245

Cornet, B., Bonmatin, J-M., Hetru, C., Hoffman, J., Ptak, M. and Vovelle, F. (1995), Refined three dimensional structure of insect defensin A. Structure, 3:435-448

Driessen, J.M., van den Hooven, H., Kuiper, W., van de Kamp, M., Sahl, H-G., Konings, R. and Konings, W., (1995), Mechanistic studies of lantibiotic-induced permeabilization of phospholipid vesicles. Biochemistry, 34: 1606-1614

Duché, D., Baty, D., Chartier, M. and Letellier, L. (1994) Unfolding of colicin A during its translocation as demonstrated by disulfide bond engineering J. Biol. Chem., 269: 6332-6339

Duché, D., Letellier, L., Géli, V., Bénédetti, H. and Baty, D. (1995), quantification of group A import sites , J. Bacteriol., 177: 4935-4939

Duché, D., Parker, M.W., Gonzalez-Manas, J.M., Pattus, F. and Baty, D. (1994), Uncoupled steps of the colicin A pore formation demonstrated by disulfide bond engineering J. Biol. Chem, 269: 6332-6339

Duclohier, H., Molle, G. and Spach, G (1989), Antimicrobial peptides magainin 1 from Xenopus skin forms anion-permeable channels in planar lipid bilayers. Biophys. J., 56: 1017-1021

Durell, S.R., Raghunathan, G., and Guy, H.R. (1992), Modeling the ion channel structure of cecropin. Biophys. J., 63: 1623-1631

Ganz, T. and Lehrer R.(1994), Defensins: Current Opinion in Immunology, 6 : 584-589

Gao, F., Abee, T., and Konings, W. (1991), Mechanism of action of the peptide antibiotic nisin in liposome and cytochrome C oxidase containing proteoliposomes. Appl. and Environ.Microbiol., 57: 2164-2170

Garcera, M., Elferink, M., Driessen, A., and Konings, W. (1993), *In vitro* pore-forming activity of the lantibiotic nisin: role of protonmotive force and lipid composition, Eur. J. Biochem., 212: 417-422

Grant, Jr., E. , Beeler, T.J., Taylor, K.M.T., Gable, K., and Roseman, M.A. (1992), Mechanism of magainin 2a induced permeabilization of phospholipid vesicles. Biochemistry, 31: 9912-9918

Guihard, G., Bénédetti, H., Besnard, M. and Letellier, L. (1993), Phosphate efflux through the channels formed by colicins and phage T5 in *E.coli* cells is responsible for the fall in cytoplasmic ATP; J. Biol. Chem., 268: 17775-17780

Guihard, G., P. Boulanger, H. Benedetti, R. Lloubès, M. Besnard and Letellier L. (1994) Colicin A and the tol proteins involved in its translocation are preferrentialy located in contact sites between the inner and outer membranes of Escherichia coli cells. J. Biol. Chem. 269: 5874-5880

Hoffmann, J.A. and Hetru, C. (1992), Insect defensins: inducible antibacterial peptides Immunol. Today, 13: 411-415

Holak, T.A., Engström, A., Kraulis, P., Lindeberg, G., Bennich, H., Jones, T., Gronenborn, A. and Clore, G.,(1988), The solution conformation of the antibacterial peptide cecropin A: a NMR and dynamic stimulated annealing study; Biochemistry, 27: 7520-7629

Jack, R.W., Tagg, J.R., and Ray, B. (1995), Bacteriocins of Gram-positive bacteria, Microbiol. Rev., 59: 171-200

Juretic, D., Hendler, R.W., Kamp, F., Caughey, W.S., Zasloff, M., and Westerhoff, H. (1994) magainin oligomers reversibly dissipate the electrochemical gradient of protons in cytochrome oxidase liposomes. Biochemistry, 33: 4562-4570

Kagan, B.L., Selsted, M.E., Ganz, T. and Lehrer, R.I. (1990), Antimicrobial defensin peptides form voltage-dependent ion-permeable channels in lipid bilayer membranes. Proc. Natl.Acad. Sci.USA, 87: 210-214

Kordel, M., Benz, R. and Sahl, H.G. (1988), Mode of action of the staphylococcin like peptide Pep5: voltage-dependent depolarization of bacterial and artificial membranes. J. Bacteriol., 170: 84-88

Kusters, R., W. Dowhan, and de Kruijff, B, (1991) Negatively charged phospholipids restore prePhoE translocation across phosphatidylglycerol-depleted *Escherichia coli* inner membranes. J. Biol. Chem. 266: 8659-8662.

Lazdunski, C., D. Baty, V. Geli, D. Cavard, J. Morlon, R.H.P. Lloubès, M. Knibiehler, M. Chartier, S. Varenne, M. Frenette, J.L. Dasseux, and Pattus.F.(1988). The membrane channel-forming colicin A: synthesis, secretion, structure, action and immunity. Biochim. Biophys. Acta 947: 44-464.

Lehrer, RI., Barton, A., Daher, KA., Harwig, S., Ganz, T. and Selsted, M. (1989), Interaction of human defensins with *E.coli*. Mechanism of bactericidal activity. J. Cli. Invest.84: 553-561

Matsuzaki, K., Murase, O., Fujii, N., and Miyajima, K. (1995), Translocation of a channel-forming antimicrobial peptide, magainin 2, across bilayers by forming a pore. Biochemistry, 34: 6521-6526

Parker, M.W. , J.P. Postma, F. Pattus, A.D. Tucker, and Tsernoglou, D. (1992) Refined structure of the pore-forming domain of colicin A at 2.4 Å resolution. J.Mol.Biol. 224: 639-657

Pattus, F., Massotte, D., Wilmsen, H.U., Lakey, J., Tsernoglou, D., Tucker, A., and Parker, M.W.(1990), Colicins: prokaryotic killer pores. Experientia, 46: 180-192

Sackmann, B., and Neher, E. Eds (1983), Single channel recording, Plenum Press, New York

Sahl, H-J., Jack, R.W. and Bierbaum, G.(1995), Biosynthesis and biological activities of lantibiotics with unique post-translational modifications . Eur. J. Biochem., 230: 827-853

Schein, S.J., Kagan, B.L., and Finkelstein, A. (1978). Colicin K acts by forming voltage-dependent channels in phospholipid bilayer membranes. Nature, 276: 159-163

Slatin, S., Qiu, X-Q., Jakes, K. and Finkelstein, A. (1994), Identification of a translocated protein segment in a voltage-dependent channel; Nature, 371: 158-161

Slatin, S.L. 1988. Colicin E1 in planar lipid bilayers. Int. J. Biochem. 20: 737-744

van der Goot, F.G., N. Didat, F. Pattus, W. Dowhan, and Letellier, L. 1993. Role of acidic lipids in the translocation and channel activity of colicins A and N in *Escherichia coli* cells. Eur. J. Biochem. 213: 217-221.

Viljanen, P., Koski, P. and Vaara, M. (1988), Effect of small cationic leukocyte peptides (defensins) on the permeability barrier of the outer membrane; Infection and immunity, 56: 2324-2329

Webster, R.E. (1991), The Tol gene products and the import of macromolecules into *E.coli*; Molecular Microbiol., 5: 1005-1011

Westerhoff, H., Juretic, D., Hendler, R.W., and Zasloff, M. (1989), Magainins and the disruption of membrane-linked free energy transduction. Proc. Natl. Acad. Sci. USA, 86: 6597-6601

White, S.H., Wimley, W.C. and Selsted, M.E. (1995), Structure, function, and membrane integration of defensins, Curr. Opinion Struct.Biol., 5: 521-527

Wiener, M., Freymann, D., Ghosh, P. and Stroud, R. (1997), Crystal structure of colicin Ia; Nature, 385: 461-464
Wimley, W.C., Selsted, M.E., and White, S.H. (1994), Interactions between human defensins and lipid bilayers: evidence for the formation of multimeric pores. Protein Sci., 3: 1362-1373.
Zasloff, M. (1987), Magainins, a class of antibacterial peptides from Xenopus skin: isolation, characterization of two active forms, and partial cDNA sequence of precursor; Proc. Natl. Acad. Sci., 84: 5449-5453

Potassium Channels in *E. Coli*

M J MACLEAN, L S NESS, S MILLER & I R BOOTH

Department of Molecular and Cell Biology, Institute of Medical Sciences, University of Aberdeen, Foresterhill, Aberdeen, AB25 2ZD, SCOTLAND.

1 Transport in bacteria

Bacterial cells are surrounded by a cell envelope, which keeps the cell distinct from its environment. This envelope consists of a cytoplasmic membrane, a peptidoglycan layer and a cell capsule. In Gram negative bacteria there is also an outer membrane and periplasm, because the peptidoglycan layer is much thinner than in Gram positive bacteria. Eubacteria cell membranes are composed of phospholipids and proteins. Any solutes wanting to enter or leave the cell have to cross this lipid barrier, therefore, several different transport mechanisms have evolved to facilitate movement across the cytoplasmic membrane and the outer membrane. Hydrophilic molecules enter and exit the cell via protein carriers embedded in the membrane, either by active transport or passive transport.

1.1 Active transport

Active transport is energy dependent and is divided into 3 different classes.

(i) PRIMARY TRANSPORT, which uses energy derived from the direct hydrolysis of a high energy chemical bond, usually ATP, eg. the Kdp-ATPase

(ii) SECONDARY TRANSPORT, which uses the electrochemical gradient of ions to drive transport of solutes.

Uniports transport a solute down its own electrochemical gradient, eg. ammonium ions.

Symports transport 2 substrates simultaneously in the same direction. An ion is translocated down its concentration gradient and this is the driving force for co-transport of the other molecule, eg. the lactose-proton symport.

Antiports transport the solute and coupling ion in opposite directions using an electrochemical gradient, eg. the sodium-proton antiport.

(iii) GROUP TRANSLOCATION is a process whereby the solute is chemically modified during transport. Therefore, no transmembrane gradient of the solute is established, eg. transport of sugars by the phosphoenol-pyruvate (PEP)-dependent transport system.

NATO ASI Series, Vol. H 106
Lipid and Protein Traffic
Pathways and Molecular Mechanisms
Edited by Jos A. F. Op den Kamp

1.2 Passive transport

Passive transport systems do not result in solute accumulation inside the cell, but instead equilibrate the solute across the membrane. Therefore, energy is not used to transport solutes. Passive transport can be divided into simple diffusion and facilitated diffusion.

(i) Simple diffusion does not require specific transport proteins. Small uncharged solutes and gases can pass through lipid bilayers by simple diffusion down their concentration gradient, eg. water, carbon dioxide.

(ii) Facilitated diffusion occurs via a transport protein embedded in the membrane, ie a pore or a channel.

1.2.1 Membrane channels in bacteria. Several types of channels have been characterised in bacteria:

- Porins
- Mechanosensitive
- Glutathione-gated
- Inwardly rectified potassium
- Shaker-like

Channels are different from other transporters in that the stoichiometry of solutes transported per conformational cycle of the channel is not fixed. The binding of ions to a channel is much weaker and, therefore, when the channel is open a large number of ions can move through. Most work, to date, has been done on eucaryotic channels. Patch clamp technology was originally developed to analyse ion channels in eukaryotic membranes, but the method has now been applied to bacterial cells.

Gram negative bacteria have many porin channels in the outer membrane (10^5 per cell), which can allow molecules up to 600 Da to cross in and out of the periplasm. Porins are not very selective for the solutes they transport. Other non-selective channels are the mechanosensitive channels in the inner membrane. They open in response to high turgor and allow the release of solutes, to relieve pressure.

Our lab is mainly concerned with ligand-gated potassium channels.

2 Potassium channels

Potassium is the major intracellular cation in the growing bacterial cell. It has several important roles within the cell, including regulation of cell turgor, pH homeostasis and activation of many cytoplasmic enzymes. Several different potassium uptake and efflux systems have been discovered in bacteria, of which the *E. coli* systems are the best studied.

2.1 Potassium uptake

At least 5 potassium uptake systems exist in *E. coli*. Kdp is an inducible system with a high affinity for potassium ions. ATP hydrolysis drives the uptake of potassium into the cell. TrkG and TrkH are constitutive, low affinity potassium uptake systems. They transport potassium at a high rate in response to hyperosmotic shock. Kup (TrkD) is constitutive, with a low affinity for potassium. Kup may be involved in potassium uptake in low osmolarity medium or when the external pH is low. TrkF is a low rate, non-saturable potassium uptake system of which little else is understood.

2.2 Potassium efflux

Potassium is lost from *E. coli* cells by addition of thiol reagents, high turgor, alkalinisation of the cytoplasm and addition of 2,4-dinitrophenol (DNP). KefA is thought to be the efflux system activated during increased cell turgor. KefB and KefC are responsible for potassium efflux after the addition of electrophiles, and will be discussed later. KefD is a homologue of KefA. Kch is a putative potassium channel, based on homology to the eukaryotic shaker channel. Not much is known about the function of Kch. DNP can elicit potassium efflux fom *E. coli* but the route of efflux is unknown. Thus another efflux system, Kef X, possibly exists, but as yet is unidentified.

2.3 Channel or antiporter?

The presence of channels in the plasma membrane of bacteria has been disputed by those who believe that the opening of ion channels would dissipate the proton motive force. Thus KefC was originally assumed to be an antiporter because potassium efflux is accompanied by an influx of protons and sodium ions and the N terminal membrane spanning domain has homology to sodium-proton antiporters in other bacteria. KefC is now thought to be a channel for several reasons:

- KefB and KefC are oligmeric
- Ligand-gated regulation by GSH and GSH adducts
- Sequence in the N terminal region of KefC is homologous to the voltage gate region in eukaryotic potassium channels; KefC residues 200-216 [RXXX]5, followed by leucine residues at every 7th interval.
- Interaction with a regulatory protein in the cytoplasm (YabF)
- Preliminary patch clamp data

Thus the channel status of the potassium efflux systems KefB and KefC is now fairly well established.

Kyte and Doolittle Hydropathicity Plots

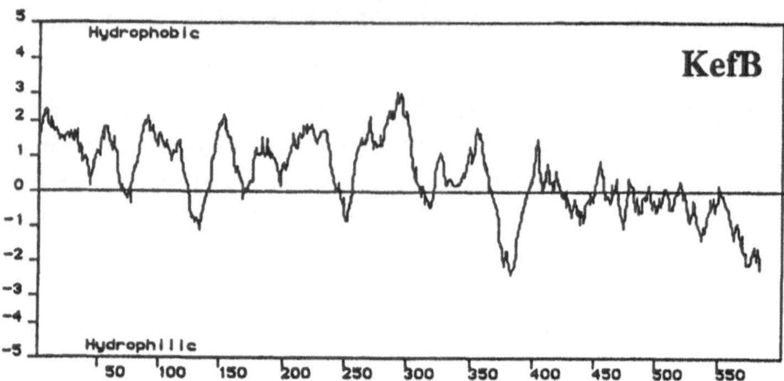

KefB and KefC have 45 % identity and 70 % similarity overall, similarity breaksdown in the C terminus

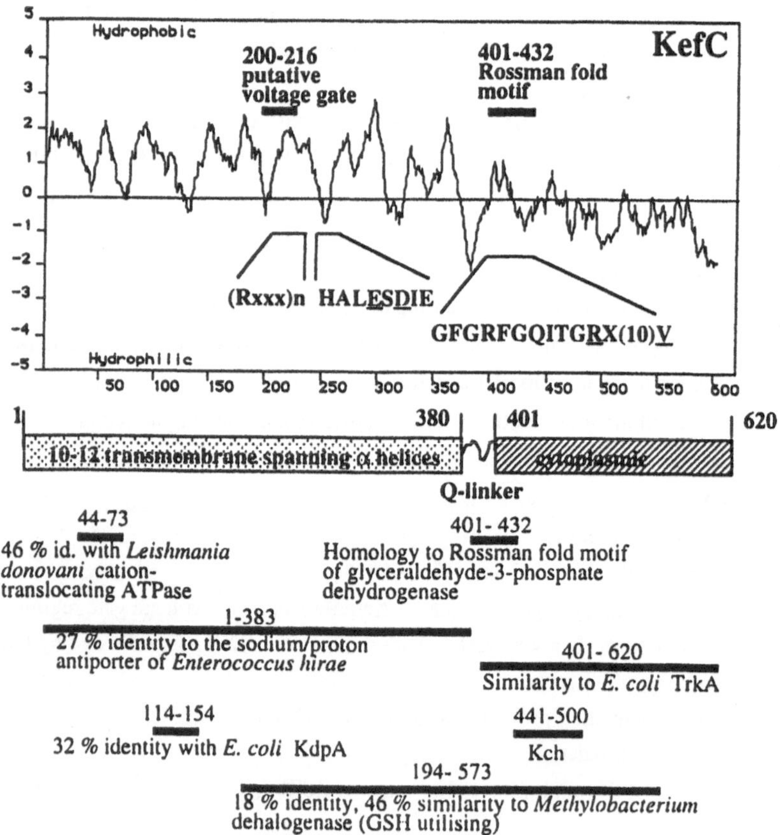

Figure 1: Hydropathy plots of KefB and KefC. Positive scores represent hydrophobic regions of the protein, and correspond with putative transmembrane spans. The membrane domain and cytoplasmic domains are separated by a Q linker, which is very hydrophilic and allows flexibility between domains. Regions of KefC that show homology to other proteins is shown below.

3 KefB and KefC

During analysis of potassium transport mutants the *kef*B and *kef*C loci were identified. Mutations at these loci evoked rapid potassium loss from the cells. These genes have since been cloned and mutated, and a physiological role assigned. KefB and KefC are highly homologous proteins, and can exist in 4 states: closed, open, partially activated or fully activated. The tripeptide glutathione, or its analogue ophthalmic acid, can keep the channels inactive (closed conformation). In the absence of glutathione KefB and KefC open and there is a small amount of potassium loss. Electrophilic compounds, such as methylglyoxal or N-ethylmaleimide (NEM), can directly activate the channels into opening with the subsequent loss of potassium. Direct activation by NEM can also be observed in glutathione deficient cells (eg. *gsh*A mutants). Direct activation by an electrophile only partially opens the channel but the precise mechanism of this activation is not understood. In *E. coli* when an electrophile is synthesised or enters the cell it will spontaneously conjugate with glutathione, through the sulphydryl group, to form a glutathione adduct. These glutathione adducts can completely activate KefB and KefC and elicit the maximum amount of potassium efflux.

These channels are present at very low levels in the membrane. The *kef*B and *kef*C genes use a high frequency of rare codons and the ribosome binding site for KefB and KefC is in the middle of a very stable stem-loop structure in the mRNA, this may account for the low level of translation.

3.1 KefB

The *kef*B gene is at 73-74 minutes on the *E. coli* chromosome and encodes KefB (TrkB), a protein of 601 amino acids. The N terminus of KefB was found to be very hydrophobic and is presumed to contain 6-12 transmembrane spans. The C terminus has a mix of hydrophilic and hydrophobic residues, believed to form a cytoplasmic domain. A Q linker separates the two domains. Strains with mutations in *kef*B will spontaneously leak potassium. Transformation of a leaky strain with wild type *kef*B will partially suppress the leak. This suggests that KefB forms a multimeric channel in the cytoplasmic membrane.

3.2 KefC

KefC (TrkC) is a 620 amino acid protein with a predicted molecular mass of 70 kDa, encoded by the *kef* C gene. The *kef* C gene lies at 1 minute on the *E. coli* chromosome. The amino acid sequence shows KefC to be composed of 2 domains, a hydrophobic N terminus and a cytoplasmic C terminus, joined together by a Q linker sequence. As for KefB, there is genetic evidence that KefC is an oligomer: cloned KefC suppresses the phenotype of potassium leak in

KefC mutants. Suppression is specific to KefC, with no cross over between KefB and KefC, suggesting the channels are **homo**oligomers.

3.3 YhaH and YabF

Upstream of *kef*B is a 555 bp open reading frame (ORF), which overlaps *kef*B by 1 bp. This ORF encodes a 184 amino acid protein which is called YhaH, and is required for full KefB activity, but not for KefC activity. Upstream of *kef*C, and overlapping it by 8 bp, is a 528 bp ORF, *yab*F, which codes for a 176 amino acid protein YabF. YabF is required for full KefC activity. YabF and YhaH show sequence similarity to quinone oxidoreductases.

KefB and KefC can be activated by electrophiles in the absence of YhaH and YabF, but activity is greatly reduced. In the presence of YabF, YhaH slightly inhibits KefC activity, suggesting that YhaH is interacting somehow with either YabF or KefC. YhaH and YabF show 58 % sequence similarity, and appear to be standard cytoplasmic soluble proteins. These regulatory subunits increase channel activity but are not absolutely essential for channel activation. This extra level of regulation reflects the complex nature of the control mechanisms required by KefB and KefC.

3.4 Regulation of KefB and KefC

Analysis of *kef*C mutants that spontaneously leak potassium has resulted in the identification of two regions in KefC that are involved in regulation of potassium efflux; the HALESDIE motif and the Rossman fold. Missense mutations in either of these regions causes a potassium leak.

The HALESDIE motif lies adjacent to the voltage sensor and is predicted to form an acidic patch between two transmembrane spans of the amino terminal region at the cytoplasmic face of the membrane. Strains containing cloned KefB (copy number of ~30) will spontaneously leak 40 % potassium, but cells with a cloned KefC are potassium tight. KefB has a HELETAID motif in place of a HALESDIE motif and site directed mutagenesis is being used to ascertain the role of this motif in potassium leak. Mutagenesis of A262D in KefB only decreases the spontaneous leak to 35 %, and decreases the **rate** of efflux elicited by electrophiles. Therefore the leak seen with KefB is not due to this residue. The HALESDIE/HELETAID motifs are presently being mutated to discover what their role is in control of channel opening. The differences in these motifs may account for the differences in kinetics between the channels.

Mutations in the Rossman fold motif result in an alteration of normal regulation by glutathione. Mutants R416S/S420A, where the conformation of the Rossman fold is incorrect, can leak potassium in the presence of glutathione. Thus the carboxy-terminal domain of KefC may play a major role in regulating activity and may be the location of the glutathione-binding site.

Mutagenesis of the voltage gate region of KefC has been performed at R212A (KefB has A at this site) and a decrease in the rate and extent of NEM elicited efflux was observed. KefB is more poorly activated by NEM than KefC, and in the absence of YhaH NEM cannot elicit **any** potassium efflux from KefB (in a delta YhaH strain YabF cannot compensate). This mutation (R212A) in KefC has resulted in a potassium efflux profile more similar to that seen in KefB.

3.5 Physiological role for KefB and KefC

These glutathione metabolite-gated transport systems are widespread among Gram-negative bacteria. It has been shown in our laboratory that possession of KefB and KefC aids survival of cells exposed to various electrophiles. During detoxification of electrophilic compounds formation of glutathione metabolites is often the first step. These glutathione adducts are required to activate the channels and elicit potassium efflux. In the absence of glutathione the electrophilic compounds cannot conjugate to form glutathione adducts and channel activation is drastically decreased. Thus glutathione deficient cells cannot fully protect themselves against electrophiles even in the presence of KefB and KefC.

Potassium efflux is accompanied by an equivalent influx of protons, to balance the charge within the cell. This increase in protons serves to decrease the internal pH. It is the acidification of the cytoplasm that protects the cells against electrophilic damage. Artificially lowering the internal pH with weak acids will afford protection against NEM and methylglyoxal. Thus survival after exposure to electrophiles relies on cytoplasmic acidification and not the potassium loss. How this decrease in internal pH aids survival is presently unknown. A model put forward proposes that the lowering of pH may activate macromolecule repair enzymes or decrease the rate at which electrophiles can cause damage within the cell. This model is under investigation within our laboratory.

4 Conclusion

KefB and KefC are ligand-gated potassium channels embedded in the inner membrane of *E. coli*. Glutathione is required to keep the channels inactive. Glutathione adducts can completely activate the channels and elicit maximal potassium efflux. Direct activation by the electrophile will only partially open KefB and KefC. Cytoplasmic proteins YabF and YhaH are required for regulation of channel activity.

The proton influx that accompanies potassium efflux is required for protection against electrophiles. How this acidification of the cytoplasm aids survival is unknown.

THIS WORK HAS BEEN SUPPORTED BY THE WELLCOME TRUST

A Link Between Fatty Acid Synthesis and Organelle Structure in Yeast

M. Lampl, S. Eder, R. Schneiter and S.D. Kohlwein

SFB Biomembrane Research Center
Institut für Biochemie und Lebensmittelchemie
Technische Universität Graz
Petersgasse 12/II
8010 Graz, Austria

1 Background

Fatty acids are essential components of all cellular membranes. They represent an efficient storage form for metabolic energy and serve structural functions when esterified in membrane-forming phospholipids. Furthermore, specific fatty acids act as second messengers in signal transduction pathways and membrane-association of proteins can be modulated by post-translational acyl-modification (for reviews see Kohlwein et al., 1996; Bhatnagar and Gordon, 1997).

Biosynthesis and assembly of membrane components is of central interest to our understanding of the cell. How the lipid moiety of the membrane, mainly consisting of phospholipids and sterols, is synthesized and assembled, is only poorly understood. Furthermore, membrane properties are strongly influenced and dependent on the nature of the fatty acids esterified in phospholipids. Considerable progress has been made in elucidating the regulatory mechanisms that control expression of genes involved in lipid biosynthesis. However, the precise role of specific lipid molecular species remains to be established.

Yeast contains a number of specialized subcellular structures, including mitochondria, the endoplasmic reticulum, vacuoles, secretory vesicles, nuclear and plasma membranes and, under certain growth conditions, peroxisomes. All of these membranes contain a similar spectrum of phospholipids, yet their relative distribution varies considerably (Zinser et al., 1991). Mitochondria for instance, contain a specific lipid, cardiolipin, which accounts for up to 15% of the lipids of the inner mitochondrial membrane. Cardiolipin contains relatively high amounts of unsaturated fatty acids, mainly palmitoleic acid ($C_{16:1}$) and oleic acid ($C_{18:1}$). The physiological

NATO ASI Series, Vol. H 106
Lipid and Protein Traffic
Pathways and Molecular Mechanisms
Edited by Jos A. F. Op den Kamp

relevance of a predominant occurance of unsaturated fatty acids esterified in the sn-2 position of phospholipids is not clear at present (Wagner and Paltauf, 1994).

In yeast, the spectrum of fatty acids is rather simple consisting predominantly of C_{16} and C_{18} fatty acids, both saturated and cis-delta^9-unsaturated. The relative distribution of the different fatty acids esterified in phospho- and neutral lipids (ergosteryl esters, triacylglycerols) has not yet been analyzed in detail. Myristic acid (C_{14}) is only a minor component of cellular lipids, as are very-long-chain fatty acids, C_{26} (1-2% of total fatty acids). Myristic acid may not play a significant role in membrane lipid architecture, but is essential for protein modification (Bhatnagar and Gordon, 1997). Very-long-chain fatty acids, on the other hand, are an important structural component of sphingolipids (Wells and Lester, 1983) and GPI-anchored proteins.

The general pathway of fatty acid uptake, biosynthesis and degradation in yeast is outlined in Figure 1. The synthesis of saturated long-chain fatty acids (LCFAs; C_{14}-C_{18}) is central to their subsequent desaturation, to their elongation to very-long-chain fatty acids (VLCFAs; >C_{22}), to peroxisomal degradation through beta-oxidation, and to their function in protein acylation. The biosynthesis of LCFAs requires the concerted action of four different enzyme systems: acetyl-CoA carboxylase (Acc1p), fatty acid synthetase (Fas1p/Fas2p), elongase (Elo1p), and fatty acid desaturase (Ole1p). The rate limiting step of the de novo synthesis of LCFA is under the control of the first of these, acetyl-CoA carboxylase. This biotinylated enzyme catalyzes the ATP-dependent carboxylation of acetyl-CoA to malonyl-CoA, which then serves as the two-carbon-unit donor for acyl chain elongation by fatty acid synthetase and elongase. Palmitoyl- and stearoyl-CoA are the end products of this reaction. Fatty acid uptake and activation requires at least four acyl-CoA synthetases (Faa1p-Faa4p). Transport of LCFAs across the peroxisomal membrane depends on two peroxisomal ATP-binding cassette transporters, Pat1p and Pat2p, while peroxisomal beta-oxidation of medium chain fatty acids depends on their activation inside peroxisomes by Faa2p. Myristoylation of membrane-associated proteins is catalyzed by Nmt1p.

Finally, little is known about how fatty acids of different chain length and degree of unsaturation contribute to membrane structure and function. Studies on two conditional yeast mutants that affect key steps of fatty acid synthesis revealed that impaired fatty acid biosynthesis is associated with severe morphological defects. Characteristically, these mutants were isolated in screens unrelated to lipid metabolism (for review see Schneiter and Kohlwein, 1997). A conditional mutant defective in the inital reaction of fatty acid synthesis, acetyl-CoA carboxylase (*mtr7*), exhibits a specific morphological defect of the nuclear envelope, resulting in a disturbed mRNA export from the nucleus. Secondly, a yeast mutant impaired in fatty acid desaturation, *mdm2*, shows a defect in mitochondrial morphology and is unable to transmit mitochondria into the daughter cell upon cell divison. In this report we present experimental strategies aimed to unveil the role of unsaturated and very-long-chain fatty acids in membrane assembly and function in the yeast *Saccharomyces cerevisiae*.

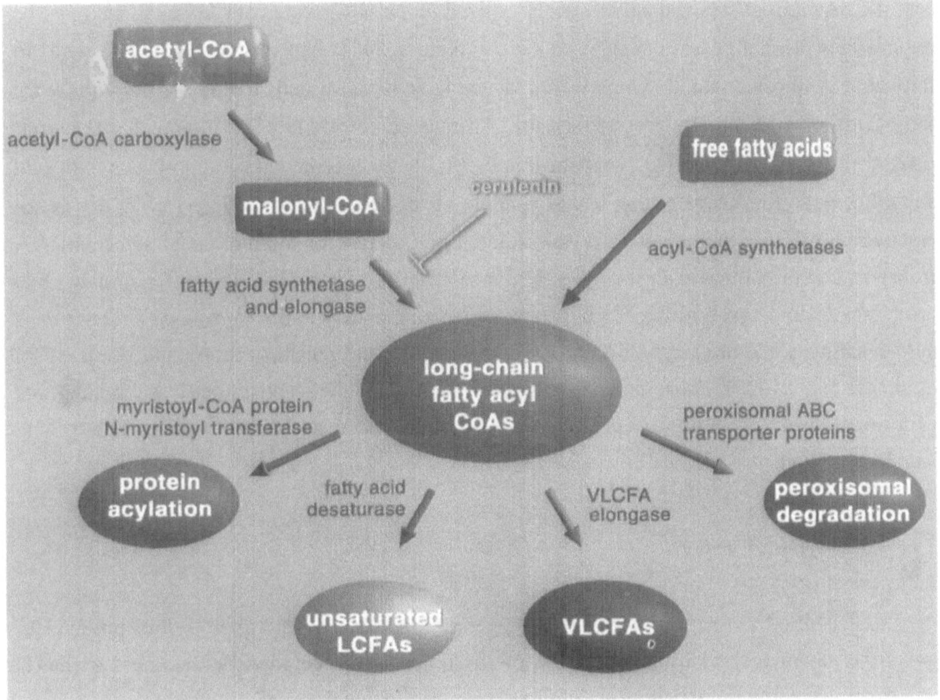

Figure 1. Pathways of fatty acid synthesis and degradation in *Saccharomyces cerevisiae*. For details see text.

2 Acetyl-CoA carboxylase and nuclear function

2.1 Conditional, non-fatty acid auxotrophic allels of *acc1*

Studies on acetyl-CoA carboxylase, the initial and rate limiting step of fatty acid biosynthesis, unveiled that mutants lacking this enzyme are inviable, even in the presence of exogenously supplied long-chain fatty acids (Hasslacher et al., 1993). Thus, malonyl-CoA synthesized by acetyl-CoA carboxylase and subsequently utilized in the fatty acid synthase-reactions must be required for other processes besides *de novo* synthesis. This observation stands in contrast to fatty acid synthesis mutants which were isolated in screens for cells that can grow only when fatty acids are added to the growth medium. All mutants selected by this approach, *ole1*, *fas1*, *fas2*, *acc1* and *acc2* (defective in apoenzyme-biotin ligase) are rescued by appropriate fatty acid supplementation. A temperature-sensitive allele of acetyl-CoA carboxylase (*mtr7*) that is not rescued by LCFA supplementation was isolated in a screen for mutants defective in *m*RNA *tr*ansport from the nucleus to the cytosol (Schneiter et al., 1995). In a screen for mutants that are synthetically lethal with *hpr1*, a hyper-recombination mutant of *Saccharomyces cerevisiae*, a cold-sensitive allele of acetyl-CoA carboxylase ($acc1^{cs}$) was isolated (Guerra and Klein, 1995; Schneiter et al., 1997). Thus, together with *mtr7*, the $acc1^{cs}$ allele constitutes a second member of a novel class of conditional *ACC1* alleles that are not rescued by long-chain fatty acid supplementation.

2.2 Phenotype of *mtr7*

mtr7 cells shifted to the non-permissive temperature reveal a severe alteration of the nuclear envelope, characterized by a separation of the inner and outer nuclear membranes that is accompanied by the formation of vesicle-like structures within the newly formed intermembrane space (Figure 2). At the same time, accumulation of poly(A)$^+$ RNA is observed in a ring-like structure at the nuclear perimeter. Neither nuclear pore complex mutants nor the fatty acid auxotrophic acetyl-CoA carboxylase mutants display this striking phenotype (Schneiter et al., 1996). Similar to *mtr7*, the cs-allele affects nuclear export of mRNA and hence displays a weak mtr-phenotype. These findings suggest that the nonsupplementable function of acetyl-CoA carboxylase is directly or indirectly related to the structure and function of the nuclear envelope/nuclear pore complex.

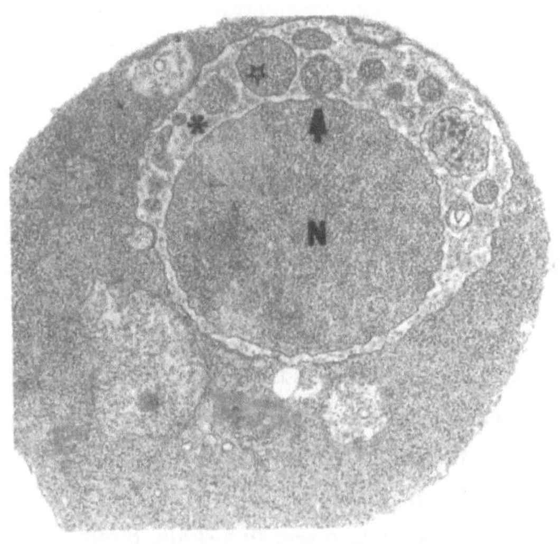

Figure 2. Morphological phenotype of *mtr7*. Separation of inner and outer nuclear membranes (asterisk) and formation of vesicle-like structures (open star) in the inter-membrane space are shown. The arrow indicates a nuclear pore complex.

2.3 *mtr7* and *acc1^cs^* affect fatty acid chain length distribution

Since fatty acid auxotrophic acetyl-CoA carboxylase mutants can elongate short chain fatty acids (C_{12}-C_{14}) to LCFA, it was proposed that in yeast, unlike in other organisms (Cinti et al., 1992), malonyl-CoA is not required for acyl chain elongation (Roggenkamp et al., 1980). This view has been challenged by the observed alteration in the fatty acid composition of wild type, *mtr7* and *acc1^{cs}* mutants shifted to non-permissive conditions. In comparison to wild type, *mtr7* and *acc1^{cs}* accumulate a significant amount of myristic acid (C_{14}) and palmitoleic acid ($C_{16:1}$) even at permissive conditions. At non-permissive conditions C_{26} drops by approximatly 40%, i.e. to about 30% of wild-type level in both mutants. Fatty acid chain length distribution is thus drastically impaired in these conditional *acc1* alleles, indicating that malonyl-CoA is required for acyl chain elongation. The degree of desaturation is not affected in the mutants. In mammalian cells a microsomal, malonyl-CoA-dependent chain elongation activity has been identified. Its protein components, however, resisted all purification and characterization attempts (Cinti et al., 1992).

2.4 *mtr7* suppressor mutants

To identify components required for malonyl-CoA dependent chain-elongation we initiated a genetic approach to isolate mutants that interact with *mtr7*. 14 mutants that suppress the temperature-sensitive phenotype of *mtr7* were isolated. 4 of these displayed a cold-sensitive phenotype. A wild-type gene complementing the cold-sensitivity was cloned. Identical inserts containing 5 candidate open reading frames were recovered from two different suppressor-strains. The fatty acid pattern of the suppressor-mutants displayed no restoration of the wild-type fatty acid distribution, suggesting that suppression of the *mtr7* phenotype does not require the reversion to a wild-type fatty acid composition.

2.5 Fate of Acc1p in *mtr7* and *acc1^{cs}* mutants

Western blot analysis, enzyme activity measurements and immunofluorescence microscopy were performed to investigate the stability and subcellular distribution of Acc1p in *mtr7* and *acc1^{cs}* mutants. The cold-sensitive mutant shows no visibly change in the amount of Acc1p after a 4h shift to the non-permissive temperature (17°C). The abundance of the Mtr7p protein, on the other hand, decreases to undetectable levels within the same time period when cells are shifted to 37°C (Figure 3). Immunofluorescence microscopic analysis revealed similar results. No visible alteration in the subcellular distribution of Acc1p was observed in *acc1^{cs}* cells shifted to the non-permissive temperature. In the temperature-sensitive strain, on the other hand, the intensity of the fluorescence signal was greatly reduced after shifting to non-permissive conditions. Enzyme activity of acetyl-CoA carboxylase measured in extracts from the two conditional *acc1* mutants is reduced approximatly 6 fold under permissive conditions and drops to undedectable levels after a shift to non-permissive temperature. Fatty acid synthase activity, on the other hand, is wild type and not affected by a temperature-shift.

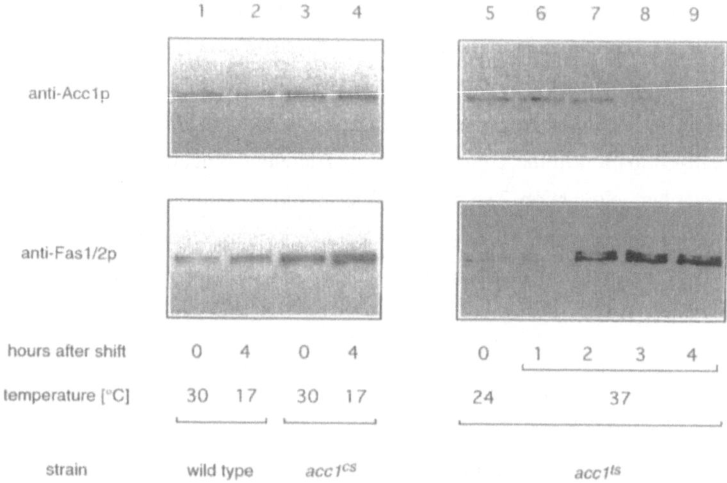

Figure 3. Western blot analysis to detect Acc1p and Fas1/2p in wild type (lanes 1 and 2), *acc1^{cs}* (lanes 3 and 4) and *mtr7* (lanes 5-9) mutant strains at the indicated permissive and non-permissive temperatures.

2.6 Very-long-chain fatty acids and the nuclear envelope

In yeast, hexacosanoic acid (C_{26}) is by far the most abundant VLCFA and comprises about 1-2% of the total fatty acid fraction (Welch and Burlingame, 1973). VLCFAs are essential since they form a structurally important part of the ceramide moiety of sphingolipids and the lipid domain of glycosylphosphatidylinositol (GPI)-anchored proteins. Mutants defective in sphingolipid synthesis were isolated in a screen for *l*ong-*c*hain *b*ase auxotrophs (i.e. *lcb1* and *lcb2*, encoding serine palmitoyl transferase). These mutants die if deprived of the long chain base, indicating that sphingolipid/GPI synthesis is essential. However, strains which suppress the long chain base requirement are able to grow without synthesizing sphingolipids. These mutants produce novel C_{26} fatty acid-substituted inositol glycerophospholipids that structurally mimic sphingolipids, an observation that emphasizes the important structural role of the very long acyl chain in sphingolipid function (Lester et al., 1993). Thus, not only the hydrophilic part of the lipid is essential but also the very long-chain fatty acid associated with it. Since sphingolipids follow the secretory pathway and are ultimately destined to the plasma membrane it is difficult to imagine how a defect in sphingolipid synthesis could affect the nuclear envelope.

2.7 Working model

Based on biochemical and morphological evidence we propose a direct involvement of VLCFA-substituted lipids in the interaction between the nuclear pore complex and the nuclear envelope. Our working model postulates a structural function of these lipids in stabilizing the 180° turn of the nuclear membrane at the interface between membrane and pore complex. This model is based on the observation that the surface of the lipid headgroup which is exposed to the aqueous phase relative to the volume and length of the hydrocarbon portion determine the molecular shape of the membrane formed (Israelachvili et al., 1980). Further support for this model is given by biophysical studies on artifical membrane systems which indicate that very-long-chain fatty acids stabilize highly curved membrane structures by occupying free volume in the hydrophobic core of the bilayer. If supply of VLCFA becomes limiting, as is observed in *mtr7*, one might predict that the membrane bend becomes more and more unstable, eventually breaks apart, thereby separating the two nuclear membranes, as seen in *mtr7* (Schneiter et al., 1996).

3 Unsaturated fatty acids and mitochondrial inheritance

In a screen to isolate mutants defective in mitochondrial morphology and distribution to the daughter cell upon cell divison, a temperature sensitive mutant, *mdm2* (*m*itochondrial *d*istribution and *m*orphology), was isolated (Stewart and Yaffe, 1991). Under restrictive conditions, enlarged mitochondria accumulate in the mother cell. Although cells are still able to bud, transmission of mitochondria to the bud is impaired, resulting in a growth arrest. Vacuoles are faithfully transmitted to developing buds; secretion and nuclear division also proceed normally.

mdm2 mutants are defective in the only fatty acid desaturase (a delta^9-desaturase) in yeast encoded by *OLE1* (Stuckey et al., 1989). When *mdm2* cells are shifted to the nonpermissive temperature, the levels of unsaturated fatty acids ($C_{16:1}$ and $C_{18:1}$) decline about 2.5-fold and the levels of precursors ($C_{16:0}$ and $C_{18:0}$ fatty acids) increase. Addition of unsaturated fatty acids, palmitoleic acid and oleic acid, to the growth medium complements the temperature-sensitive growth and mitochondrial distribution defect of *mdm2*. Thus, unsaturated fatty acids are required for maintenance of mitochondrial structure and distribution of mitochondria to the daughter cell. While the unsaturated fatty acids seem to play an essential role in a yet undefined, probably membrane-associated process, the molecular details and control of this mitochondrial partitioning, however, have yet to be described.

We have started to investigate morphological alterations of mitochondria in conditional mutants lacking fatty acid desaturase, using high resolution light microscopy of living cells. Mitochondria were labelled using potential sensitive fluorescent dyes and by introducing a GFP (green fluorescent protein)-tagged mitochondrial protein. Time lapse studies suggest that the dynamic equilibrium between the mitochondrial reticular structure and vesiculation is impaired in the mutants, resulting in disassembly of mitochondria in the mother cell.

References

Bhatnagar, R. S., and Gordon, J. I. (1997). Understanding covalent modifications of proteins by lipids: Where cell biology and biophysics mingle. Tr. Cell Biol. *7*, 14-20.

Cinti, D.L., Cook, L., Nagi, M. N., and Suneja, S.K. (1992). The fatty acid chain elongation system of mammalian endoplasmic reticulum. Prog. Lipid Res. *31*, 1-51.

Guerra, C. E., and Klein, H. L. (1995). Mapping of the *ACC1/FAS3* gene to the right arm of chromosome XIV of *Saccharomyces cerevisiae*. Yeast *11*, 697-700.

Hasslacher, M., Ivessa, A. S., Paltauf, F., and Kohlwein, S. D. (1993). Acetyl-CoA carboxylase from yeast is an essential enzyme and is regulated by factors that control phospholipid metabolism. J. Biol. Chem. *268*, 10946-10952.

Israelachvili, J. N., Marcelja, S., and Horn, R. G. (1980). Physical principles of membrane organization. Q. Rev. Biophys. *13*, 121-200.

Kohlwein, S. D., Daum, G., Schneiter, R., and Paltauf, F. (1996). Phospholipids: synthesis, sorting, subcellular traffic - the yeast approach. Tr. Cell Biol. *6*, 260-266.

Lester, R. L., Wells, G. B., Oxford, G., and Dickson, R. C. (1993). Mutant strains of *Saccharomyces cerevisiae* lacking sphingolipids synthesize novel inositol glycerolipids that mimic sphingolipid structure. J. Biol. Chem. *268*, 845-856.

Roggenkamp, R., Numa, S., and Schweizer, E. (1980). Fatty acid-requiring mutant of *Saccharomyces cerevisiae* defective in acetyl-CoA carboxylase. Proc. Natl. Acad. Sci. USA *77*, 1814-1817.

Schneiter, R., and Kohlwein, S. D. (1997). Organelle structure, function and inheritance in yeast: a role for fatty acid synthesis? Cell *88*, 431-434.

Schneiter, R., Hitomi, M., Ivessa, A. S., Fasch, E. V., Kohlwein, S. D., and Tartakoff, A. M. (1996). A yeast acetyl coenzymeA carboxylase mutant links very-long-chain fatty acid synthesis to the structure and function of the nuclear membrane-pore complex. Mol. Cell. Biol. *16*, 7161-7172.

Schneiter, R., Guerra, C.E., Lampl, M., Kohlwein, S.D., and Klein, H.L. (1997). Characterization of a novel cold-sensitive allele of acetyl-CoA carboxylase ($acc1^{cs}$) that displays intragenic complementation and is lethal with the *hpr1* mutant of *Saccharomyces cerevisiae*. Submitted.

Stewart, L. C., and Yaffe, M. P. (1991). A role for unsaturated fatty acids in mitochondrial movement and inheritance. J. Cell Biol. *115*, 1249-1257.

Stuckey, J. E., McDonough, V. M., and Martin, C. E. (1989). Isolation and characterization of *OLE1*, a gene affecting fatty acid desaturation from *Saccharomyces cerevisiae*. J. Biol. Chem. *264*, 16537-16544.

Wagner, S., and Paltauf, F. (1994). Generation of glycerophospholipid molecular species in the yeast *Saccharomyces cerevisiae*. Fatty acid pattern of phospholipid classes and selective acyl turnover at sn-1 and sn-2 positions. Yeast *10*, 1429-1437.

Welch, J. W., and Burlingame, A. L. (1973). Very long-chain fatty acids in yeast. J. Bacteriol. *115*, 464-466.

Zinser, E., Sperka-Gottlieb, C. D. M., Fasch, E.-V., Kohlwein, S. D., Paltauf, F., and Daum, G. (1991). Phospholipid synthesis and lipid composition of subcellular membranes in the unicellular eukaryote *Saccharomyces cerevisiae*. J. Bacteriol. *173*, 2026-2034.

Acknowledgements

This work was supported by the Fonds zur Förderung der wissenschaftlichen Forschung in Österreich (project 11731 to S.D.K.) and by the Swiss National Science Foundation (823A-046702 to R.S.). We would like to thank Michael Yaffe for providing *mdm2* mutants and Rob Jensen for providing GFP-constructs.

Lipid Translocation from the Cytosolic Leaflet of the Plasma Membrane to the Cell Surface by MultidrugTransporters

Gerrit van Meer[1*], René Raggers[1], Hein Sprong[1] and Ardy van Helvoort[2]
[1] Laboratory of Cell Biology and Histology, Academic Medical Center, University of Amsterdam, P.O. Box 22700, 1100 DE Amsterdam, The Netherlands
[2] The Netherlands Cancer Institute, Plesmanlaan 121, 1066 CX Amsterdam, The Netherlands

Keywords: sphingomyelin, glucosylceramide, glycosphingolipid, MDR P-glycoprotein, flippase

1 Lipids on the cell surface

Many molecules that are exposed on the outside of the cell are involved in recognition and signaling events. This is true for receptor proteins but also for various classes of membrane lipids like the phospholipid phosphatidylserine and various glycosphingolipids. To obtain a better understanding of the function of lipids in these processes we and others have studied the general mechanisms by which lipid molecules reach the cell surface, i.e. the outer leaflet of the plasma membrane. Based on the pioneering work by Pagano and colleagues (Pagano et al., 1983; Lipsky and Pagano, 1985) we have extensively used short-chain analogs of membrane lipids. More specifically, we have added to the cells a ceramide or a diacylglycerol with a C_6-NBD-fluorescent fatty acyl chain at the C2 position of the sphingosine and at the *sn*2 position of the diacylglycerol (Figure 1). The general idea is that these precursors will be converted to (short-chain) membrane lipids in the cell and will be transported to the plasma membrane. Because of the short-acyl chain these products will display a higher off-rate from membranes than the regular membrane lipids. Conveniently, as an assay for their appearance in the outer leaflet of the plasma membrane, they can be extracted from that surface by albumin in the bathing medium. Obviously, we hoped that the short-chain analogs would follow the same transport pathways as their natural counterparts. However, it has been clear from the start that any finding obtained by the use of short-chain lipids must finally be confirmed for the natural lipids, which generally possess two long acyl chains.

---

* This research was carried out in the Dept. of Cell Biology of Utrecht University, and was supported by the Netherlands Foundation for Chemical Research (SON) with financial aid from the Netherlands Organization for Scientific Research (NWO), by the Deutsche Forschungsgemeinschaft, and by E.C. contract B102-CT93-0348.

NATO ASI Series, Vol. H 106
Lipid and Protein Traffic
Pathways and Molecular Mechanisms
Edited by Jos A. F. Op den Kamp

2 Sphingomyelin transport to the cell surface

As a first subject of study we selected sphingomyelin (SM). The bulk of the sphingomyelin is synthesized in the lumen of the Golgi and we showed previously (Jeckel et al., 1992) that newly synthesized C_6-NBD-SM is sequestered in the Golgi lumen and is unable to translocate towards the cytosolic side. Because most of the cellular SM is exposed on the cell surface, it would seem that SM transport from the Golgi to the plasma membrane occurs exclusively on the inside of carrier vesicles (Figure 2). We decided to test this by interfering with vesicular traffic from the Golgi to the plasma membrane and seeing whether transport of newly synthesized C_6-NBD-SM was indeed inhibited. Addition of C_6-NBD-Cer to HepG2 cells or CHO cells resulted in synthesis of both C_6-NBD-SM and C_6-NBD-GlcCer. This shows that C_6-NBD-Cer after insertion into the plasma membrane was able to translocate to the cytosolic leaflet (most likely because of the absence of a polar headgroup), to cross the cytosol to the Golgi (because of its short acyl chain) and to translocate to the Golgi lumen where SM synthase is located. At 37°C C_6-NBD-SM became accessible for albumin extraction on the outside of the cell. Unexpectedly, when we inhibited vesicular traffic to the cell surface by adding the drug brefeldin A (BFA), transport of C_6-NBD-SM to the cell surface was unchanged (Table 1). To see whether this was true for native SM, we radiolabeled newly synthesized SM and measured transport by accessibility of newly synthesized SM to exogenously added bacterial SMase (where after a 37°C transport incubation the SMase was added in the cold). While under control conditions newly synthesized SM equilibrated with surface SM as shown by the fact that 60% of each was degradable by SMase after 1.5 h of transport, BFA nearly completely inhibited this equilibration. While native SM is transported by a BFA-sensitive vesicular pathway, C_6-NBD-SM appears to be transported by a BFA-insensitive pathway. Alternatively, BFA induced a new transport pathway for C_6-NBD-SM but not for native SM.

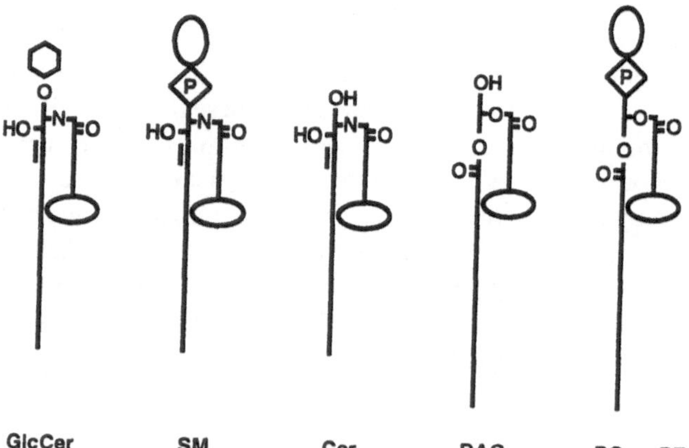

Figure 1. While the C_6-analog of diacylglycerol (DAG) is converted by the cell to triacylglycerol (not shown), phosphatidylcholine (PC) and phosphatidylethanolamine (PE), ceramide (Cer) is converted to the phosphosphingolipid sphingomyelin (SM) and the glycosphingolipid glucosylceramide (GlcCer).

A BFA-insensitive vesicular pathway from ER to the cell surface has been suggested to exist from studies on cholesterol transport (Urbani and Simoni, 1990). However, BFA is known to inhibit vesicle budding at the ER and also to result in fusion between the Golgi cisternae and the ER. Indeed, we could show that SM is now synthesized in the ER-Golgi fusion compartment (van Helvoort et al., 1997). So from the BFA-induced fusion compartment a vesicular route would persist that would exclude proteins (Klausner et al., 1992) and a natural lipid, SM, but would efficiently transport cholesterol and an analog of SM. We thought this level of lipid sorting unlikely, and, rather, considered an alternative possibility. Interestingly, it has been observed that short-chain SM can translocate across the ER membrane (Herrmann et al., 1990). After BFA treatment, C_6-NBD-SM would thus have access to the cytosolic surface of the ER and, because of its short chain, would be able to transfer as a monomer to the cytosolic leaflet of the plasma membrane. The only hypothetical step would be a translocation step across the plasma membrane (Figure 2, route 3).

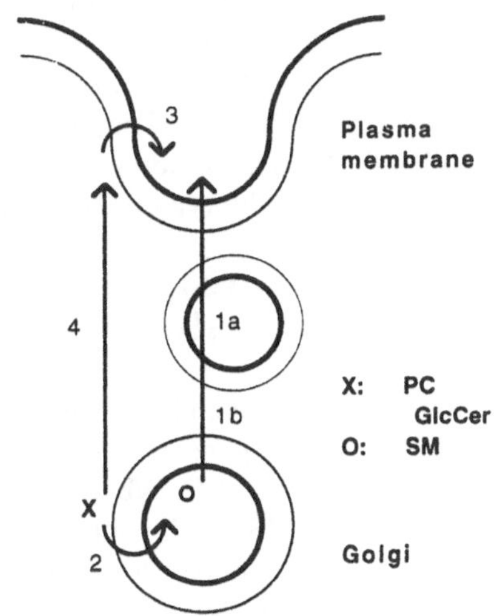

Figure 2. After synthesis in the lumenal leaflet of the Golgi SM follows the vesicular route to the cell surface (1). In contrast, newly synthesized GlcCer and PC must translocate across a membrane to reach the cell surface. Translocation could occur across the membrane of Golgi or ER (2), or across the plasma membrane (3). In the latter case GlcCer or PC can reach the cytoplasmic leaflet of the plasma membrane either on the cytosolic surface of transport vesicles (1b), or by a monomeric transfer process through the cytosol (4).

The only protein known to be able to translocate a phospholipid from the cytosolic leaflet across the plasma membrane is the MDR3 P-glycoprotein, which is present in the bile canalicular surface of the hepatocyte and is required for the presence of the phosphatidylcholine in the bile (Smit et al., 1993; Ruetz and Gros, 1994; Smith et al., 1994). Since also the multidrug transporter MDR1 P-glycoprotein has been suggested to be a flippase of amphipathic molecules (Higgins and Gottesman, 1992), we tested the effect of MDR-inhibitors on the observed transport of C_6-NBD-SM to the cell surface in the presence of BFA. The MDR-inhibitor cyclosporin A and its analog PSC 833 dramatically inhibited the transport (Table 1), suggesting that an MDR-like activity was responsible for translocating C_6-NBD-SM across the plasma membrane of CHO cells. Interestingly, in the absence of BFA, where C_6-NBD-SM has no access to the cytosolic surface (Jeckel et al., 1992), transport to the cell surface was insensitive to the MDR-inhibitors, in line with a vesicular pathway. In contrast to C_6-NBD-SM, C_6-NBD-GlcCer is synthesized on the cytosolic surface of the Golgi

Table 1. MDR1-inhibitors affect non-vesicular transport of C_6-NBD-SM and C_6-NBD-GlcCer to the CHO cell surface

	% of in BSA-medium after 1 h 37°C			
Addition	– brefeldin A		+ brefeldin A (1μg/ml)	
Lipid	C_6-NBD-SM	C_6-NBD-GlcCer	C_6-NBD-SM	C_6-NBD-GlcCer
None	73 ± 8	86 ± 2	70 ± 3	86 ± 3
Cyclosporin A[a]	62 ± 3	12 ± 2	10 ± 1	8 ± 3
PSC 833	59 ± 7	29 ± 1	14 ± 1	21 ± 1
Progesterone	53 ± 1	47 ± 2	32 ± 0	40 ± 1
Verapamil	66 ± 2	49 ± 1	34 ± 1	44 ± 1

[a] Cyclosporin A and PSC 833: 10 μM; Progesterone and verapamil: 20 μM.

(see Jeckel et al., 1992). Transport of C_6-NBD-GlcCer to the outer leaflet of the plasma membrane turned out to be sensitive to MDR-inhibitors, even in the absence of BFA (Table 1), providing the first evidence for a potential physiological role of this translocator activity (van Helvoort et al., 1997).

3 MDR1 P-glycoprotein can actually translocate C_6-NBD-SM and C_6-NBD-GlcCer

MDR3 P-glycoprotein has been shown to possess the ability to translocate C_6-NBD-PC across a membrane, whereas MDR1 P-glycoprotein did not display this property (Ruetz and Gros, 1994). We decided to test whether MDR1 and/or MDR3 can translocate C_6-NBD-sphingolipids. As a model we used two pig kidney epithelial cells lines (LLC PK1) transfected with the human MDR1 or MDR3 P-glycoprotein (In collaboration with the group of Piet Borst, Netherlands Cancer Institute, Amsterdam). We reasoned that the overexpression of a lipid translocator might result in the ability to transport C_6-NBD-lipids to the cell surface even at low temperature (≤15°C) where such traffic normally does not occur (van Helvoort et al., 1994). After synthesis of C_6-NBD-GlcCer on the cytosolic surface of the Golgi, which has been shown to occur at least down to 10°C (van Helvoort et al., 1994), it should freely equilibrate with the cytosolic leaflet of the plasma membrane because monomeric exchange of this lipid is rapid even below 10°C (Lipsky and Pagano, 1985; van Meer et al., 1987). An experiment at 15°C, a temperature where vesicular traffic to the cell surface is blocked, might thus allow us to measure translocation without interference of vesicular traffic. As both MDR1 and MDR3 are apical proteins in the epithelial cells, translocation should occur selectively across the apical membrane.

The results of the incubation with C_6-NBD-Cer were unequivocal (Figure 3). C_6-NBD-GlcCer was translocated to the apical surface of MDR1-transfected cells, but not to the basolateral surface, nor did it reach the surface of untransfected or MDR3-transfected cells. When cells were pretreated with BFA to give the newly synthesized

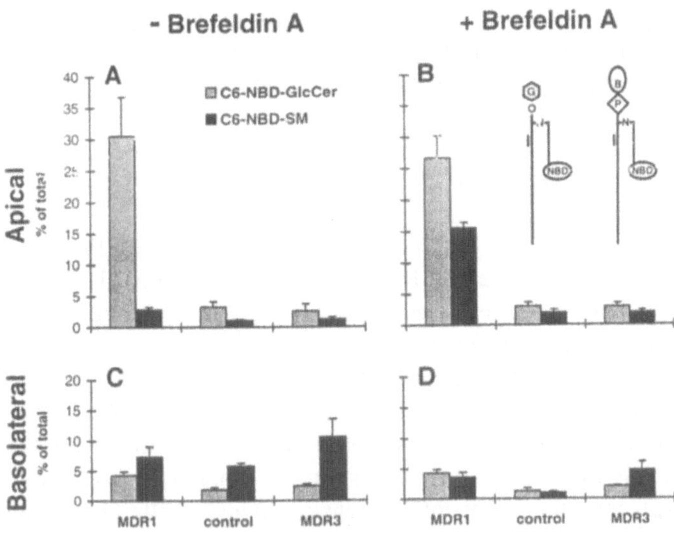

Figure 3. Translocation of C_6-NBD-SM and C_6-NBD-GlcCer across the apical plasma membrane of MDR1-transfected cells selectively. The C_6-NBD-SM signal on the basolateral surface of all three cell lines reflects synthesis on the surface by a cell surface SM-synthase (van Helvoort et al., 1994). Structure: B, choline in SM; G, glucose in GlcCer; NBD, fluorescent group; P, phosphate.

C_6-NBD-SM access to the cytosolic surface of the plasma membrane (see above), indeed, also C_6-NBD-SM appeared selectively on the apical surface of MDR1-transfected cells. In a series of experiments, we demonstrated that sphingolipids carrying a C_6-chain without the NBD-moiety are recognized and translocated by MDR1. Because it had been reported that MDR3 but not MDR1 can translocate C_6-NBD-PC, we incubated the various cell lines with C_6-NBD-PA, which results in the synthesis of C_6-NBD-PC and -PE. Indeed, C_6-NBD-PC but not C_6-NBD-PE was translocated by MDR3. However, both lipids were translocated by MDR1! In contrast to MDR3, MDR1 also transported C_8C_8-PC and C_8C_8-PE. From this we concluded that MDR1 P-glycoprotein is a lipid translocase of broad specificity, while MDR3 P-glycoprotein specifically translocates phosphatidylcholine (van Helvoort et al., 1996). The observation that MDR3 did not translocate C_8C_8-PC is difficult to understand at present.

Interestingly, in the various experiments a fair amount of the newly synthesized C_6-NBD-SM and C_6-NBD-PC was found on the basolateral cell surface of each cell line. We have provided evidence that this is due to the presence of a SM-synthase activity on the basolateral surface of epithelial cells (van Helvoort et al., 1994). The MDR-study proves that the SM-synthase is on the cell surface and not on the inside of the basolateral plasma membrane. In that case, C_6-NBD-SM and C_6-NBD-PC synthesized by that enzyme would have had the opportunity to diffuse to the cytosolic surface of the apical membrane where in the MDR1-cells they would have been recognized by the MDR1 and translocated towards the apical cell surface. No difference in the amounts of C_6-NBD-SM and C_6-NBD-PC on the basolateral surface was observed between the MDR1-cells and the untransfected cells.

4 Implications for lipid transport.

We have observed an MDR1 P-glycoprotein inhibitor-sensitive translocation of C_6-NBD-GlcCer in all cell types tested: CHO, HepG2, HeLa, Caco-2, LLC-PK1, and MDCK cells. In the latter three cell types the activity is limited to the apical plasma membrane domain (K.N.J. Burger et al., unpublished observations) and it has indeed been reported that transport of C_6-NBD-SM continued in BFA-treated Caco-2 cells, but that it now appeared on the apical rather than the basolateral surface (van Meer and van 't Hof, 1993). Most epithelial cells express the MDR1 P-glycoprotein at their apical surface and also many tumor cells express this drug transporter. The observation that human MDR1 P-glycoprotein is capable of translocating C_6-NBD-lipids, suggests the MDR1 P-glycoprotein as a candidate for the translocator activity observed in the non-transfected cell lines.

The finding that C_6-NBD-GlcCer can be translocated across the plasma membrane (van Helvoort et al., 1996; 1997) forces us to reconsider the interpretation of our ealier data on the transport of short-chain sphingolipids to the surface of epithelial cells transport data published before (summarized in van der Bijl et al., 1996). In preliminary studies in which we have addressed the effect of MDR-inhibitors in this system (K.N.J. Burger, R. Raggers, and G. van Meer, unpublished), we observed that the transport of C_6-NBD-GlcCer to the apical surface was largely mediated by MDR. In contrast, this was not the case for another short-chain GlcCer analog, C_6-(αOH)-GlcCer. Its transport to the cell surface was insensitive to MDR-inhibitors, suggesting that it was transported on the lumenal side of carrier vesicles. Transport of both C_6-NBD- and C_6-(αOH)-SM to the cell surface was vesicular as it was insensitive to MDR-inhibitors. So, the enrichment of C_6-NBD-GlcCer over C_6-NBD-SM on the apical surface of both MDCK and Caco-2 cells (see van der Bijl et al., 1996) could be largely ascribed to the fact that C_6-NBD-GlcCer was translocated across the apical surface whereas C_6-NBD-SM was exclusively transported by vesicles. The apical enrichment of C_6-(αOH)-GlcCer over C_6-(αOH)-SM may be due to an apical translocator of different specificity, or, alternatively as we suggested before (Simons and van Meer, 1988), to a lateral segregation of C_6-(αOH)-GlcCer from -SM in the lumen of the sorting compartment, the TGN. We are presently developing assays to study whether one or both mechanisms are involved in the establishment and maintenance of the surface polarity of the native lipids. Interesting in this respect is the purification and sequencing of a monohexosylceramide transfer protein from cytosol (Abe, 1990), which may be responsible for monomeric transport of GlcCer from the Golgi to the plasma membrane that was found to be BFA-insensitive (Warnock et al., 1994).

Potentially, in all earlier (Karrenbauer et al., 1990; summarized in van der Bijl et al., 1996) or later (van IJzendoorn et al., 1997) studies on transport of short-chain lipids, they were translocated by MDR-type proteins. Control experiments will have to be performed in all these cases to allow an interpretation of the data in terms of possible transport and sorting mechanisms. Finally, further studies on transport of short-chain lipids should be published only when including a test of the effect of MDR-inhibitors. However, the essential question remaining is whether endogenous GlcCer or other endogenous long-chain lipids can be translocated by the MDR1 P-glycoprotein. Further questions are whether MDR1 could be responsible for the release from cells of the short-chain PC platelet activating factor (PAF) or other short-chain lipids, like oxidized lipids, or lysophospholipids, like lysophosphatidic acid, LPA. Finally, other multidrug transporters (like MRP1) or other ABC-transporters may be translocators of the native membrane lipids.

References

Abe A (1990) Primary structure of glycolipid transfer protein from pig brain. J. Biol. Chem. 265: 9634-9637

Herrmann A, Zachowski A and Devaux PF (1990) Protein-mediated phospholipid translocation in the endoplasmic reticulum with a low lipid specificity. Biochemistry 29: 2023-2027

Higgins CF and Gottesman MM (1992) Is the multidrug transporter a flippase? Trends Biochem. Sci. 17: 18-21

Jeckel D, Karrenbauer A, Burger KNJ, van Meer G and Wieland F (1992) Glucosylceramide is synthesized at the cytosolic surface of various Golgi subfractions. J. Cell Biol. 117: 259-267

Karrenbauer A, Jeckel D, Just W, Birk R, Schmidt RR, Rothman JE and Wieland FT (1990) The rate of bulk flow from the Golgi to the plasma membrane. Cell 63: 259-267

Klausner RD, Donaldson JG and Lippincott-Schwartz J (1992) Brefeldin A: Insights into the control of membrane traffic and organelle structure. J. Cell Biol. 116: 1071-1080

Lipsky NG and Pagano RE (1985) Intracellular translocation of fluorescent sphingolipids in cultured fibroblasts: Endogenously synthesized sphingomyelin and glucocerebroside analogues pass through the Golgi apparatus en route to the plasma membrane. J. Cell Biol. 100: 27-34

Pagano RE, Longmuir KJ and Martin OC (1983) Intracellular translocation and metabolism of a fluorescent phosphatidic acid analogue in cultured fibroblasts. J. Biol. Chem. 258: 2034-2040

Ruetz S and Gros P (1994) Phosphatidylcholine translocase: A physiological role for the *mdr2* gene. Cell 77: 1-20

Simons K and van Meer G (1988) Lipid sorting in epithelial cells. Biochemistry 27: 6197-6202

Smit JJM, Schinkel AH, Oude Elferink RPJ, Groen AK, Wagenaar E, van Deemter L,

Mol CAAM, Ottenhoff R, van der Lugt NMT, van Roon MA, van der Valk MA, Offerhaus GJA, Berns AJM and Borst P (1993) Homozygous disruption of the murine *mdr2* P-glycoprotein gene leads to a complete absence of phospholipid from bile and to liver disease. Cell 75: 451-462

Smith AJ, Timmermans-Hereijgers JLPM, Roelofsen B, Wirtz KWA, van Blitterswijk WJ, Smit JJM, Schinkel AH and Borst P (1994) The human MDR3 P-glycoprotein promotes translocation of phosphatidylcholine through the plasma membrane of fibroblasts from transgenic mice. FEBS Lett. 354: 263-266

Urbani L and Simoni RD (1990) Cholesterol and Vesicular Stomatitis virus G protein take separate routes from the Endoplasmic Reticulum to the plasma membrane. J. Biol. Chem. 265: 1919-1923

van der Bijl P, Lopes-Cardozo M and van Meer G (1996) Sorting of newly synthesized galactosphingolipids to the two surface domains of epithelial cells. J. Cell Biol. 132: 813-821

van Helvoort A, Giudici ML, Thielemans M and van Meer G (1997) Transport of sphingomyelin to the cell surface is inhibited by brefeldin A and in mitosis, where C_6-NBD-sphingomyelin is translocated across the plasma membrane by a multidrug transporter activity. J. Cell Sci. 110: 75-83

van Helvoort A, Smith AJ, Sprong H, Fritzsche I, Schinkel AH, Borst P and van Meer G (1996) MDR1 P-glycoprotein is a lipid translocase of broad specificity, while MDR3 P-glycoprotein specifically translocates phosphatidylcholine. Cell 87: 507-517

van Helvoort A, van 't Hof W, Ritsema T, Sandra A and van Meer G (1994) Conversion of diacylglycerol to phosphatidylcholine on the basolateral surface of epithelial (Madin-Darby canine kidney) cells. Evidence for the reverse action of a sphingomyelin synthase. J. Biol. Chem. 269: 1763-1769

van IJzendoorn SVD, Zegers MMP, Kok JW and Hoekstra D (1997) Segregation of glucosylceramide and sphingomyelin occurs in the apical to basolateral transcytotic route in HepG2 cells. J. Cell Biol. 137: 347-357

van Meer G, Stelzer EHK, Wijnaendts-van-Resandt RW and Simons K (1987) Sorting of sphingolipids in epithelial (Madin-Darby canine kidney) cells. J. Cell Biol. 105: 1623-1635

van Meer G and van 't Hof W (1993) Epithelial sphingolipid sorting is insensitive to reorganization of the Golgi by nocodazole, but is abolished by monensin in MDCK cells and by brefeldin A in Caco-2 cells. J. Cell Sci. 104: 833-842

Warnock DE, Lutz MS, Blackburn WA, Young WWJr and Baenziger JU (1994) Transport of newly synthesized glucosylceramide to the plasma membrane by a non-Golgi pathway. Proc. Natl. Acad. Sci. USA 91: 2708-2712

Mycrobacterial Glycopeptidolipid Interactions with Membranes: An Air-Water Monolayer Study by FTIR Spectroscopy

Isabelle Vergne[1], Bernard Desbat[2], Jean-François Tocanne[1] and Gilbert Lanéelle[1]

[1] Institut de Pharmacologie et de Biologie Structurale : CNRS, Université Paul Sabatier. 31062 Toulouse. France.

[2] Laboratoire de Spectroscopie Moléculaire et Cristalline. Université Bordeaux I. 33405 Talence. France.

Pathogenic mycobacteria multiply within macrophagic phagosomes. They produce abundantly glycopeptidolipids (GPLs) in their envelope. GPL suspensions increased membrane permeability and uncoupled oxidative phosphorylation in mitochondria. Disturbances of the macrophage membrane may take part in intracellular survival.

GPLs with light glycosylation are more active than heavily glycosylated ones. The former inserted less efficiently in phospholipid monolayers than the latter. Infrared studies at the air-water interface indicate that GPL with one or two sugars increased disorder in phospholipid acyl chains while GPL with three sugars ordered them. Highly glycosylated GPLs self-assembled in antiparallel beta-sheets that segregated in a phospholipid monolayer. Glycosylation has opposite effects, leading to an optimum activity with one or two sugar residues.

Key Words : glycopeptidolipids, air-water monolayer, infrared spectroscopy.

NATO ASI Series, Vol. H 106
Lipid and Protein Traffic
Pathways and Molecular Mechanisms
Edited by Jos A. F. Op den Kamp

Organisms belonging to the *Mycobacterium avium* complex are the most common cause of bacteremia in patients with AIDS. These opportunistic pathogenic mycobacteria are obligate intracellular parasites, which survive and multiply within the host's macrophages. After phagocytosis, bacteria reside in a phagosomal vacuole and circumvent the defences of the macrophage like the hydrolytic enzymes resulting from fusion with lysosomes (Frehel et al., 1986) and the decrease of phagosomal pH (Sturgill-Koxzycki et al., 1994). We investigated the possible involvement of the outermost constituents of cell envelopes in maintenance of bacteria in phagosome.

Glycopeptidolipids are profusely produced by *Mycobacterium avium* (Tereletsky et al., 1983; Rulong et al., 1991) and, as they are not covalently linked to the bacterial wall, these amphiphilic molecules can interact with macrophagic membranes and disturb their properties. Membrane associated-functions are essential for the phagocytosis process in particular during phagosome-lysosome fusion and acidification of the phagosome.

Our studies examined the interaction of mycobacterial glycopeptidolipids (GPL) with membranes. These molecules share a common peptidolipid core with a thirty-carbon acyl chain and four amino-acid residues; carbohydrates are linked to the peptidic moiety. The number and the nature of the sugars are species and serovar-dependent (Figure 1).

Acyl - D Phe - D allo Thr - D Ala - L Alaninol
(D allo Thr bears OX_1; L Alaninol bears OX_2)

Acyl : fatty acid with 30 carbon atoms

X : 6 - deoxysugars or hydrogen

Fig. 1. Chemical structure of the glycopeptidolipids and lipopeptide. A number following GPL corresponds to the number of carbohydrate residues linked to the peptidic moiety. GPL0 : without sugar; GPL1 : one sugar, GPL3 : three sugars (Vergne et al., 1995), GPL2 : two sugars; GPL5 : five sugars (Lopez-Marin et al., 1994). GPL1 was obtained by beta-elimination of the sugar residue on Thr. This introduced a double bond on C2 of the Thr.

Firstly, we tested the effects of GPLs differing in their carbohydrate moiety on isolated mitochondria and on liposomes. Then, we tried to explain the disturbances of membrane properties induced by these molecules, using monolayers at the air-water interface.

1. Effects of glycopeptidolipids on isolated mitochondria and on liposomes (Lopez-Marin et al., 1994)

1.1 Oxidative phosphorylation. In order to test the effects of mycobacterial glycopeptidolipids on a natural membrane, isolated rat liver mitochondria were used and oxidative phosphorylation was followed after addition of different GPL suspensions. Five species of GPLs were tested which have zero, one, two, three or five sugars on their peptidic moieties. It was shown that in the presence of GPL, controlled respiration increased and phosphorylation efficiency decreased (ADP/O ratio) (Fig. 2.A).

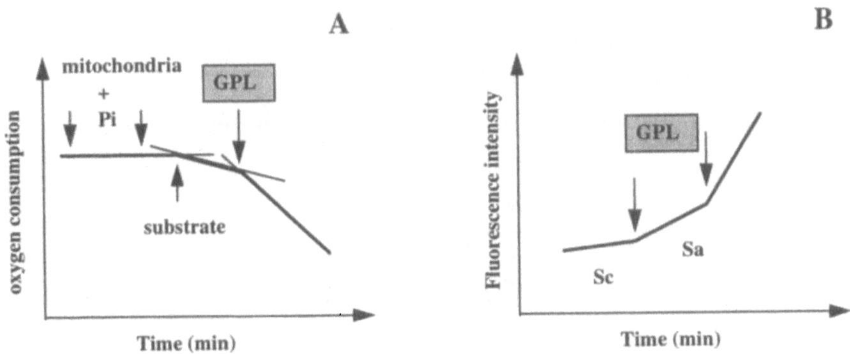

Fig. 2. (A): Typical experiment of oxidative phosphorylation measurement. Oxygen consumption is followed after addition of mitochondria and substrate. Arrows indicate successive additions.
(B): Release of carboxyfluorescein from small unilamellar vesicles (PC/PS/cholesterol, 2/1/1). Fluorescence intensity was followed. GPL activities were compared through their respective ratio assay slope (Sa) / control slope (Sc).

1.2 Membrane permeability. Release of carboxyfluorescein from liposomes after addition of GPL was measured (Fig. 2.B). The results showed that GPL increased membrane permeability. This suggests that GPL-induced uncoupling of oxidative phosphorylation could result from proton leaks through the inner mitochondrial membrane. By comparing their dose-dependent effects, the GPLs were seen to have the following decreasing efficiency : GPL1 > GPL2 >> GPL0 ~ GPL3 ~ GPL5.

To understand the origin of membrane perturbation and of the GPL efficiency (i) GPL insertion in a phospholipid monolayer at the air-water interface was followed, (ii) an analysis of conformation and orientation of GPL and phospholipid acyl chain order by Fourier Transformed InfraRed spectroscopy (FTIR) was performed.

2. GPL insertion into phospholipid monolayers. GPL insertion into a phospholipidic film was detected by following the surface pressure increase with time (Vergne et al., 1995). Briefly, a GPL suspension was injected through a hole underneath an egg-phosphatidylcholine monolayer at the air-water interface. As the monolayer surface area was constant, insertion of GPL resulted in an increase of surface pressure. Insertion kinetics at an initial pressure of 10 mN/m showed that GPL0 molecules did not insert into the film, in constrat with GPL1 and GPL3 (Fig. 3.A). Surface pressure variation versus initial pressure (Fig. 3.B) indicated that whatever the initial pressure, GPL0 did not significantly insert into the monolayer explaining its very low effect on membranes. On the other hand, GPL3 inserted more efficiently than GPL1, but the former was less active than the latter. Thus, we looked for differences in conformation and/or organization between GPL1 and GPL3.

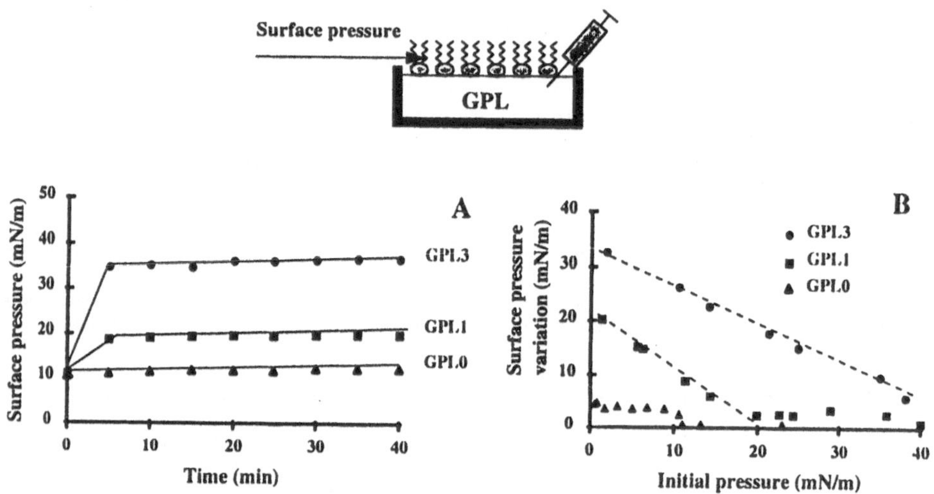

Fig. 3. Insertion of GPL molecules into an egg-phosphatidylcholine monolayer at the air-water interface.
(A): Insertion kinetics (initial pressure 10mN/m).
(B) : Surface pressure variation versus initial pressure.

3. Organization of GPL-phospholipid monolayers : an infrared study

The conformation of GPL and phospholipid molecules in mixed monolayers was investigated by Fourier Transformed InfraRed spectroscopy (FTIR). This technique has two main advantages, (i) it does not require probe molecules, (ii) it allows structural information on peptide and lipid molecules to be obtained (Arrondo and Goñi, 1993), since they have separate characteristic infrared absorption bands, at frequencies which depend on the molecular conformation.

Infrared studies were done at the air-water interface, in order to determine the number of GPL molecules present in the phospholipidic film and the surface pressure. Conventional FTIR is not suitable for performing air-water interfacial studies, due to the small number of molecules excited and to strong absorption of water molecules. These difficulties were overcome by differential reflectivity measurements by polarization modulation of the incident light, which allows isotropic absorption in the spectrum to be eliminated (Blaudez et al., 1993). Recently, this technique of Polarization Modulated Infra Red Reflection Absorption Spectroscopy (PMIRRAS) proved to be an efficient method to determine the conformation and orientation of the bee venom peptide melittin in a dimyristoylphosphatidylcholine monolayer at the air-water interface (Cornut et al., 1996).

3.1 Principle of PMIRRAS. A detailed description of PMIRRAS as well as the experimental procedure has been published (Blaudez et al., 1994). Briefly, this method uses a rapid modulation of the electric field polarization between polarization in the plane of incidence (p) and polarization perpendicular (s) to this plane (Figure 4). The polarized infrared beams were focused on the interface and the corresponding reflectivities were detected. The PMIRRAS signal is essentially conditioned by the differential reflectivity signal : (Rp-Rs)/(Rp+Ps). Monolayer bands were obtained by normalizing monolayer spectra with that of the subphase. The direction of the bands versus the baseline is an indication of the orientation of the transition moment relative to the interface, and therefore of molecule orientation. In this way, three main situations can occur : (i) an upward-oriented band indicates a transition moment preferentially parallel to the interface; (ii) a downward-oriented band indicates a transition moment preferentially perpendicular to the interface; (iii) a disappeared-band indicates that either the angle between the transition absorption moment and the interface is 40°, or that the transition moments are weak and isotropic (Figure 4). Frequencies and band area analysis give information on the conformation and orientation of molecules at the interface.

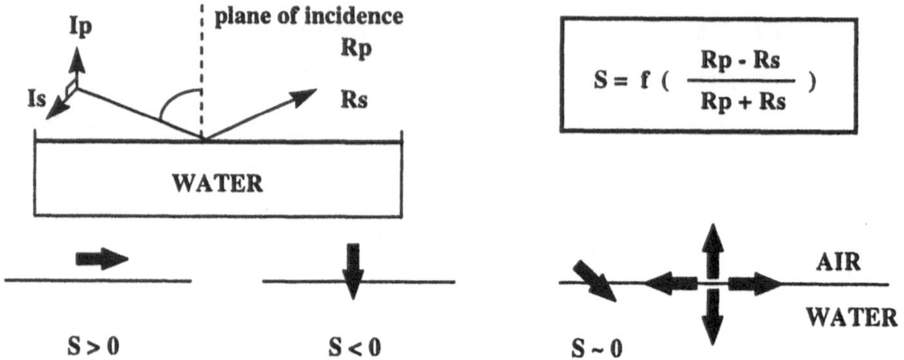

Fig. 4. Principle of Polarization Modulated Infra Red Reflection Absorption Spectroscopy.
I = incident beam. R = reflected beam. (p) is polarization in the plane of incidence. (s) is polarization perpendicular to this plane. On the bottom of the scheme, broad arrows indicate transition moments.
S = PMIRRAS signal.

3.2 GPL and DPPC-d62 monolayers

3.2.1 GPL monolayers. Three types of GPL were studied, which differ in their respective number of sugars on the peptidic core : GPL1, GPL2, and GPL3. The spectra of the corresponding monolayers at an initial pressure 25 mN/m are presented in figure 5.

C-H stretching vibrations of the acyl chain give rise to bands in the 3100-2800 cm^{-1} region : antisymmetric and symmetric CH_2 bands have frequencies which are gauche-trans conformation sensitive. On GPL2 and GPL3 PMIRRAS spectra, we noted the absence of these bands, although they were present and strong on absorption spectra of solid anhydrous GPL (data not shown). The GPL1 spectrum showed two broad bands at 2922 cm^{-1} and 2854 cm^{-1} corresponding to antisymmetric and symmetric vibrations respectively. These results suggest mostly a gauche conformation of the acyl chain for the three GPL species and are indicative of a highly disordered hydrocarbon chains (Mitchell and Dluhy, 1988).

The peptidic moiety mainly give rise to a strong band corresponding to C=O stretching (Amide I) with a conformation-sensitive frequency and to a weaker band corresponding to N-H bending (Amide II). The GPL1 PMIRRAS spectrum shows a broad weak band at 1640 cm-1, while that of GPL2 and GPL3 exhibit an intense sharp peak at 1625 cm-1 (Amide I. pi, 0) and a weaker band at 1690 cm-1 (Amide I. 0, pi). This latter results suggests the formation of antiparallel beta-sheets in GPL2 and GPL3 monolayers, while GPL1 molecules were in an unordered structure (Arrondo and Goñi, 1993; Haris and Chapman, 1995). Because of the small size of the peptide (four amino-acid residues), an single-molecule antiparallel beta-sheet can be excluded. We propose an intermolecular antiparallel beta-sheet arrangement of GPL2 and GPL3 molecules in monolayers. Band area analysis indicates that the beta-sheet was preferentially parallel to the air-water interface (Miyazawa, 1960).

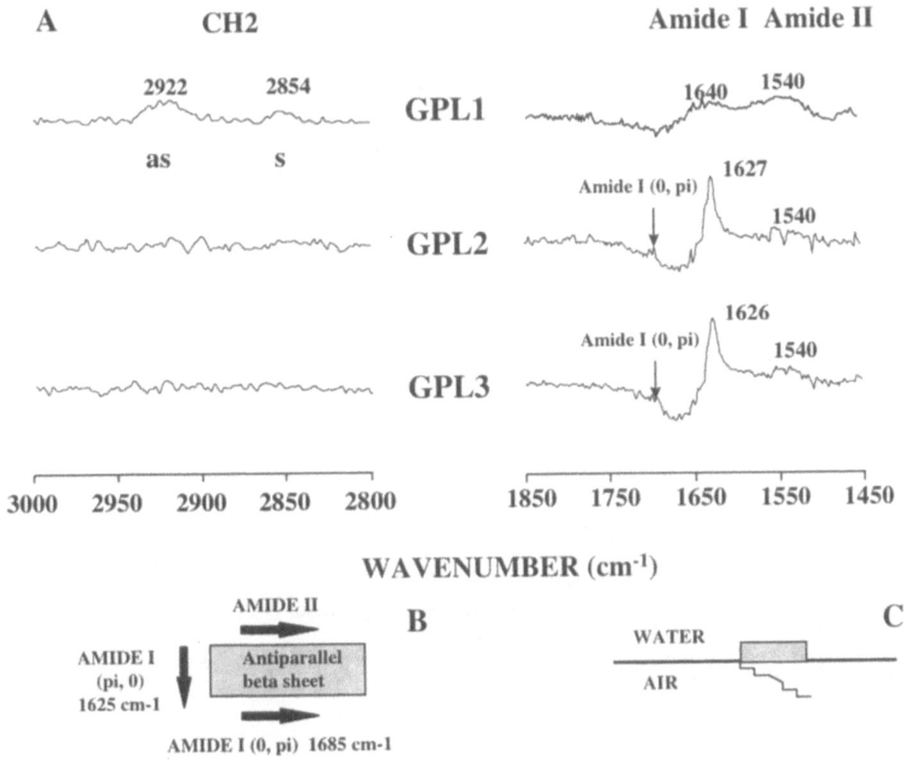

Fig.5. (A): GPL monolayer PMIRRAS spectra at 25 mN/m.
CH_2: C-H stretching vibrations of acyl chains. (as): antisymmetric; (s) : symmetric.
Amide I : C=O stretching . Amide II : N-H bending.
(B): Transition moment orientation in antiparallel beta-sheet.
(C): Schematic representation of GPL molecule orientation at air-water interface.

3.2.2 DPPC d62-film. In order to follow conformational changes of phospholipid acyl chains in the presence of GPL, investigations were performed with dipalmitoyl phosphatidyl choline with perdeuterated acyl chains (DPPC-d62). Figure 6 presents the spectrum of the DPPC-d62 monolayer at 25mN/m. The CD_2 stretching frequencies were observed at 2194.5 cm^{-1} and 2089 cm^{-1} for antisymmetric and symmetric vibrations respectively. This indicates predominantly trans conformation for the acyl chains. The high intensity of these bands suggests a favorable orientation of the C-D bonds in the plane of the monolayer and ordered hydrocarbon chains (Mitchell and Dluhy, 1988).

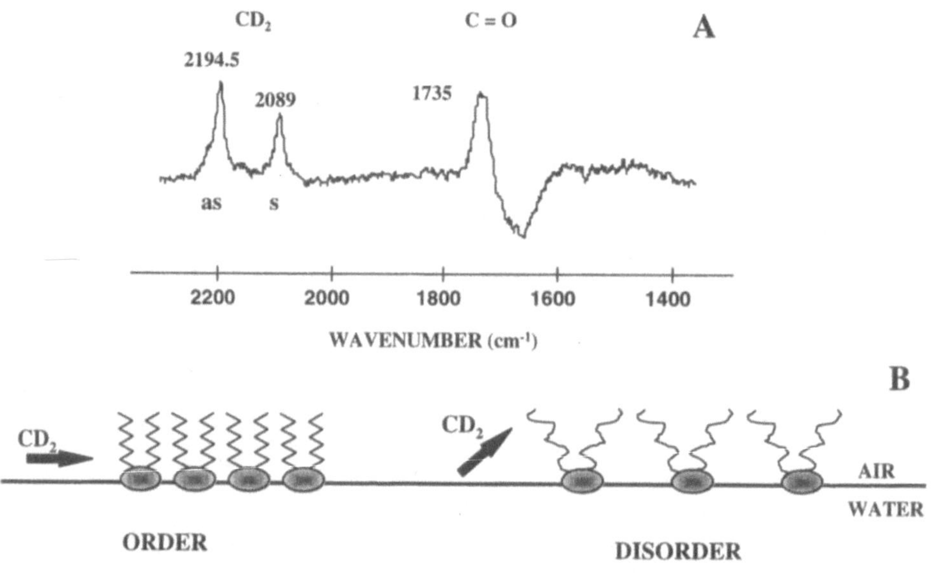

Fig. 6. (A): DPPC-d62 monolayer PMIRRAS spectrum at 25mN/m.
CD_2 : C-D stretching vibration of acyl chains. (as) : antisymmetric. (s) : symmetric.
C=O : C=O stretching of ester bonds.
(B): Schematic representation of phospholipid molecules at the air-water interface. Arrows indicate the transition moment orientation of C-D stretching vibrations.

3.3 Mixed films. GPL-phospholipid interactions were monitored by forming mixed monolayers with 40%GPL at 25mN/m. This proportion of GPL allowed signals of both phospholipid and of glycopeptidolipid molecules to be detected. On GPL2 and GPL3 mixed film spectra (Figure 7), amide I at 1625 cm^{-1} was noted suggesting that even when mixed with phospholipids, GPL molecules self-associated to form beta-sheet structures. Data of CD_2 band analysis are reported in Table I. Integrated intensity were corrected for surface density of DPPC-d62 molecules of one molecule per nm^2. In the presence of GPL1, CD_2 stretching frequencies increased from 2194.6 to 2195.6 cm^{-1} for antisymmetric and from 2090.3 to 2094.5 cm^{-1} for symmetric bands. The integrated intensity decreased by almost a factor of two, average transition moment withdrawing from the interface. In contrast, a decrease of CD_2 frequencies and an increase of integrated intensity per molecule were noted with GPL3. Finally, GPL2 induced a decrease of the band area. This results suggests that GPL2, and more intensely GPL1 molecules, increased acyl chain disorder. On the contrary, GPL3 ordered the phospholipid acyl chains.

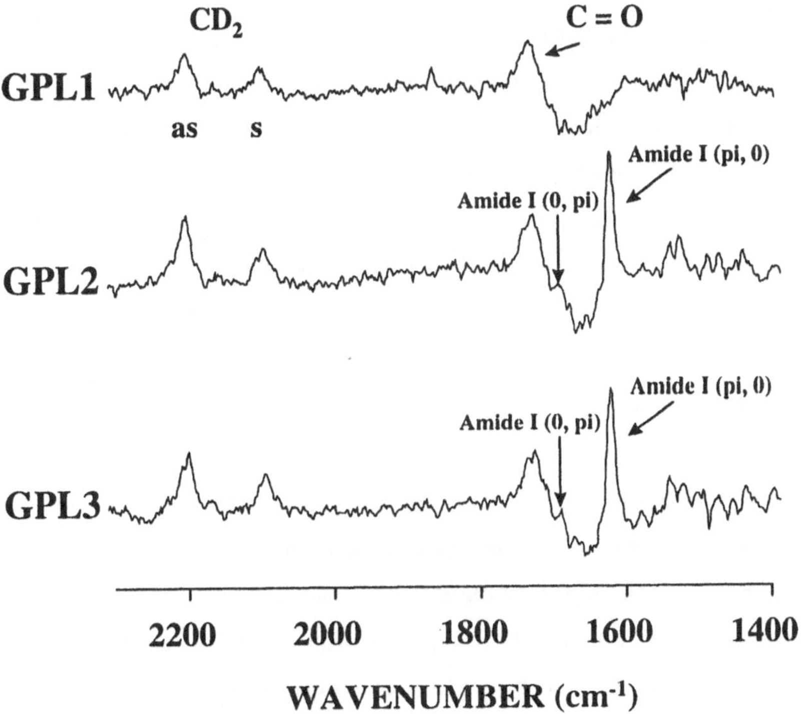

Fig. 7. Mixed GPL/DPPC-d62 monolayer PMIRRAS spectra at 25mN/m.
CD_2: C-D stretching vibration of phospholipid acyl chains. (as) : antisymmetric. (s) : symmetric.
C=0: C=0 ester bond stretching.
Amide I: C=O amide bond stretching.

	Antisymmetrical Frequencies cm-1	CD_2 stretching Integrated Intensities	Symmetrical Frequencies cm-1	CD_2 stretching Integrated Intensities
DPPC d62	2194.6	2.5	2090.3	1.6
40% GPL1	2195.6	1.3	2094.5	0.8
40% GPL2	2194.2	1.4	2090.7	1.1
40% GPL3	2193.8	3.5	2089.2	2.7

Table 1. Data of CD_2 band analysis of DPPC-d62 and mixed film PMIRRAS spectrum at 25 mN/m
Integrated intensities correspond to corrected integrated intensities calculated for surface density of DPPC-d62 of one molecule per nm^2.

PMIRRAS results suggest that GPL increased membrane permeability by increasing phospholipid acyl chain disorder. Indeed, GPL1 is more active than GPL3, and it disordered the hydrocarbon chains whereas GPL3 ordered them. But how can we explain the different behavior of the GPLs? GPL1 molecules do not self-associate probably due to the presence of a double bond which introduces an ethylenic carbon atom in the peptidic backbone. Thus, its molecules could freely diffuse in the phospholipid phase and disturb it. Both GPL2 and GPL3 molecules were organized in beta-sheet structures in the presence of phospholipids, but the former increased disorder while the latter did not. We can propose at least two hypotheses, either the beta-structures have different orientations relative to the interface and do not interact in the same way with phospholipids, or some GPL2 molecules do not participate in a beta-structure and they are free to mix with phospholipids. Integrated intensities do not allow these two possibilities to be distinguished.

Finally, the importance of the sugars linked to the peptidic chain seems to affect both insertion of the molecules and their organization within the phospholipid film. Carbohydrates increase the hydrophilic/ hydrophobic balance, favoring molecule insertion, but they may facilitate molecule self-association, extending the beta-sheet, inducing lateral phase separation between GPL and phospholipids. Thus, the most efficient GPL seems to be the one that has sufficient sugars to insert in membranes, but no enough to form an extended beta-structure in the presence of phospholipids.

References:

Arrondo JLR and Goñi FM (1993) Infrared spectroscopies studies of lipid-protein interactions in membranes. In Protein-Lipid Interactions (Watts A, ed.) pp 321-349, Elsevier Science Publishers BV

Blaudez D, Buffeteau T, Cornut JC, Desbat B, Escafre N, Pezolet M, and Turlet JM (1993) Polarization-modulated FT-IR spectroscopy of spread monolayer at the air-water interface. Appl. Spectrosc. 47: 869-874

Blaudez D, Buffeteau T, Cornut JC, Desbat B, Escafre N, Pezolet M, and Turlet JM (1994) Polarization modulation FTIR spectroscopy at the air-water interface. Thin Solid Films. 242: 146-150

Cornut I, Desbat B, Turlet JM and Dufourcq J (1996) In situ study by polarization modulated Fourier Transform Infrared Spectroscopy of the structure and orientation of lipids and amphipathic peptides at the air-water interface. Biophys. J. 70: 305-312

Flach CR, Brauner JW, Taylor JW, Baldwin RC and Mendelsohn R (1994) External reflection FTIR of peptide monolayer films in situ at the air-water interface: experimental design, spectra-structure correlations and effects of hydrogen-deuterium exchange. Biophys. J. 67: 402-410

Frehel C, De Chastellier C, Lang T, and Rastogi N (1986) Evidence for inhibition of fusion of lysosomal and prelysosomal compartments with phagosomes in macrophages infected with pathogenic *Mycobacterium avium*. Infect. Immun. 52: 252-262

Haris PI and Chapman D (1995) The conformational analysis of peptides using Fourier-transform IR spectroscopy. Biopolymers. 37: 251-263

Lopez-Marin LM, Quesada D, Lakhdar-Ghazal F, Tocanne JF, and Laneelle G (1994) Interactions of mycobacterial glycopeptidolipids with membranes: influence of carbohydrate on induced alterations. Biochemistry. 33: 7056-7061

Mitchell ML and Dluhy RA (1988) In situ FT-IR investigation of phospholipid monolayer phase transitions at the air-water interface. J. Am. Chem. Soc. 110: 712-718

Miyazawa T (1960) Perturbation treatment of the characteristic vibrations of the polypeptide chains in various configurations. J. Chem. Phys. 32: 1647-1652

Rulong S, Aguas AP, Da Silva PP, and Silva MT (1991) Intramacrophagic *Mycobacterium avium* bacilli coated by multiple lamellar structure : freeze fracture analysis of infected mouse liver. Infect. Immun. 59:3895-3902

Sturgill-Koxzycki S, Schlesinger PH, Chakraborty P, Haddix PL, Collins HL, Fok AK, Allen RD, Gluck SL, Heuser J and Russell DG (1994) Lack of acidification in *Mycobacterium* phagosomes produced by exclusion of the vesicular proton-ATPase. Science. 263: 678-681

Tereletsky MJ, and Barrow WW (1983) Postphagocytic detection of glycopeptidolipids associated with the superficial L1 Layer of *M. intracellulare*. Infect. Immun. 4: 312-1321

Vergne I, Prats M, Tocanne JF, and Laneelle G (1995) Mycobacterial glycopeptidolipid interactions with membranes: a monolayer study. FEBS Lett. 375: 254-258

Consequences of Hydrophobic Matching on the Lateral Distribution of Lipids Around Bacteriorhodopsin Reconstituted in DLPC/DSPC Mixtures

Fabrice DUMAS[1], Maria-Chantal LEBRUN[1], Maria M. SPEROTTO[2], Ole.G. MOURITSEN[2] and Jean-François TOCANNE[1].

[1] Institut de Pharmacologie et de Biologie Structurale, 118 Route de Narbonne, F-31062 Toulouse Cedex, France
[2] Memphys, Department of Chemistry, The technical University of Denmark Building 206, DK-2800 Lyngby, Denmark.

ABSTRACT To test the hypothesis that differences in lipid bilayer and membrane protein hydrophobic thicknesses can brings about a mechanism of lipid molecular sorting by proteins, computer simulations and Fluorescence Resonance Energy Transfer experiments (between NBD-labeled lipids as donor and retinal as acceptor) were carried out on Bacteriorhodopsin reconstituted in DLPC/DSPC mixtures. This study took advantage of the non-ideal mixing behavior of DLPC and DSPC and the fact that the average lipid acyl-chain length depends on temperature. At low temperature, in the gel-gel coexistence region, BR is found associated with the short-chain lipid DLPC. At moderate temperature, in the fluid-gel coexistence region, BR still shows preference for DLPC but now stands at the fluid-gel boundary. At high temperature, in the fluid-fluid phase, the theoretical data shows preference of BR for the long-chain DSPC at the expense of the short-chain DLPC molecules.

Biological membranes are complex assemblies composed of a wide variety of lipid and protein molecular species. One of the most challenging problem in membrane biology is to define the suprastructure of biological membranes. In the Fluid Mosaic Model of membranes, lipids are organized in the form of a bilayer supporting peripheral and integral proteins. This model considers the lipid bilayer as a two dimensional fluid in which lipids and proteins are free to diffuse. As a direct consequence, both types of molecules would be expected to be randomly distributed within the membrane. In fact, membrane organization is certainly more complex and evidence is accumulating to indicate the occurrence of both a transverse (Devaux, 1993) and lateral (Tocanne, 1992; Glaser, 1993) compartmentation of membranes which can be described in terms of lipid and protein macro- and microdomains. Gel/fluid phase lateral separations have been suggested as being responsible for lateral heterogeneities in membranes in specific cases like in ram sperm plasma (Wolf and al, 1990). However, in many cases, the lateral heterogeneities in lipid distribution detected in natural membranes are no longer observed in the corresponding protein-free lipid bilayers (Tocanne and al, 1994). With respect to proteins, one current opinion is to consider that their confinement in domains is due principally to interactions of their cytosolic part with membrane skeleton and/or cytoskeleton elements (Sheets and al, 1995; Kusumi and al, 1996).

Abbreviations used: BR, bacteriorhodopsin; eggPC, egg yolk phosphatydilcholine; DLPC, dilauroylphosphatydilcholine; DSPC, distearoylphosphatydilcholine; PE, phosphatydilethanolamine obtained from egg yolk phosphatydilcholine ; NBD, 7-nitrobenz-2-oxa-1,3-diazol-4-yl; N-NBD-DLPE, N-NBD-dilauroylphosphatydilethanolamine; N-NBD-DSPE, distearoylphosphatydilethanolamine.

NATO ASI Series, Vol. H 106
Lipid and Protein Traffic
Pathways and Molecular Mechanisms
Edited by Jos A. F. Op den Kamp

Another very attractive possibility is to consider the role of interactions between lipids and transmembrane proteins. It is generally agreed that interactions with lipids modulate the function of at least some membrane proteins. However despite extensive studies on the structure of biological membranes, we still have no direct information on the molecular organization of lipids in contact with or far from the proteins, nor on the potential specificity of lipid distribution around membrane proteins. A related question is that of the existence of lipid domains on the nanometer-scale (1-100 nm) (Bergelson et al; 1995) and the relationship which can exist between lipid-domain formation and the functional properties of membrane-associated proteins. Do lipids organize themselves into nonrandom clusters or domains? If domain exist, have they any functional importance? Are they intimately involved, for example, in basic membrane processes such as fusion, transport, recognition and catalysis?

With respect to the molecular mechanisms involved in lipid-protein interactions, matching of the lipid-bilayer and integral membrane proteins hydrophobic thicknesses has been shown to be an important parameter for the coupling between lipid and integral membrane proteins structure. It may also contribute to the formation of lipid microdomains around intrinsic membrane proteins with fluidity and/or composition different from that of the bulk of the lipid phase (Sperotto and Mouritsen, 1991a; Mouritsen and Biltonen, 1992)

From studies carried out with reconstituted lipid-protein systems, it is clear that interactions with lipids modulate the function of at least some membrane proteins. It is tempting to speculate on whether the hydrophobic mismatch can play a role in the regulation of the functions of membrane proteins. Thus, for the Ca^{2+}-ATPase from reticulum sarcoplasmique and for (Na^{+}-K^{+})-ATPase, optimal activity occurs over a fairly narrow range of bilayer thicknesses (Caffrey and Fergenson, 1981; Johannson et al 1981; Moore et al 1981; Starling and al, 1993; Cornea and al 1993; Johannsson et al., 1981a). In the case of gramicidin in solvent-containing black lipid membranes, the activity was found to decrease as the bilayer thickness was increased (Haydon et Hladky, 1972). It has been suggest that hydrophobic lipid-protein interactions will influence the tension on the protein which in some cases may introduce conformational changes in the protein and therefore couple indirectly to the protein function as has recently been found in the case of the Meta-I to Meta-II transition in Rhodopsin (Brown, 1994).

Although it is generally recognized that hydrophobic interactions are key factors in membrane stability, their possible role in influencing the planar organization of membrane components has to be examined thoroughly. The consequences of hydrophobic matching on the lateral organization of lipids in membrane has been rarely investigated experimentally. Imperfect matching of protein and lipid bilayer surfaces may result in aggregation of transmembrane proteins in order to minimize the exposure of hydrophobic surfaces to polar environments (Israelachvili 1977). Recently, bilayer thickness adaptations have been shown to occur. Thus, the reaction center and the light-harvesting antenna from *Rhodobacter sphaeroides* and bacteriorhodopsin from *Halobacterium halobium* have been reconstitued with phosphatidylcholines with various acyl chain length and shown to modify the gel/liquid phase transition temperature of the lipid bilayer (Riegler and Möhwald, 1986; Peschke et al 1987, Piknova et al, 1993).

Consequences of hydrophobic matching between lipids and proteins have also been studied from a theoretical point of view (Mouritsen and Bloom, 1984; Sperotto and Mouritsen, 1988, 1991a, 1991b).

In the case of a binary lipid system consisting of two lipid species with different acyl chain lengths, the hydrophobic matching principle has been proposed to act as a mechanism of lipid sorting at the lipid-protein interface because the protein will, on a statistical basis, prefer to be associated with the lipid species that is hydrophobically most well matched (Sperotto and Mouritsen, 1993),

In the present work we have explored the possibility of molecular sorting of lipids by proteins. We have analyzed the potential specificity of lipid distribution around membrane protein, using a fully defined system in which the hydrophobic thickness of both protein and lipids are known. Reconstituted vesicles were obtained from delipidated bacteriorhodopsin, a protein whose the structure is known to 7 Å resolution (Henderson and al, 1990) and a binary DLPC/DSPC lipid mixture. The large difference in hydrophobic length of the two lipid species implies a strongly non-ideal mixing behavior and phase equilibria that involve large gel-fluid and gel-gel phase coexistence regions. Depending on their physical state, these lipids can match or mismatch the hydrophobic length of the protein. Thus depending on the temperature two situations are expected : either a random distribution of protein molecules within the bilayer with an adaptation of the phospholipid chain length in contact with BR as shown by Piknova and al. (1993), or a preference of the protein for one or the other lipid species via its specific hydrophobic length. One would expect that the energy required for conformational changes should be quite greater.

In this paper, a theoretical calculation based on computer simulation of the specific molecular interaction model of the BR/DLPC/DSPC system is presented and results are compared with the data obtained through fluorescence transfer measurements. This approach brings strong support to the concept of lipid sorting by membrane proteins.

The theoretical model

The theoretical model used describes a two-component bilayer incorporated with an immobile and very large protein, i.e. a protein whose cross-sectional diameter is much larger than the cross-sectional diameter of a lipid-acyl chain. The DLPC/DSPC binary mixture and the ternary BR/DLPC/DSPC have been chosen for this study.

This model is built on the ten-state Pink model (Pink et al, 1980) to describe the gel-fluid transition for each of the pure lipid components in the fully hydrated state. This model has proved to describe a wealth of thermodynamic, thermomechanic, and spectroscopic data for a variety of phospholipid membranes (Mouritsen, 1990; 1991; Dammann et al.,1995). Within the Pink model, the bilayer is considered as two independent monolayers, each represented by a triangular lattice on which the lipid chains are arrayed, one chain at each site. Within the lattice formulation, a protein molecule can occupy a certain number of sites and its hydrophobic part is assumed to be smooth, rod-like and characterized only by a cross-sectional area and length of its hydrophobic transmembrane domain. These data have been chosen in order to model a protein size corresponding to that of BR (Henderson and Unwin, 1975).

The lipid-protein interactions have been incorporated into the microscopic Pink model in the form of attractive van der Waals-like interactions and by identifying part of the interaction parameters in terms of a hydrophobic matching between the hydrophobic length of the lipid and that of the protein (Mouritsen and Bloom 1984; 1993).

The microscopic and the thermodynamic properties of the model have been calculated by standard Metropolis Monte Carlo simulation techniques (Mouritsen, 1990), both in absence and in the presence of protein.

The results refer to the DLPC/DSPC (25:75 mol/mol) mixture used for the fluorescence energy transfer experiments and to a BR-to-lipid molar ratios of 1/514 (the closest possible to the experimental ratio of 1/590 when a lattice of 40×40 sites is used). The simulation data are presented in the form of the microconfigurations shown in Fig. 1.

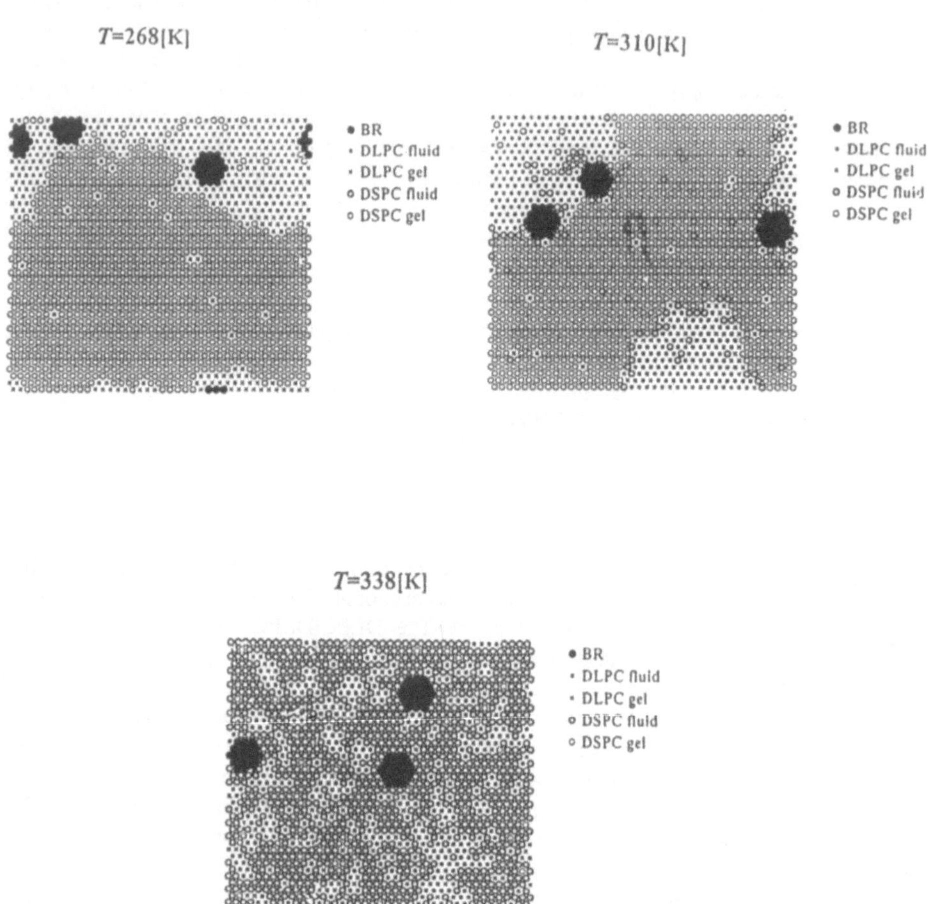

FIGURE 1: Microconfigurations obtained for DLPC/DSPC (25:75 mol/mol) mixtures in presence of BR with a protein-to-lipid ratio of 1/514. The data refers to the following temperatures: T=268K (a), T=310K (b), and T=338K (c). The symbols indicate the following: black=BR, cross=DLPC, circle=DSPC), small symbols = gel, large symbols = fluid. The data refer to simulations on a lattice with 40×40 sites.

At low temperature (T= 268K), there is a massive gel-gel phase separation as expected from the phase diagram of the DLPC/DSPC mixture (Marbrey and Sturtevant, 1976) and the protein is predominantly found in the DLPC-rich gel phase. At T= 310K, in the gel-fluid coexistence region, the protein tends to be located at the phase boundary with a shell of fluid DSPC around it. At T = 338K, in the fluid phase, a dictinct local demixing of the lipid mixture takes place and the protein shows preference for DSPC at the expense of DLPC.

Thus, the distribution of lipids around the protein molecules show there is an enrichment of one of the species and a depletion of the other in a systematic fashion reflecting the hydrophobic matching.

Experimental studies

Fluorescence energy transfer is a well established technique for measuring distances in biological systems. Successful application of this approach requires proper choice of lipid probes, in particular to ensure that the probe molecules reflect faithfully the behavior of the unlabeled species whose property are of studied. The emission spectrum of N-NBD-PE and the absorption spectrum of the retinal group of Bacteriorhodopsin show a significant overlap which enable fluorescence energy transfer measurements to be carried out with this donor-acceptor couple. Furthermore, when attached to the polar headgroup of phosphatidylethanolamine, the NBD group is located at the water/lipid interface and has no influence on the lipid molecular packing (Mazères et al., 1996). Thus, for studying the consequences of hydrophobic lipid/protein interactions in the DLPC/DSPC binary mixture, the probes N-NBD-DSPE and N-NBD-DLPE were used.

To test the feasibility of fluorescence energy transfer in studying the spatial relationship between NBD-labeled phospholipids and the retinal group of bacteriorhodopsin we employed egg-PC as host phospholipids and N-NBD-eggPE as probe. The extent of fluorescence quenching was determined by comparing the fluorescence I of vesicles containing the probe and Bacteriorhodopsin to the fluorescence I_0 of vesicles containing the same amount of the probe alone. The efficiency of energy transfer (quenching of the fluorescence of the donor) is defined by the relationship (Fung and Stryer, 1978):

$$E = 1 - I/I_0$$

The efficiency of transfer was evaluated experimentally for vesicles containing increasing amounts of egg-PC molecules in order to decrease the protein surface density, σ, i.e. to increase the distance between the donor and the acceptor. The dependence of energy transfer efficiency on the acceptor surface density was evaluated at different temperatures as shown in Fig.2.

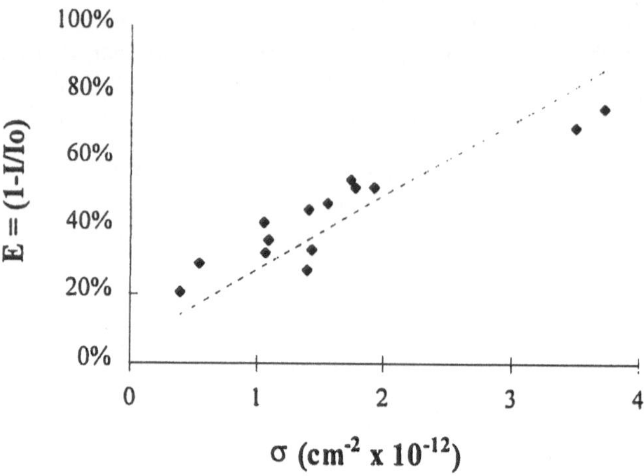

FIGURE 2: Influence of Bacteriorhodopsin surface density σ (molecules/cm²) on fluorescence energy transfer efficiency $E = 1 - I / I_o$. Bacteriorhodopsin was reconstitued in increasing amounts of egg-PC in presence of N-NBD-eggPE with a constant BR-to-probe ratio of 1/1. The reconstitution procedure used was described by Rigaud et al. (1995). Fluorescence emission spectra of liposomes (I_o) and proteoliposomes (I) were recorded at a 293K. σ was calculated assuming a molecular area of 6,90 nm² for BR in the monomeric form (Henderson et al 1990) and a molecular area of 0,63 nm² for egg-PC (Nagle, 1993).

As expected, E decreased when increasing σ. In addition, fluorescence emissions of lipid vesicles containing the donor only and of vesicles containing the donor and bleached Bacteriorhodopsin were identical. Thus, the fluorescence quenching observed in the presence of BR was due to a resonance energy transfer mechanism and the data shown in Fig.2 can be used to evaluate the surface densities of BR.

To explore the possibility of a lipid sorting by Bacteriorhodopsin in DLPC/DSPC mixtures, the protein was reconstituted in DLPC/DSPC (25:75 mol/mol) mixtures at a protein/lipid ratio of 1/590 with either N-NBD-DLPE or N-NBD-DSPE as probes. The Fluorescence energy transfer was measured over the temperature range 268K-338K.

As can be seen in Fig.3 the fluorescence energy transfer with the probe N-NBD-DLPE was high (60%) and remained unchanged up to 306K. Upon further temperature increase, it decreased rapidly down to a value less than 20%. An opposite effect was observed with the probe N-NBD-DSPE. In this case, the changes in energy transfer were weak and varied from 5% to 20%.

From this data, the apparent surface density of BR was calculated at each temperature and the results are shown in Fig.4.

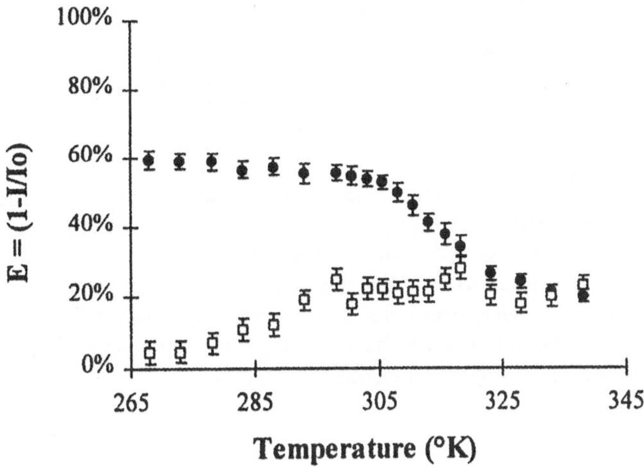

FIGURE 3: Temperature dependence of the fluorescence energy transfer efficiency for bacteriorhodopsin reconstitued in DLPC/DSPC (25:75 mol/mol) vesicles with a lipid-to-protein ratio of 1/590, in presence of N-NBD-DLPE (•) or N-NBD-DSPE (□) (prot:probe =1/1). Verticals bars indicate the dispersion of the data.

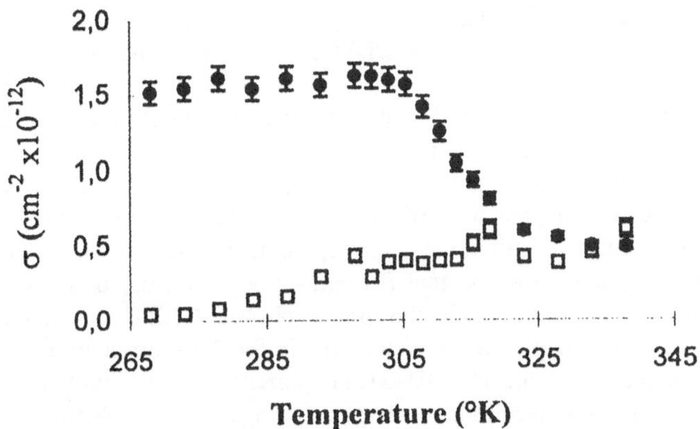

FIGURE 4: Temperature dependence of the apparent surface density σ (molecules/cm^2) for bacteriorhodopsin reconstitued in DLPC/DSPC. For each temperature tested, the fluorescence energy transfer values given in the Fig.3 are expressed in terms of σ using the data shown in the Fig. 2 (Dumas et al 1997).

If random distribution of BR in the DLPC-DSPC mixture occurred, the transfer efficiency would have been expected to be the same as that obtained with proteoliposomes of egg-PC at the protein to lipid ratio of 1/590. This corresponds to a value of $\sigma = 0{,}7\ 10^{12}$ and $0{,}52\ 10^{12}$ BR/cm^2 in the gel and the fluid phases respectively. However, for the probe N-NBD-C12-PE, σ_{app} remained unchanged at the high value of $1{,}6 \times 10^{12}$ BR/cm^2 for temperatures above 306K. Then, it decreased to a lower value of $0{,}5 \times 10^{12}$ BR/cm^2 for temperatures above 325K. For the probe N-NBD-C18-PE, σ progressively increased from a low value of $0{,}05 \times 10^{12}$ BR/cm^2 at 268K up to a value of $0{,}6 \times 10^{12}$ BR/cm^2 at 338K.

The data obtained shows that at low temperature the donor-acceptor pair distribution was not uniform. Close proximity between BR and the short-chain lipid existed in the gel phase and inversely BR was far from the long-chain lipid. At high temperature in the fluid phase the lipid-protein distribution was similar to that obtained with egg-PC for protein to lipid ratio of 1/590, which may correspond to an apparent random distribution of the various protein and lipid molecular species. .

Discussion and conclusion

Since most biological membranes contain a wide variety of lipid species with different acyl chain length, one can expected from a structural point of view, a transmembrane protein is solvated by the lipid species which better match its hydrophobic surface. For a few proteins, namely the reaction center and the light-harvesting antenna from *Rhodobacter sphaeroides* and bacteriorhodopsin from *Halobacterium halobium* reconstitued in phosphatidylcholines with different chain lengths, upward or downward shifts in the phase transition temperature of the lipids were detected depending on the relative protein/lipid hydrophobic thickness. On this ground, the consequences of hydrophobic matching on the lateral organization of lipids was studied in the ternary Bacteriorhodopsin/DLPC/DSPC mixture by combining a theoretical and an experimental approach.

From the phase diagram of DLPC/DSPC mixtures (Marbrey and Sturtevant, 1976; Dumas et al 1997) and fluorescence polarization data of bacteriorhodopsin vesicles (Dumas et al 1997), as well as from the computer-simulation results, one can assume that at 268K all the lipids species are in the gel state, that at 293K the DLPC molecules are in the fluid state while most of the DSPC molecules are still in the gel state, and that at 338K all the lipids are in the fluid state. At low temperature, the apparent BR surface densities, σ_{app} of 1.51×10^{12} BR/cm^2 and 0.05 10^{12} BR/cm^2 measured with the probe N-NBD-C12PE and N-NBD-C18PE respectively suggest a marked preference of the protein for the short-chain lipids in the gel phase. The data obtained with the N-NBD-C12PE corresponds to a protein/lipid ratio of 1/257. This number of lipid molecules is too large as compared to the 147 DLPC molecules (25%) associated on the average with one BR molecule. At this temperature and as shown by the computer simulations, the two lipid species are not fully separated since there is a small but finite solubility of DLPC in the DSPC-rich domains far from the protein. When taking account precisely the lipid distributions obtained from the computer simulation, one can recalculate an apparent surface density, σ_{cal}.. Good agreement between experimental (σ_{app}) and calculated (σ_{cal}) values is observed.

When the temperature was increased, σ_{app} was observed to decrease and increase for the probes N-NBD-C12PE and N-NBD-C18PE respectively. It should be remembered that the specific and fixed hydrophobic length of BR of ~ 30Å better matches the thickness of the DSPC bilayer in the fluid state (~ 30Å) or of the DLPC bilayer in the gel state (~ 29Å) as compared to DSPC in the gel state (~ 45Å) or DLPC in the fluid state (~20Å). The progressive increase in σ_{app} observed with N-NBD-DSPE may be explained by the fact that the melted DSPC molecules, which progressively enter the DLPC-rich phase, are selected by the protein and that BR displays a tendency to position itself at the interface between the gel and fluid lipid domains as shown by computer simulations. At 338K, in the fluid phase, the apparent surface density measured account for a random distribution. The computer simulations showed that in the fluid state, at least the first layer around BR is enriched in DSPC at the expense of DLPC and thus only small deviations from a random distribution of the two lipids species are expected. Taking account that at this temperatures the fluorescence approach reaches its limits of resolution, the experimental data do not allow to conclude definitely that in the fluid phase bacteriorhodopsin prefers DSPC as compared to DLPC. Nevertheless, the good agreement between the experimental and theoretical data observed in the gel-gel and gel-fluid coexistence regions gives strong support to the theoretical prediction that such a lipid sorting takes place.

It is likely that the molecular sorting mechanism described in the present paper in the case of Bacteriorhodopsin is operative for a large class of integral membrane proteins. Another example is provided by the hydrophobic pulmonary surfactant protein, SP-C, which was recently shown to prefer palmitoyl lipids according to the hydrophobic matching principle (Horowitz and al, 1995). A particular question is related to the overall effect of the organization of lipid-protein assemblies on the function of the protein. The function of BR is believed not to be associated with any conformational changes that alter the hydrophobic length of the membrane domain. However, this may be the case for other proteins, whose function therefore can be controlled by the lipid bilayer structure near the protein. It would be of interest to extend this study to proteins whose the function depends on the surrounding lipids. Furthermore, it would be of interest to investigate whether structured lipid concentration profiles around proteins might influence their state of segregation and/or aggregation in the case of proteins whose biological activity depends on their aggregation state (Kaprelyants, 1988; Andersen, 1989).

References

Andersen, A. S. 1989. Reception and transmission. Nature 337:12.

Bergelson, L. O., Gawrisch, J.A. Feretti, and R. Blumenthal (eds.). 1995. Special Issue on Domain Organization in Biological Membranes. Mol. Memb Biol. 12:1-162.

Brown, M. F. 1994. Modulation of rhodopsin function by properties of the membrane bilayer. Chem. Phys. Lipids 73:159-180.

Caffrey, M., and G.W. Feigenson. 1981. Fluorescence quenching in model membranes. Relationship between calcium adenosinetriphosphate enzyme activity of the protein forphosphatidylcholines with different acyl chain characteristics. Biochemistry 20:1949-1961.

Cornea, R.L. and Thomas, D.D. 1994. Effects of membrane thickness on the molecular dynamics and enzymatic activity of reconstituted Ca-ATPase. Biochemistry 33: 2912-2920.

Devaux, P.F. 1993. Lipid transmembrane asymmetry and flip-flop in biological membranes and lipid bilayers. Curr. Opin. Struct. Biol. 3: 489-494.

Dumas, F., Sperotto, M.M., Lebrun, C., Tocanne J.F. and Mouritsen, O.G. 1997. Molecular sorting of lipids by bacteriorhodopsin in DLPC/DSPC lipid bilayers. Biophys. J. In the press.

Glaser, M. 1993. Lipid domains in biological membranes. Curr. Opin. Struct. Biol. 3: 475-481.

Haydon. D. A. and Hladky. S. B. (1972). Quart. Rev. Biophys. 5:187-282.

Henderson, R., J.M. Baldwin, T.A. Ceska, F. Zemlin, E. Beckman, and K.H. Downing. 1990. Model of the structure of Bacteriorhodopsin based on high-resolution electron cryo-microscopy. J. Mol. Biol. 213:899-929.

Henderson, R., and P.N.T. Unwin. 1975. Three-dimensional of purple membrane obtained by electron microscopy. Nature 257:28-32.

Horowitz, A. D. 1995. Exclusion of SP-C, but not SP-B, by gel phase palmitoyl lipids. Chem. Phys. Lipids. 76: 27-39.

Israelachvili, J. 1992. Intermolecular and Surface Forces. Academic Press, London.

Johannsson, A., G. A. Smith, and J.C. Metcalfe 1981a. The effect of bilayer thickness on the activity of $(Na^{+}-K^{+})$-ATPase. Biochim. Biophys. Acta 641:416-421.

Johannsson, A., C. A. Keithley, G.A. Smith, C.D. Richards, T.R. Hesketh, and J.C. Metcalfe. 1981b. The effect of bilayer thickness and n-alkanes on the activity of the $(Ca^{2+}-Mg^{2+})$dependent ATPase of Sarcoplasmic Reticulum. J. Biol. Chem. 256:1643-1650.

Kaprelyants, A.S. 1988. Dynamic spatial distribution of proteins in the cell. TIBS 13:43-46.

Kusumi, A. and Sako, Y. 1996. Cell surface organization by the membrane skeleton. Curr. Opin. Cell. Biol. 8: 566-574

Marbrey, S., and J.M. Sturtevant. 1976. Investigation of phase transitions of lipids and lipid mixtures by high sensitivity differential scanning calorimetry. Proc. Natl. Acad. Sci. USA 76:3862-3866.

Mazères, S, V.Schram, J-F. Tocanne, and A. Lopez. 1996. 7-nitrobene-2-oxa-1,3-diazole-4-yl labeled phospholipids in lipid membranes: differences in fluorescence behaviour. Biophys. J. 71:327-335.

Moore. B. M., Lentz. B.R., Hoechli. M. and Meissner. G. (1981). Biochemistry. 20:6810-6817.

Mouritsen, O. G. and Bloom, M. 1984. Mattress model of lipid-protein interactions in membranes. Biophys. J. 46: 141-153.

Mouritsen, O.G. 1990. Computer simulation of cooperative phenomena in lipid membranes. In moleculmar Description of Biological Membrane Components by Computer Aided conformational Analysis. R. Brasseur, editor. CRC Press, Boca Raton, FL. 3-83.

Mouritsen, O. G., and R. L. Biltonen. 1993. Protein-lipid interactions and membrane heterogeneity. In: Protein-Lipid Interactions (Wattsd, A., Ed.), pp. 1-39, Elsevier Science, Amsterdam.

Mouritsen, O.G., and M.Bloom. 1993. Models of Lipid-Protein Interactions in Membranes. Ann. Rev. Biophys. Biomol. Struct. 22:145-171.

Peschke. J., Riegler, J. and Möhwald, H. 1987. Quantitative analysis of membrane distorsions induced by mismatch of protein and lipid hydrophobic thickness. Eur. Biophys. J. 14: 385-391.

Piknova, B., Perochon, E. and Tocanne, J.F. 1993. Hydrophobic mismatch and long-range protein-lipid interactions in bacteriorhodopsin/phosphatidylcholine vesicles. Eur. J.Biochem. 218:385-396

Pink, D.A., J. G. Green, a,nd D. Chapman. 1980. Raman scattering in bilayers of saturated phosphatidylcholines. Experiment and theory. Biochemistry 19:349-356.

Riegler, J. and Möhwald, H. 1986. Elastic interactions of photosynthetic reaction center proteins affecting phase transitions and protein distributions. Biophys. J. 49: 1111-1118.

Rigaud, J.L., Pitard, and D. Levy. 1995. Reconstitution of membrane proteins into liposomes: Application to energy transducing membrane proteins. Biochim. Biophys. Acta. 1231:223-246.

Sheets, E.D., Simson, R. and Jacobson, K. 1995. New insights into membrane dynamics from the anlysis of cell surface interactions by physical methods. Curr. Opin. Cell. Biol. 7: 707-714.

Sperotto, M. M., and O. G. Mouritsen. 1991a. Mean-field and Monte Carlo simulation studies of the lateral distribution of proteins in membranes. Eur. Biophys. J. 19:157-168.

Sperotto, M. M., and O. G. Mouritsen. 1991b. Moye Carlo simulation studies of lipid order parameter profiles near integralm membra,nes proteins. Biophys. J. 59:261-270.

Sperotto, M. and Mouritsen, O. 1993. Lipid enrichment and selectivity of integral membrane proteins in two component lipid bilayers. Eur. Biophys. J. 22: 323-328.

Starling, A.P., East, J.M. and Lee, A.G. 1993. Effects of phosphatidylcholine fatty acyl chain length on calcium binding and other functions of the (Ca^{++} - Mg^{++})-ATPase. Biochemistry 32: 1593-1600.

Tocanne, J.F. 1992. Detection of lipid domains in biological membranes. Comments Mol. Cell. Biophys. 8: 53-72.

Tocanne, J.F., Dupou-Cézanne, L., Lopez, A., Perochon, E., Piknova, B., Schram, V., Tournier, J.F. and Welby, M. 1994. Lipid domains and lipid/protein interactions in biological membranes. Chem. Phys. Lipids. 73, 139-158

Wolf, D.E., Maynard, V.M., McKinnon, C.A. and Melchior, D.L. 1990. Lipid domains in the ram sperm plasmamembrane demonstrated by differential scanning calorimetry. Proc. Natl. Acad. Sci. USA. 87: 6893-6896

Translocation of Phospholipids Between the Endoplasmic Reticulum and Mitochondria in Yeast

G. Tuller, G. Achleitner, B. Gaigg, A. Krasser, E. Kainersdorfer, S.D. Kohlwein*, A. Perktold§, G. Zellnig§ and G. Daum*

*SFB Biomembrane Research Center
Institut für Biochemie und Lebensmittelchemie
Technische Universität, Petersgasse 12/2
A-8010 Graz, Austria
§Institut für Pflanzenphysiologie
Karl-Franzens Universität
Schubertstr. 51
A-8010 Graz, Austria

Several mechanisms have been proposed by which lipids are translocated between membranes of eukaryotic cells: (i) transport of lipid monomers with or without the aid of soluble carrier proteins; (ii) vesicle flux; and (iii) migration of lipids between donor and acceptor through membrane contact (for a recent review see Trotter and Voelker, 1994). The aim of the present study was to characterize the transport of aminoglycerophospholipids between the endoplasmic reticulum and the mitochondria of the yeast, *Saccharomyces cerevisiae*. This process is important because synthesis of phosphatidylethanolamine catalyzed by the mitochondrial phosphatidylserine decarboxylase, Psd1p (Trotter et al., 1993), requires import into mitochondria of phosphatidylserine formed by phosphatidylserine synthase in the endoplasmic reticulum (Zinser et al., 1991). In intact mammalian cells the rate of phosphatidylserine transport from microsomes to mitochondria is reduced by the energy blockers azide and fluoride (Voelker, 1988) and enhanced by ATP in permeabilized cells (Voelker, 1989b; 1990; 1993). Interorganelle translocation of phosphatidylserine was shown to be independent of cytosolic factors in permeabilized mammalian cells (Voelker, 1989b) and in a cell free system (Voelker, 1989a).

NATO ASI Series, Vol. H 106
Lipid and Protein Traffic
Pathways and Molecular Mechanisms
Edited by Jos A. F. Op den Kamp

Recently, phosphatidylserine import into mitochondria was characterized in permeabilized yeast cells (Achleitner et al., 1995). Translocation of phosphatidylserine from its site of synthesis to Psd1p located in the inner mitochondrial membrane was neither affected by cytosolic proteins nor ATP. In contrast, export of phosphatidylethanolamine from mitochondria to the endoplasmic reticulum is ATP-dependent. Membrane contact between the endoplasmic reticulum and mitochondria was suggested as a prerequisite for lipid transport between the two organelles.

Similar to mammalian cells (Vance, 1990) a phospholipid-synthesizing mitochondria associated membrane (MAM) was isolated from yeast (Gaigg et al., 1995). MAM is a specialized subfraction of the endoplasmic reticulum which contains phosphatidylserine and phosphatidylinositol synthase at high specific activity. Physical contact between MAM and mitochondria leads to functional reconstitution of synthesis, transport and decarboxylation of phosphatidylserine. These results are in line with findings that structural elements coupling a donor membrane compartment to the mitochondria are required for phosphatidylserine transport to mitochondria (Voelker, 1993; Achleitner et al., 1995).

METHODOLOGY

The wild-type yeast strain *Saccharomyces cerevisiae* D273-10B, and a revertant strain of a phosphatidylserine synthase deficient *cho1* mutant, SDK03-1AR (Sperka-Gottlieb et al., 1990) were used throughout these studies. D273-10B was grown on 2% lactate medium (Daum et al., 1982), and SDK03-1AR was cultivated on YPD-medium with 3% glucose as the carbon source. Both strains were cultivated under aerobic conditions at 30°C and harvested in the logarithmic phase.

For three-dimensional image reconstruction of yeast cells serial ultrathin sections (80-90 nm) were transferred to single slot grids, stained with lead citrate and viewed with a Philips CM 10 electron microscope. Images visualized by Transmission Electron Microscopy were stored and processed using Optimas 4.02 software. Tracing of structures was carried out with program Corel 6.0. Parallel projection was used for 3D reconstruction.

Yeast organelles were isolated by published procedures (see Zinser and Daum, 1995). A mitochondria associated membrane (MAM) was prepared as described by Gaigg et al.(1995). For import assays of phosphatidylserine from MAM into mitochondria *in vitro* MAM isolated from the wild-type yeast strain D273-10B (final protein concentration 0.1 mg/ml) was mixed with density gradient-purified mitochondria from the *cho1* revertant SDK03-1AR in 0.6 M mannitol, 10 mM Tris-HCl, pH 7.4, 0.6 mM $MnCl_2$, 0.2 mM CDP-diacylglycerol.

Mitochondria of SDK03-1AR do not exhibit phosphatidylserine synthase activity. This is important because all phosphatidylserine formed in this assay can be assigned to synthesis in the MAM fraction. Radioactive phosphatidylserine was synthesized in the assay mixture using [^{3}H]serine (30 µCi; 170 mCi/mmol) as a precursor. After 10 or 20 min of incubation EDTA was added (final concentration 5 mM) to stimulate the formation of phosphatidylethanolamine. The cation-sensitive phosphatidylserine decarboxylase was used as the reporter enzyme for the import of phosphatidylserine into mitochondria. During the third phase of the assay synthesis of [^{3}H]phosphatidylcholine in the MAM fraction from [^{3}H]phosphatidylethanolamine newly formed in mitochondria was monitored after addition of 8 mM $MgCl_2$ and 0.2 mM SAM (final concentration, each) to the incubation mixture. Mg^{++} stimulates the phosphatidylethanolamine *N*-methyltransferase, and SAM is the methyl-donor for the synthesis of phosphatidylcholine.

To distinguish between co- and post-synthetic import of phosphatidylserine into mitochondria MAM were first labeled with [^{3}H]serine (see above) and re-isolated. Then pre-labeled MAM and mitochondria were mixed, and labeling was continued in the presence of both organelles with [^{14}C]serine for 10 min. Finally 5 mM EDTA was added to convert phosphatidylserine imported into mitochondria to phosphatidylethanolamine.

Lipids were extracted according to the method of Folch et al. (1957), and individual phospholipids were separated by thin-layer chromatography with chloroform/ methanol/25 % NH_3 (50:25:6; per vol.) as a developing solvent. Spots were visualized by iodine vapor, scraped off from the plate, and radioactivity was measured by liquid scintillation counting.

RESULTS

In this study we addressed the question as to the transport of aminoglycerophospholipids between MAM and mitochondria and its dependence on metabolic energy, cytosolic factors, organelle surface proteins, and ongoing lipid biosynthesis. Translocation of these lipids between the two organelles can be monitored by the appearance of metabolic products with phosphatidylethanolamine serving as a marker for phosphatidylserine import into mitochondria, and phosphatidylcholine as a marker for phosphatidylethanolamine export to the endoplasmic reticulum.

To visualize physical association of the endoplasmic reticulum with mitochondria we used a computer-supported method of 3D-image reconstruction of yeast cells from serial ultrathin-sections. Fig. 1 shows that parts of the endoplasmic reticulum network are located in juxtaposition to mitochondria. These regions of the endoplasmic reticulum may be identical to

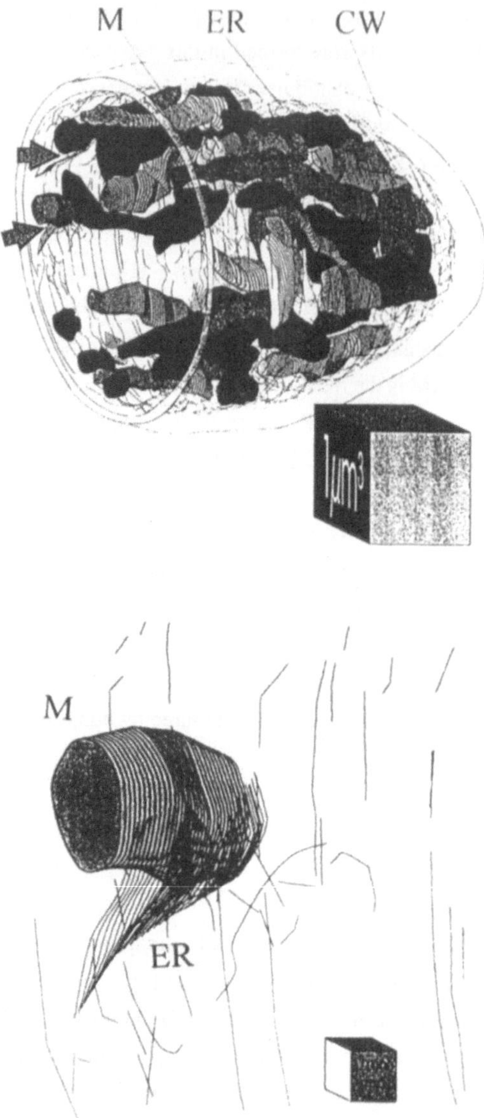

Fig. 1 Three-dimensional imaging of a fixed yeast cell. The upper panel shows cell wall (CW), endoplasmic reticulum (ER) and mitochondria (M) of a whole cell. Other organelles have been faded out. The lower panel shows a region of contact between endoplasmic reticulum and mitochondria in more detail.

MAM characterized before in our laboratory (Gaigg et al., 1995) and serve as bridges for lipid translocation between the endoplasmic reticulum and mitochondria.

Re-association of MAM with mitochondria leads to reconstitution of the import of phosphatidylserine into mitochondria and its conversion to phosphatidylethanolamine by the mitochondrial phosphatidylserine decarboxylase (Fig. 2). Conversion of [^{3}H]phosphatidylserine to phosphatidylethanolamine served as a measure for the import of phosphatidylserine to the inner mitochondrial membrane.

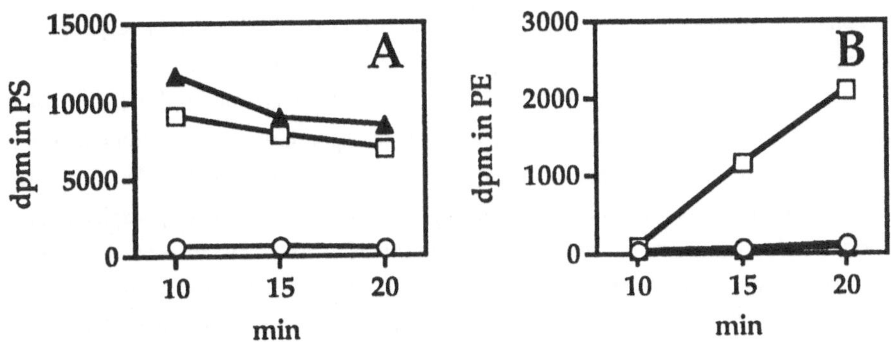

Fig. 2 Reconstitution of phosphatidylserine import from MAM into mitochondria. [^{3}H]Phosphatidylserine (A) was synthesized in an assay containing MAM (▲--▲) of wild-type cells or mitochondria (O--O) of the *cho1* revertant SDK03-1AR, or a mixture (□--□) of both organelles. After 10 min of incubation, EDTA was added and formation of phosphatidylethanolamine (B) was monitored.

To characterize translocation of phosphatidylserine from MAM to mitochondria as a result of membrane association we tested the role of cytosolic and organelle surface protein and the requirement of energy for this process. A yeast phosphatidylserine transfer protein (Lafer et al., 1991) significantly enhanced the transport of phosphatidylserine from unilamellar phospholipid vesicles to mitochondria (Simbeni et al., 1993). No such enhancement could be observed in the reconstituted system of MAM and mitochondria. Under standard conditions approximately 20 % of newly formed phosphatidylserine were converted to phosphatidylethanolamine. When cytosolic proteins were present in the assay mixture, the synthesis of phosphatidylserine dropped to 70% of the control without cytosol, and 12% of phosphatidylserine formed were subsequently decarboxylated. Thus, cytosolic proteins rather inhibited than stimulated the transfer of phosphatidylserine between MAM and mitochondria *in vitro*.

In permeabilized yeast cells the import of phosphatidylserine into mitochondria is energy-independent (Achleitner et al., 1995). Similar results were obtained with the MAM-mitochondria reconstituted system (data not shown). Neither addition of ATP to the assay mixture nor energy depletion by addition of apyrase, CCCP, F^- and N_3^- resulted in marked changes of phosphatidylserine import from MAM into mitochondria.

Ongoing synthesis of phosphatidylserine was tested as another possible driving force for the import of the phospholipid into mitochondria. A double labeling experiment (see Methods section) was designed to distinguish between post- and co-synthetic transport of phosphatidylserine from MAM to mitochondria. Phosphatidylserine either pre-formed in MAM or synthesized during contact of MAM with mitochondria reached the inner mitochondrial membrane at similar rates. This result demonstrates that synthesis of the phospholipid and interorganelle translocation can occur independently of each other *in vitro*.

To test whether proteins of mitochondria and/or MAM are involved in the association between the two organelles and translocation of phosphatidylserine, proteins of the two fractions were inactivated either by treatment with protease or heat (Fig. 3). Proteinase K treated mitochondria exhibited a slightly but significantly reduced rate of phosphatidylserine import. Similar results were obtained with trypsin (not shown). Inactivation of MAM proteins by heat also resulted in a decrease of phosphatidylserine import. When surface proteins of both MAM and mitochondria were inactivated with proteinase K the import of phosphatidylserine decreased to approximately 50 % of the control.

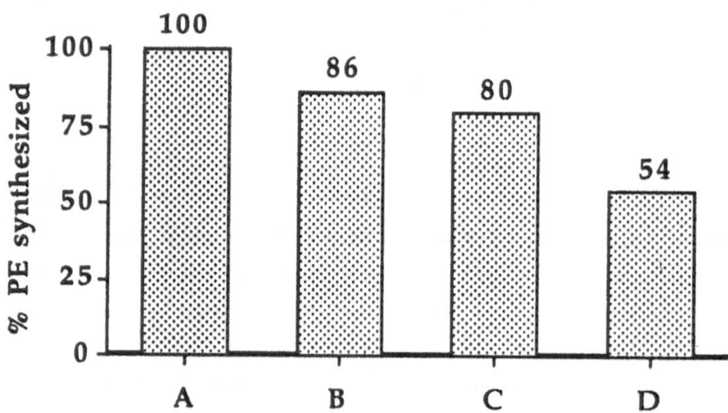

Fig. 3 Proteins of MAM and mitochondria influence phosphatidylserine transport between the two organelles. Translocation assays were carried out as described in the Methods section. A: control; B: MAM treated at 90°C for 15 min; C: mitochondria treated with proteinase K; D: MAM and mitochondria treated with proteinase K.

Simbeni et al. (1990) had demonstrated that phosphatidylethanolamine newly formed in the inner mitochondrial membrane was rapidly exported to the mitochondrial surface. Further conversion to phosphatidylcholine requires translocation to the microsomal phosphatidylethanolamine- and phospholipid-*N*-methyltransferases (Zinser et al., 1991; Gaigg et al., 1995). MAM is not the compartment with the highest specific activity of phosphatidylethanolamine- and phospholipid-*N*-methyltransferase, but contains significant levels of both enzymes. Thus we were able to tested whether association between MAM and mitochondria is sufficient for the export of phosphatidylethanolamine and further conversion to phosphatidylcholine. These experiments were carried out with "crude" mitochondria which have MAM still attached to the cytosolic side of the outer mitochondrial membrane. As can be seen from Fig. 4 the functional unit of mitochondria and MAM has not only the capacity to synthesize phosphatidylserine, import it into mitochondria, and convert it to phosphatidylethanolamine, but also to export phosphatidylethanolamine which serves as a substrate for phosphatidylcholine synthesis in MAM in the presence of the methyl donor S-adenosyl methionine. A three-step assay had to be designed for the sequential formation of phosphatidylserine, phosphatidylethanolamine and phosphatidylcholine because of the different biochemical requirements of each step involved (see Fig. 4 and Methods section). Conversion of phosphatidylethanolamine to phosphatidylcholine is not very efficient, and the intermediate methylated product, phosphatidyldimethyl-

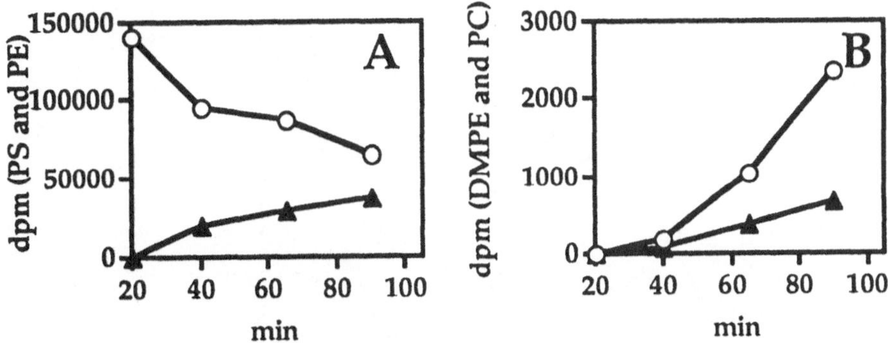

Fig. 4 Export of phosphatidylethanolamine from mitochondria to MAM *in vitro*. Mitochondria with MAM attached were incubated with [^{3}H]serine as a precursor for aminoglycerophospholipid synthesis in a three-step assay as described in the Methods section. Mn^{++} was present during the first phase (not shown; 0-20 min), EDTA during the second phase (20-40 min), and Mg^{++} and SAM during the third phase (40-90 min) of the assay. A: phosphatidylserine (O--O); phosphatidylethanolamine (▲--▲); B: phosphatidyldimethylethanolamine (O--O); phosphatidylcholine (▲--▲).

ethanolamine, was always found to a significant amount. The ratio of the three aminoglycerophospholipids formed in this *in vitro* assay was similar to that obtained with permeabilized yeast cells (Achleitner et al., 1995).

In permeabilized yeast cells export of phosphatidylethanolamine from mitochondria to the endoplasmic reticulum did not require cytosolic proteins, but was ATP-dependent (Achleitner et al., 1995). These results are in agreement with previous findings obtained with intact yeast cells (Gnamusch et al., 1992). Preliminary results indicate that export of phosphatidylethanolamine from isolated mitochondria to the endoplasmic reticulum may also require ATP in the reconstituted system presented in this paper.

DISCUSSION

A mitochondria associated membrane (MAM) of the yeast, *Saccharomyces cerevisiae*, purified by density gradient centrifugation (Gaigg et al., 1995) can be re-associated with highly

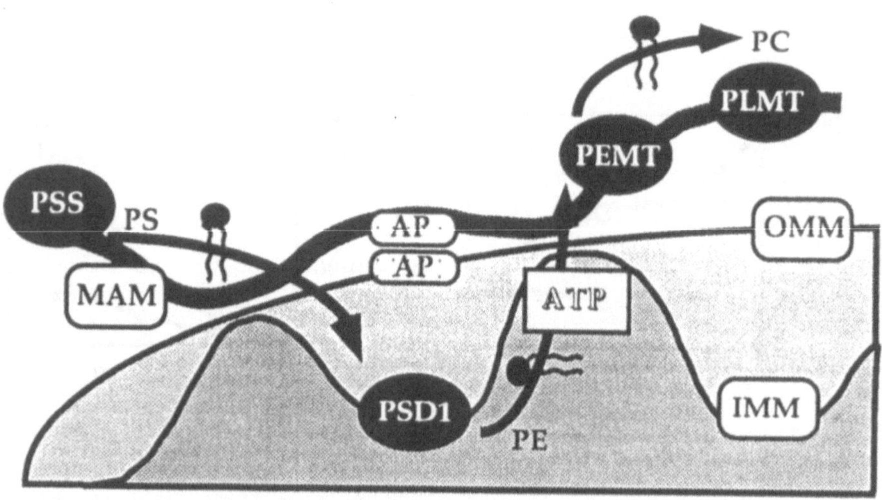

Fig. 5 Hypothetical scheme of transport of phosphatidylserine (PS) synthesized by phosphatidylserine synthase (PSS) in the MAM (mitochondria associated membrane) fraction to phosphatidylserine decarboxylase 1 (PSD1), and export of phosphatidylethanolamine (PE) for conversion to phosphatidylcholine (PC). OMM: outer mitochondrial membrane: IMM: inner mitochondrial membrane; PEMT: phosphatidylethanolamine-*N*-methyltransferase; PLMT: phospholipid-*N*-methyltransferase; AP: hypothetical association protein.

purified mitochondria. The physical contact of MAM, which is a subfraction of the endoplasmic reticulum, with mitochondria restores the capability to synthesize, translocate and decarboxylate phosphatidylserine (Fig. 5). Stimulation of phosphatidylserine transfer by membrane surface proteins of organelles in a vesicle/organelle transport assay has been shown before by Gaigg et al. (1993). We assume that proteins present on the surface of MAM and mitochondria (named AP in Fig. 5) stimulate the association and thus translocation of phospholipids between the two compartments. Fusogenic proteins similar to those described recently by Rakowska et al. (1994) may be good candidates to enhance membrane association. In contrast, cytosolic phospholipid transfer proteins do not increase the translocation rate of phosphatidylserine between yeast MAM and mitochondria *in vitro*. These findings are in line with previous observations from our laboratory with permeabilized yeast (Achleitner et al., 1995). Similarly, translocation of phosphatidylserine between isolated endoplasmic reticulum and mitochondria of mammalian cells was not stimulated by cytosolic proteins (Voelker, 1989a).

In the reconstituted system of yeast MAM with mitochondria neither addition of ATP nor ATP depletion had a significant effect on the import of phosphatidylserine into mitochondria which is in good agreement with previous results obtained with permeabilized yeast cells (Achleitner et al., 1995). These findings are in contrast to observations made with permeabilized mammalian cells which require ATP for translocation of phosphatidylserine from the endoplasmic reticulum to mitochondria (Voelker, 1989b, 1990, 1993). It is very likely that this discrepancy between the mammalian and the yeast system is due to the difference in the enzymatic formation of phosphatidylserine. In mammalian cells phosphatidylserine synthesis is a base-exchange reaction and coupled to the action of a Ca^{++} sequestering ATPase of the endoplasmic reticulum whereas in yeast phosphatidylserine is synthesized from CDP-diacylglycerol and serine (see Paltauf et al., 1992). In yeast, requirement of ATP for phosphatidylserine biosynthesis is indirect because formation of CDP-diacylglycerol requires energy in the form of nucleotide triphosphates. Accessibility of nascent phosphatidylserine to the translocation machinery may be different in mammalian cells and yeast resulting in different requirements for intracellular transport.

ATP requirement for the export of phosphatidylethanolamine from mitochondria to the endoplasmic reticulum in yeast is not understood at the molecular level at present. We can only speculate that phosphorylation of a component participating in this process, energy-consuming steps of vesicle flux, or transmembrane movement of the phospholipid catalyzed by ATP-dependent translocases or proteins of the ABC transporter family may be the reason for this finding.

ACKNOWLEDGMENTS

This work was supported by the Fonds zur Förderung der wissenschaftlichen Forschung in Österreich (projects S-5811 and 12076 to G.D., and S-5812 and F 706 to S.D.K.).

REFERENCES

Achleitner, G., Zweytick, D. Trotter, P. J., Voelker, D. R. and Daum, G. (1995) J. Biol. Chem. 270, 29836-29842

Daum, G., Boehni, P.C. and Schatz, G. (1982) J. Biol. Chem. 257, 13028-13033

Folch, J., Lees, M. and Sloane-Stanley, G. H. (1957) J. Biol. Chem. 226, 497-509

Gaigg, B., Simbeni, R., Hrastnik, C., Paltauf, F. and Daum, G. (1995) Biochim. Biophys. Acta 1234, 214-220

Gaigg, B., Lafer, G., Paltauf, F. and Daum, G. (1993) Biochim. Biophys. Acta 1146, 301-304

Gnamusch, E., Kalaus, C., Hrastnik, C., Paltauf, F. and Daum, G. (1992) Biochim. Biophys. Acta 1111, 120-126

Lafer, G., Szolderits, G., Paltauf, F. and Daum, G. (1991) Biochim. Biophys. Acta 1069, 139-144

Paltauf, F., Kohlwein, S. D. and Henry, S. A. (1992) In The molecular and cellular biology of the yeast Saccharomyces: Gene expression. Volume II. (E. W. Jones, J. R. Pringle, J. R. Broach eds.) pp. 415-500, Cold Spring Harbor Laboratory Press

Rakowska, M., Zborowski, J. and Corazzi, L. (1994) Membr. Biol. 142, 35-42

Simbeni, R., Paltauf, F. and Daum, G. (1990) J. Biol. Chem. 265, 281-285

Simbeni, R., Tangemann, K., Schmidt, M., Ceolotto, C., Paltauf, F. and Daum, G. (1993) Biochim. Biophys. Acta 1145, 1-7

Sperka-Gottlieb, C.D.M., Fasch, E.-V., Kuchler, K., Bailis, A.M., Henry, S.A., Paltauf, F. and Kohlwein, S.D. (1990) Yeast 6, 331-343

Trotter, P. J. and Voelker, D. R. (1994) Biochim. Biophys. Acta 1213, 241-261

Trotter, P. J., Pedretti, J. and Voelker, D. R. (1993) J. Biol. Chem. 268, 21416-21424

Vance, J.E. (1990) J. Biol. Chem. 265, 7248-7256

Voelker, D.R. (1989a) J. Biol. Chem. 264, 8019-8025

Voelker, D.R. (1989b) Proc. Natl. Acad. Sci. U.S.A. 86, 9921-9925

Voelker, D.R. (1990) J. Biol. Chem. 265, 14340-14346

Voelker, D. R. (1988) in Biological Membranes: Aberrations in Membrane Structure and Function. pp. 153-164, Alan R. Liss, Inc.

Voelker, D. R. (1993) J. Biol. Chem. 268, 7069-7074

Zinser, E. and Daum, G. (1995) Yeast 11, 493-536

Zinser, E., Sperka-Gottlieb, C. D. M., Fasch, E.-V., Kohlwein, S.D., Paltauf, F. and Daum, G. (1991) J. Bacteriol. 173, 2026-2034

Lipid Peroxidation in Myocardial Cells Under Ischemia and Oxidative Stress. The Effect of Allopurinol and Oxypurinol

Jan A. Post[*], Chris Th.W.M. Schneijdenberg[*], Yvette C.M. de Hingh[+], Robert van Rooijen, Jaap R. Lahpor[++], Jos A.F. Op den Kamp[+] and Arie J. Verkleij[*].

Institute of Biomembranes; [*]Department of Molecular Cell Biology; [+]Center for Biomembranes and Lipid Enzymology, Utrecht University, The Netherlands and [++]Heart Lung Institute, University Hospital Utrecht, Utrecht, The Netherlands.[1]

Abstract

Oxygen derived free radicals (OFR) are thought to play an important role in the development of reperfusion injury of the heart after a period of ischemia. One of the targets of OFR are considered to be (phospho)lipids and one of the mechanisms of generation of OFR is the activation of xanthine oxidase (XO) during reperfusion. This study was set up to assess i) whether activation of XO can lead to cell lysis during ischemia/reperfusion and whether inhibitors of XO (allopurinol and oxypurinol) can attenuate this cell lysis and ii) whether activation of XO can lead to lipid peroxidation and whether this is affected by allopurinol and oxypurinol.

Cultured neonatal rat heart cells were subjected to simulated ischemia/reperfusion conditions in the presence and absence of (hypo)xanthine and XO. It was found that inclusion of the OFR generating system increased cell lysis five fold during the ischemia/reperfusion protocol. Under normoxic conditions the OFR generating system did not induce cell lysis. Both allopurinol and oxypurinol were able to reduce cell lysis during ischemia/reperfusion + OFR generation at concentrations of 1 μmol/l or higher.

Lipid peroxidation was studied using the fluorescent fatty acid parinaric acid (PnA), which was incorporated in a lipid matrix. Application of the OFR generating system led to a decrease of the PnA signal, directly showing lipid peroxidation by OFR generated by XO. This lipid

[1] Correspondence: Dr. Jan A. Post, Dept. Mol. Cell Biol. /EMSA, Padualaan 8, 3584 CH Utrecht, The Netherlands. E-mail: jap@emsaserv.biol.ruu.nl

NATO ASI Series, Vol. H 106
Lipid and Protein Traffic
Pathways and Molecular Mechanisms
Edited by Jos A. F. Op den Kamp

peroxidation was inhibited by both allopurinol and oxypurinol, again in the μmol/l range. Next, we tested the effect of these drugs on XO activity, giving results identical to the lipid peroxidation studies. This indicates that the effect of the drugs on lipid peroxidation is mediated by a direct effect on OFR production by the enzyme, rather than scavenging OFR.

The data show that cultured neonatal rat heart cells become susceptible to xanthine oxidase derived OFR upon ischemia, that xanthine oxidase derived OFR are capable of peroxidizing lipids and that both processes are inhibited by allopurinol and oxypurinol.

Introduction

Ischemia/reperfusion of myocardial tissue occurs under a variety of conditions, one of them being aortocoronary by-pass grafting, where it is necessary to crossclamp the aorta temporarily, followed by release of the aortic clamp, leading to reperfusion of the myocardium and associated reperfusion injury (Guanieri *et al.* 1980, Hearse *et al.* 1992, Tompson *et al.* 1986). One of the clinically important forms of reperfusion injury is stunning, which results in a measurable reduction of myocardial contractility during the first post-operative hours (Bolli *et al.* 1990, Coghlan *et al.* 1993, Flameng *et al.* 1983). Reperfusion injury can also lead to cell lysis and necrosis. Although the exact molecular mechanism(s) underlying ischemia/reperfusion injury remain to be elucidated, it is generally accepted that the different mechanisms involved lead to a disturbance of calcium homeostasis of the cardiac myocyte, subsequently leading to possible activation of phospholipases, proteases and/or affecting physico-chemical behavior of lipids and cell damage (Jennings *et al.* 1991, Post *et al.* 1995a).

As one of the mediators of reperfusion injury, oxidative stress has been put forward. Among the sources of oxygen free radicals the enzyme xanthine oxidase (EC 1.3.2.2) is thought to be a major one at the time of reperfusion (Hearse *et al.* 1986, McCord *et al.* 1985). During ischemia xanthine oxidase is formed by proteolytic conversion from xanthine dehydrogenase (McCord *et al.* 1982). Xanthine oxidase catalyzes the oxidation of (hypo)xanthine, thereby generating the superoxide anion, which can lead to the production of the very reactive hydroxyl radical.

Although in experimental animal studies the role of xanthine oxidase generated free radicals is clearly shown (Akizuki *et al.* 1985, Badylak *et al.* 1987), the role of xanthine oxidase in the pathophysiology of ischemia/reperfusion injury of the human myocardium is at the moment controversial (for review see Kooij 1994, Janssen *et al.* 1993). However, a very strong indication that xanthine oxidase derived free radicals do play a role in the development of reperfusion injury is given by the fact that allopurinol, a competitive inhibitor of xanthine oxidase, has a beneficial effect on recovery of the human myocardium upon ischemia/reperfusion during open heart surgery (Sisto *et al.* 1995, Tabayashi *et al.* 1991, Stewart *et al.* 1985, Gimpel *et al.* 1995).

To gain further insight in the mechanism of action of xanthine oxidase derived oxygen free radicals and of the possible protective effect of allopurinol, and its metabolite oxypurinol, we used in an *in vitro* cell culture model of "simulated ischemia". This model, which displays myocytal damage similar to Langendorff perfused preparations and *in vivo* situations (Musters *et al.* 1991, 1993), allows a tight control of the extracellular environment during ischemia/reperfusion. It is tested whether xanthine oxidase derived free radicals are able to induce cell lysis and whether inhibitors of xanthine oxidase can attenuate this damage. Next to this an *in vitro* system is used to specifically measure lipid peroxidation upon the production of oxygen free radicals by xanthine oxidase, and to study the effect of both drugs on this process.

Results

In the present study we used standard protocols and procedures as described earlier. Cell isolation and culturing (Musters et al., 1991; Post et al., 1988) was followed by application of ischemic conditions (Musters et al. 1993) and incubations as described in this paper. Cell integrity was followed by measuring lactate dehydrogenase release (Post et al., 1995b) and lipid peroxidation using the fluorescent probe parinaric acid (PnA) in a matrix of submitochondrial particles, as described by de Hingh (1995). The procedure to measure xanthine oxidase activity is described by DellaCorte and Stirpe, 1972). Figure 1 shows the effect of xanthine and xanthine oxidase on cell lysis induced by "simulated ischemia". Based on dose response curves for both xanthine and xanthine oxidase (results not shown) a concentration of 4.5 mIU/ml xanthine oxidase and 50 μmol/l xanthine or hypoxanthine were used. No significant differences were observed between the two substrates used. The LDH release was determined after 120 min of "simulated ischemia" and a subsequent 15 min of reoxygenation or during 135 min normoxic conditions. Under normoxic conditions the addition of xanthine and xanthine oxidase did not induce cell lysis. "Simulated ischemia" caused a significant increase in cell lysis, which was further increased five fold by the inclusion of the xanthine-xanthine oxidase system. Addition of xanthine alone during "simulated ischemia" did not affect cell lysis, whereas the addition of xanthine oxidase resulted in a small, not significant, increase in LDH release.

To study the effect of allopurinol and oxypurinol on the cell lysis induced by "simulated ischemia" the drugs were added both in the presence and absence of xanthine and xanthine oxidase and LDH release was determined. In order to compare the results of various experiments performed with cells from different cell isolations, the LDH release by cells subjected to "simulated ischemia" and xanthine and xanthine oxidase in the absence of the drugs was set at 100% and the LDH release during the other incubations were expressed relative to this.

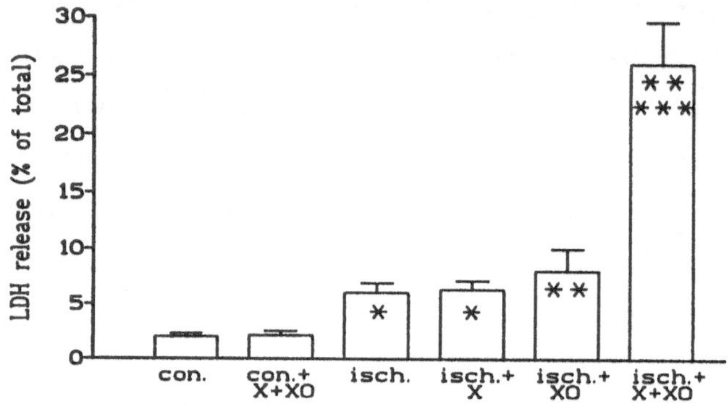

Figure 1: The effect of xanthine oxidase (XO) (4.5 mIU/ml) and/or xanthine (X) (50 μmol/l) on cultured neonatal rat heart myocytes integrity during control, normoxic conditions (con.) and simulated ischemic/reperfusion conditions (isch.). Cell integrity was assessed by LDH release and it is clear that the inclusion of X + XO during ischemia/reperfusion leads to extensive cell damage, whereas this inclusion had no effect on normoxic cells. ($4 \leq n \geq 6$, ± s.d).

Figure 2A shows the effect of allopurinol. A dose dependent attenuation of cell lysis during the "simulated ischemia/reperfusion" can be observed in the presence of xanthine and xanthine oxidase and a similar effect is obtained with oxypurinol (figure 2B). Addition of either compounds to cells subjected to "simulated ischemia/reperfusion" in the absence of xanthine and xanthine oxidase did not affect LDH release during "simulated ischemia" (not shown).

As one of the mechanisms by which xanthine / xanthine oxidase might cause cell lysis is the induction of lipid peroxidation, we focussed on this process using a procedure developed in this laboratory based on the poly-unsaturated fluorescent fatty acid parinaric acid (DeHingh *et al.* 1995, Kuypers *et al.* 1987, van den Berg 1994).

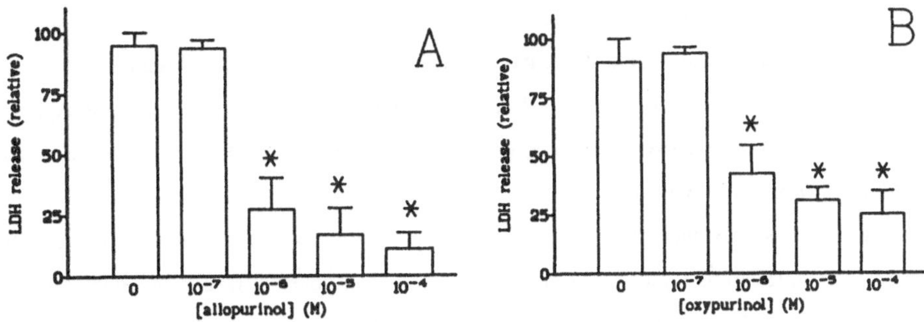

Figure 2: The effect of the inhibitors of xanthine oxidase allopurinol (panel A) and oxypurinol (panel B) on LDH release during ischemia/reperfusion. The LDH released by cells incubated in the absence of the drugs was set at 100% in order to compare different cell preparations (see text) ($4 \leq n \geq 6$, ± s.d.; * = $p<0.05$).

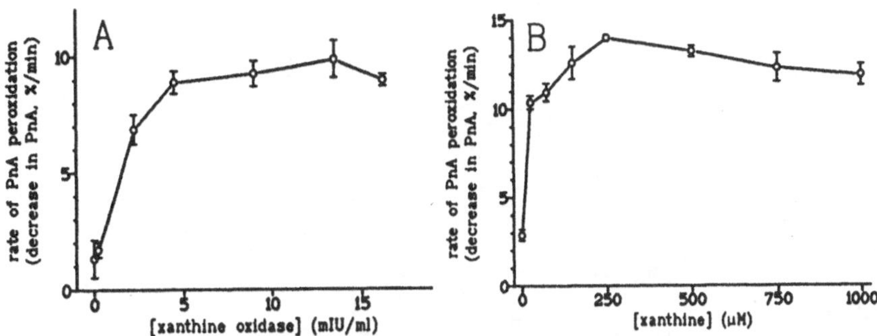

Figure 3: The effect of varying xanthine oxidase activity (panel A) at 50 µmol/l xanthine and varying xanthine concentrations (panel B) at 13.5 mIU xanthine oxidase/ml on the rate of parinaric acid peroxidation in the submitochondrial particle system (n=3, ± s.d).

In the present experimental set-up the fatty acid could, however, not be used with the intact myocardial cells because β-oxidation resulted in a fast and almost complete loss of the fluorescent signal directly after the labelling procedure (Post, Op den Kamp, unpublished observation). Therefore we performed the lipid peroxidation studies in a model system consisting of a biological membrane containing parinaric acid. As a biological membrane submitochondrial particles were used, which have been used before in lipid peroxidation studies (DeHingh *et al.* 1995). Since the experimental conditions vary between the experiments carried out with intact cells and the subsequent lipid peroxidation experiments, using an isolated lipid matrix, we set up dose response curves for both xanthine oxidase (Figure 3A) and xanthine (Figure 3B). Based on these results the following experiments designed to study the effect of allopurinol and oxypurinol on the observed lipid peroxidation were performed in the presence of 150 μmol/l xanthine and 13.5 mIU/ml xanthine oxidase.

Figure 4A shows the effect of oxypurinol on xanthine -xanthine oxidase induced lipid peroxidation. It can be seen clearly that the addition of oxypurinol results in a decrease in the rate of loss of fluorescence signal, with a 50% inhibition around 27 μmol/l oxypurinol. Figures 4B/C show the effect of the addition of allopurinol. Allopurinol has two effects on lipid peroxidation. First of all, allopurinol results in the occurrence of a delay of PnA peroxidation upon the addition of xanthine and the duration of this delay, the lag-phase, is clearly dose dependent (Figure 4C). Secondly, once the loss of fluorescence signal occurs, the rate of loss of the signal is decreased, with a 50% inhibition around 3 μmol/l (Figure 4B).

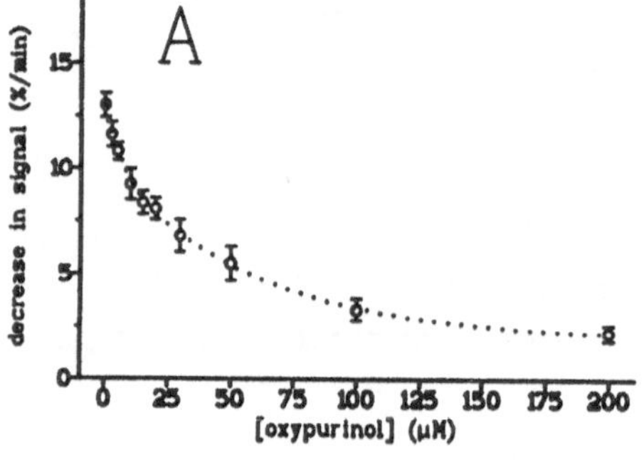

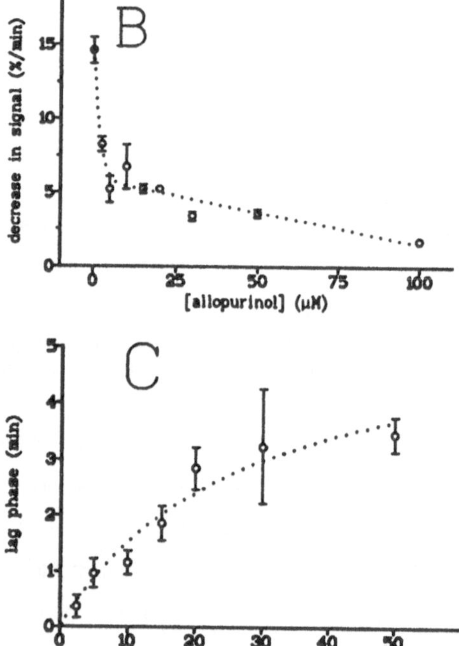

Figure 4: The effect of the inhibitors of xanthine oxidase, oxypurinol (panel A) and allopurinol (panels B and C) on parinaric acid peroxidation induced by xanthine oxidase (13.5 mIU/ml) and xanthine (150 μmol/l). The percentage of inhibition is shown. In the absence of inhibitors the rate of parinaric acid peroxidation was 12.5 % / min. Oxypurinol inhibits peroxidation with a 50% inhibition at 27 μmol/l (panel A). Allopurinol inhibits peroxidation with a 50% inhibition at 3 μmol/l (panel B) and furthermore induces a delay of parinaric acid peroxidation upon the addition of xanthine (panel C) (n=3, ± s.d). All concentrations of inhibitors showed significant inhibition of peroxidation or increase of the duration of lag phase ($p < 0.05$).

These results directly show that free radicals, generated by xanthine and xanthine oxidase, are able to cause lipid peroxidation and that this proces of peroxidation can be attenuated and almost completely inhibited by both oxypurinol and allopurinol.

To further investigate the inhibition of lipid peroxidation by the two compounds we took advantage of the possibility of the sub mitochondrial particles to generate oxygen free radicals within the lipid matrix. In the presence of rotenone and ADP-Fe(III) the conversion of NADH by complex I of the mitochondrial electron transport chain yields reactive oxygen species capable of lipid peroxidation (DeHingh *et al.* 1995). In contrast to the xanthine oxidase system, which generates free radicals in the water phase, the rotenone/ADP system generates free radicals within the membrane matrix. Comparison of both systems shows that the production of radicals within the membrane results in a stronger reduction of the PnA signal, than generation in the water phase (decrease in PnA was 23.7 ± 1.6 and 14.6 ± 0.7 % PnA/min respectively, $p < 0.001$). More importantly, the interference of allopurinol and oxypurinol with the peroxidation process was not significant in the situation were the free radicals where generated in the lipid phase, whereas it was highly significant when the free radicals were generated in the water phase (Figure 5A/B). This implies that the major action of the drugs is not at the level of the lipid peroxidation process within the bilayer.

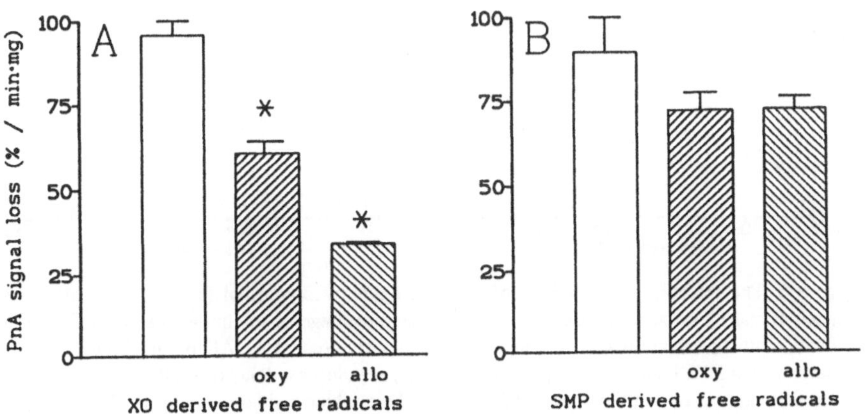

Figure 5: The effect of oxypurinol (20 μmol/l) and allopurinol (20 μmol/l) on parinaric acid peroxidation induced by free radicals generated in the aqueous phase (panel A) or within the lipid matrix (panel B). The rate of parinaric acid signal loss in the absence of the inhibitors was set at 100%. The inhibitors have a strong effect under the conditions when the free radicals are generated in the aqueous phase ($p < 0.001$); no significant inhibition is observed when the radicals are generated in the lipid phase (n=3, ± s.d.).

To answer whether the drugs attenuated lipid peroxidation by scavenging oxygen free radicals or whether they reduced the production of free radicals, we studied the effect of allopurinol and oxypurinol on the enzymatic activity of xanthine oxidase. The effect of the drugs was assessed after a five minute pre-incubation of the enzyme and the drug, identical as for the lipid peroxidation studies. Figures 6A,B and C show the dose dependency of the inhibition of the enzymatic activity and of the occurrence of the lag phase upon allopurinol incubation. For allopurinol the inhibition was 50% at 5 μmol/l, for oxypurinol 34 μmol/l. Because in all experiments using allopurinol or oxypurinol reported in this study the enzym was preincubated with the drug, we looked at the effect of adding the drug while the enzyme was active.

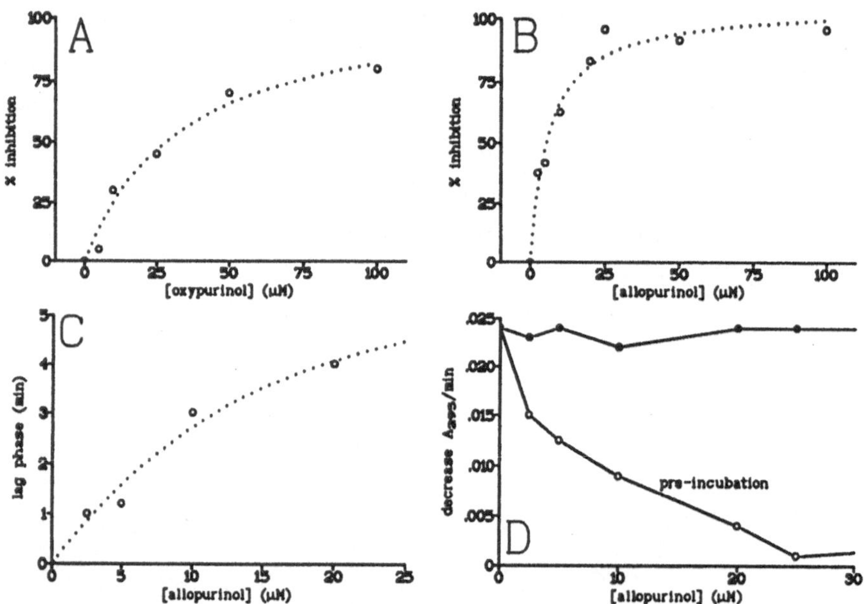

Figure 6: The effect of the inhibitors of xanthine oxidase, oxypurinol (panel A) and allopurinol (panels B and C) on xanthine oxidase activity. The percentage of inhibition is shown. Oxypurinol inhibits xanthine oxidase with a 50% inhibition at 34 μmol/l (panel A). Allopurinol inhibits xanthine oxidase with a 50% inhibition at 5 μmol/l (panel B) and induces a delay of xanthine oxidase activity upon the addition of xanthine (panel C). In panel D xanthine oxidase activity is plotted and it is shown that the effect of allopurinol, in the μmol/l range, on xanthine oxidase is absent when the compound is added when xanthine oxidase is active (closed circles), whereas after a 5 min pre-incubation of the enzyme with the inhibitor, the compound is highly effective (open circles). In all experiments xanthine oxidase activity was 13.5 mIU/ml and xanthine concentration was 150 μmol/l. (data represent the mean of two seperate experiments, which results did not vary more than 10%).

Figure 6D shows that preincubation of the enzyme with the inhibitor is far more effective than adding the inhibitor while the reaction is going on.

Discussion

Effect of xanthine oxidase on cell integrity.

The present study clearly shows that a combination of "simulated ischemia/reperfusion" and xanthine - xanthine oxidase can result in extensive damage of the plasma membrane of the myocyte. The data also show that in the presently used cell model none, or very little, xanthine oxidase is released in the incubation medium during "simulated ischemia" and reperfusion, since the inclusion of xanthine, in the absence of exogenous xanthine oxidase, did not result in an increased cell lysis (figure 1). The clinically used drug allopurinol and its metabolite oxypurinol, were shown to attenuate cell damage caused by the combination of "simulated ischemia / reperfusion" and xanthine oxidase (figure 2), indicating that xanthine oxidase activity is indeed involved in this cell lysis. Inclusion of the drugs in the absence of exogenous xanthine oxidase did not affect cell lysis during "simulated ischemia/reperfusion", which is in line with the observation that addition of xanthine alone did not affect cell lysis either.

In contrast to xanthine - xanthine oxidase induced cell lysis during "simulated ischemia/reperfusion", exposure of control cells to the xanthine - xanthine oxidase system does not result in any increase in lysis at all. This in contrast to other sources of oxygen free radicals, such as cumene hydroperoxide, which does cause cell lysis of normoxic cells (Persoon-Rothert *et al.* 1990). This indicates that during "simulated ischemia" changes take place in the myocardial cell, making the cell more vulnerable for damage induced by free radicals generated by the xanthine - xanthine oxidase system. It has indeed been shown that during ischemia a reduction in the capacity of the defence system against oxygen derived free radicals, eg. superoxide dismutase- and catalase activity, takes place (Hearse 1977, Ferrari *et al.* 1985), which most likely also occurs in the currently used model. However, this defence mechanism is cytosolic in its origin and will not affect the attack on the sarcolemma by free radicals originating from the extracellular space. This implies that during "simulated ischemia" alterations in the sarcolemma must occur, making the membrane more vulnerable to oxidative stress. Are there indications that the sarcolemma is altered during "simulated ischemia", which might cause the membrane to be more vulnerable to free radical attack? Indeed changes do occur at the sarcolemma level during ischemia. First of all, a redistribution of the sarcolemmal phospholipids take place after 60 minutes of "simulated ischemia", before any cell lysis is detectable, resulting in a net outward movement of phosphatidylethanolamine (Musters *et al.* 1993). This will result in an increase in unsaturation of the extracellular leaflet of the sarcolemma (Matos *et al.* 1990), making the sarcolemma more vulnerable for peroxidative damage. Secondly, a phase segregation of the sarcolemmal phospholipids takes place (Post *et al.* 1995a), which affects the packing of the sarcolemmal lipids, possibly making the membrane even more vulnerable for oxidative stress.

Xanthine oxidase induced lipid peroxidation.

How can the sarcolemma be damaged by xanthine oxidase derived free radicals and how can allopurinol and oxypurinol affect these processes? In general, oxygen derived free radicals can lead to damage of proteins, DNA and lipids. We focussed on lipid peroxidation using the parinaric acid technique, allowing a direct, on line, measurement of the loss of a membrane component by lipid peroxidation, rather than measuring peroxidation products. The present data clearly show that the xanthine - xanthine oxidase is capable of lipid peroxidation in a biological membrane (Figure 3). Furthermore, a clear inhibition of this lipid peroxidation by both oxypurinol and allopurinol was observed in the same concentration range at which they attenuated cell lysis in the experiments involving the intact cells. Allopurinol was more effective than oxypurinol in inhibiting lipid peroxidation, with a half maximal concentration for allopurinol of 3 μmol/l and for oxypurinol of 27 μmol/l. This difference in efficiency is not due to a difference in scavenging of membrane soluble radicals, since both compounds no significant inhibition of lipid peroxidation induced by free radicals generated within the lipid matrix (see figure 5B). The difference must either be accounted for by a difference in inhibition of the xanthine oxidase or by a different scavenging capacity of the water soluble free radicals, generated by xanthine oxidase. Both allopurinol and oxypurinol have been shown to be scavengers of the highly reactive hydroxyl radical, with a half maximum concentration of 1 to 2 mmol/l and with oxypurinol being more efficient than allopurinol (Moorhouse *et al.* 1987). In the present experiments the drugs were effective at a much lower concentrations, both in the cell studies and in the lipid peroxidation studies, indicating that the observed protection is unlikely to be due to scavenging of the hydroxyl radical. Direct measurements of the xanthine oxidase activity (figure 6) showed that allopurinol and oxypurinol inhibit the enzyme activity, with almost similar concentrations of half maximal inhibition, indicating that the effect of the compounds on lipid peroxidation is mediated by their effect on the free radical production by the enzyme.

Figure 5 shows an additional effect of allopurinol on xanthine - xanthine oxidase induced lipid peroxidation, namely the induction of a lag phase, which for instance has also been observed for vitamine E (van den Berg 1994). Induction of a lag phase suggests that the added compound scavenges free radicals, resulting in its degradation and that, at a certain timepoint, the number of savenger molecules becomes depleted and peroxidation of the reporter molecule, parinaric acid, commences and reaching a rate comparable to that observed in the absence of the scavenger (van den Berg 1994). However, this is not the case with allopurinol because: i) although lipid peroxidation commences after a while, its rate is strongly reduced for long periods of time and ii) a comparable induction of a lag phase is observed at the level of xanthine oxidase activity (figure 6).

Implications

Although the role of myocardial derived xanthine oxidase in ischemia/reperfusion damage is not settled yet (as put foreward in the introduction) another pool of xanthine oxidase, derived from distant organs, such as liver and lung, might be involved in mediating oxygen derived free radicals and thereby cause myocytal ischemia/reperfusion damage (Tabayashi *et al.* 1991). The induction of cell lysis in the present study by xanthine oxidase does not necessarily prove that xanthine oxidase is involved in cell injury during ischemia/reperfusion *in vivo*. However, it does show that activation of xanthine oxidase can result in free radical generation and cell damage, as has been suggested by others (Hearse *et al.* 1986, McCord *et al.* 1985). Furthermore, it shows directly that the activation of xanthine oxidase leads to lipid peroxidation in a membrane. Since there is still a controversy regarding the involvement of the xanthine oxidase derived oxygen free radicals in ischemia/reperfusion damage and therefore the controversy of the benificial effects of allopurinol under *in vivo* conditions is present as well. The present *in vitro* study clearly shows a protective effect of both allopurinol and oxypurinol and this observed protection is in line with several studies showing a protective effect allopurinol administration has on ischemic-reperfusion injury in man (McCord *et al.* 1985, Sisto *et al.* 1995, Tabayashi *et al.* 1991, Stewart *et al.* 1985, Gimpel *et al.* 1995).

Both oxypurinol and allopurinol reduced cell lysis during the combination of ischemia/reperfusion and xanthine - xanthine oxidase exposure at concentrations above 10^{-6} mol/l. Allopurinol and oxypurinol peak plasma concentrations have been reported to be 10 to

40 μmol/l (Murrell and Rapeport 1986), showing that the concentrations of the drugs that resulted in attenuation of cell lysis in the present study are in a pharmacological relevant range. *In vivo*, allopurinol is rapidly converted to oxypurinol by the liver, which has a biological half-life of 18 hours (Murrell and Rapeport 1986). The data presented in this paper indicate that allopurinol is more effective in inhibiting xanthine oxidase and lipid peroxidation than oxypurinol, suggesting that administration of allopurinol just before onset of the surgery might be more effective than longer treatment, this to prevent conversion of the drug to oxypurinol.

With regard to the half maximal concentration for both enzyme activity and lipid peroxidation, it can be noted that the values reported in the present study are 10 to 100 fold lower than reported in other studies (Moorhouse *et al.* 1987, Zweier *et al.* 1994). A likely explanation for this difference is shown in figure 6D, where it is clearly shown that the timepoint at which the drugs are introduced dramatically affects their effect on xanthine oxidase activity. When for instance 5 μmol/l allopurinol is preincubated with xanthine oxidase, before xanthine is added, an almost total inhibition of the enzyme is obtained. In contrast, the addition of 25 μmol/l allopurinol in the presence of xanthine, had no effect on xanthine oxidase activity. This suggests that under controlled ischemia/reperfusion conditions, such as aortocoronary by-pass grafting, the drug(s) should be present before the onset of ischemia and the formation of (hypo)xanthine.

Acknowledgements

This study was supported by the Dutch Heart Organization. The research of Dr Post has been made possible by a fellowship of the Royal Netherlands Academy of Arts and Sciences.

References:

Akizuki, S. Yoshida, S., Chambers, D.E., Eddy, L.J., Parmley, L.F., Yellon, D.M., Downey, J.M. 1985 Infarct size limitation by xanthine oxidase inhibitor, allopurinol, in closed chest dogs qith small infarcts, *Cardiovasc Res* **19**:686-692

Badylak, S.F., Simmons, A., Turek, J., Babbs, C.F. 1987 Protection from reperfusion injury in the isolated heart by post-ischemic deferoxamine and oxypurinol administration. *Cardiovasc Res* **21**:500-506

Bolli, R., Hartley, C.J., Chelly, J.E., Patel, R., Rabinovitz, R.S., Jeroudi, M.O., Roberts, R., Noon, G. 1990 An accurate nontraumatic ultrasonic method to monitor myocardial wall thickening in patients undergoing cardiac surgery. *J Am Coll Cardiol* **15**:1055-1065

Coghlan, J.G., Flitter, W.D., Clutton, S.M., Ilsley, C.D., Rees, A., Slater, T.F. 1993 Lipid peroxidation and changes in vitamin E levels during coronary artery bypass grafting. *J Thorac Cardiovasc Surg* **106**:268-274

De Hingh, Y.C.M., Meyer, J., Fischer, J., Berger, R., Smeitink, J.A.M., Op den Kamp, J.A.F.

1995 Direct measurement of lipid peroxidation in submitochondrial particles. *Biochem* **34**:12755-12760

Della Corte, E., Stirpe, F. 1972 The regulation of rat liver xanthine oxidase. *Biochemical J* **126**:739-745.

Ferrari, R., Ceconi, S., Curello, S., Cargnoni, A., Passini, E., De Guili, E., Albertini, A. and Visoli O. 1985 Oxygen-mediated myocardial damage during ischaemia and reperfusion: role of the cellular defences against oxygen toxicity. *J Mol Cell Cardiol* **17**:937-945

Flameng, W., Borgers, M., van der Vusse, G.J., Demeyere, R., Vandermeersch, E., Thoné, F., Suy, R. 1983 Cardioprotective effects of lidoflazine in extensive aorto-coronary bypass grafting. *J Thorac Cardiovasc Surg* **85**:758-768

Gimpel, J.A., Lahpor, J.R., van der Molen, A.J., Damen, J., Hitchcock, J.F. 1995 Reduction of reperfusion injury of human myocardium by allopurinol: A clinical study. *Free Rad Biol Med* **19**:251-255.

Guarnieri, C., Flamigni, F., Caldarera, C.M. 1980 Role of oxygen in the cellular damage induced by reoxygenation of hypoxic heart. *J Mol Cell Cardiol* **12**:797-808

Hearse, D.J. 1977 Reperfusion of the ischemic myocardium. *J Mol Cell Cardiol* **9**:605-616

Hearse, D.J., Manning, A.S., Downey, J.M., Yellon, D.M. 1986 Xanthine oxidase: a critical mediator of myocardial injury during ischemia and reperfusion. *Acta Physiol Scand* **548**(Suppl):65-78

Hearse, D.J., Bolli, R. 1992 Reperfusion induced injury, manifestations, mechanisms and clinical relevance. *Cardiovasc Res* **26**:101-108

Janssen, M., van der Meer, P., de Jong, J.W. 1993 Antioxydant defences in rat, pig, guinea pigs and human hearts: comparison with xanthine oxidoreductase activity. *Cardiovasc Res* **27**:2052-2057

Jennings, R.B., Reimer, K.A. 1991 The cell biology of acute myocardial ischemia. *Ann Rev Med* **42**:225-246

Kooij, A. 1994 A re-evaluation of of the tissue distribution and physiology xanthine oxidoreductase. *Histochem J* **26**:889-915

Kuypers, F.A., van den Berg, J.J.M., Schalkwijk, C., Roelofsen, B., Op den Kamp, J.A.F. 1987 Cis-parinaric acid as a fluorescent membrane probe to determine lipid peroxidation. *Biochim Biophys Acta* **921**:266-274.

Matos, M.J., Post, J.A., Roelofsen, B., Op den Kamp, J.A.F. 1990 Composition and organization of sarcolemmal fatty acids in cultured neonatal rat cardiomyocytes. *Cell Biol. Intl. Rep.* **14**:343-352

McCord, J.M., Roy, R.S. 1982 The pathophysiology of superoxide: roles in inflammation and ischemia *Can. J. Physiol. Pharmacol.* **60**:1346-1352

McCord, J.M., Roy, R.S., Schaffer, S.W. 1985 Free radicals and myocardial ischemia. The role of xanthine oxidase. *Adv Myocardiol* **5**:183-189

Moorhouse, P.C., Grootveld, M., Halliwell, B., Quinlan, J.G., Gutteridge, J.M.C. 1987 Allopurinol and oxypurinol are hydroxyl radical scavengers. *FEBS Lett* **213**:23-28

Moriwaki, Y., Yamamoto, T., Suda, M., Nasako, Y., Takahashi, S., Agbedana, O.E., Hada, T., Higashino, K. 1993 Purification and immunohistochemical tissue localization of human xanthine oxidase. *Biochim Biophys Acta* **1164**:327-330

Murrell, J.G., Rapeport, W.G. 1986 Clinical pharmacokinetics of allopurinol. *Clin Pharmacokinetics* **11**:343-353

Musters, R.J.P., Post, J.A., Verkleij, A.J. 1991 The isolated neonatal rat-cardiomyocyte used in an in vitro model for "ischemia",I: a morphological study. *Biochim. Biophys. Acta* **1091**:270-277

Musters, R.J.P., Otten, E., Biegelmann, E., Bijvelt, J., Keijzer, J.J.H., Post, J.A., Op den Kamp, J.A.F., Verkleij, A.J. 1993 Loss of asymmetric distribution of sarcolemmal phosphatidylethanolamine during simulated ischemia in the isolated neonatal rat cardiomyocyte. *Circ. Res.* **73**: 514-523

Persoon-Rothert, M., Egas-Kenniphaas, J.M., van der Valk-Kokshoorn, E.J.M., Mauve, I., van der Laarse, A. 1990 Prevention of cumene hydroperoxide induced oxidative stress in cultured neonatal rat myocytes by scavengers and enzyme inhibitors. *J Mol Cell Cardiol* **22**:1147-1155

Post, J.A., Langer, G.A., Op den Kamp, J.A.F., Verkleij, A.J. 1988 Phospholipid asymmetry in cardiac sarcolemma. Analysis of intact cells and "gas-dissected" membranes. *Biochim Biophys Acta* **943**:256-266

Post, J.A., Verkleij, A.J., Langer, G.A. 1995a Organization and function of sarcolemmal phospholipids in control and ischemic/reperfused cardiomyocytes. *J. Mol. Cell. Cardiol.* **27**:749-760

Post, J.A., Bijvelt, J.J.M., Verkleij, A.J. 1995b The role of phosphatidylethanolamine in sarcolemmal damage of cultured heart myocytes during simulated ischemia and metabolic inhibition. *Am. J. Physiol. (Heart Circ. Physiol.)* **268**:H773-H780

Sisto, T., Paajanen, H., Metsaketela, T., Harmoinen, A., Nordback, I., Tarkka, M. 1995 Pretreatment with antioxydants and allopurinol diminishes cardiac onset events in coronary artery bypass grafting. *Ann Thorac Surg* **59**:1519-1523

Stewart, J.R., Crute, S.L., Loughlin, V., Hess, M.L., Greenfield, L.J. 1985 Prevention of free radical-induced myocardial reperfusion injury with allopurinol. *J Thorac Cardiovasc Surg* **90**:68-72

Tabayashi, S., Suzuki, Y., Nagame, S., Ho, Y., Sekino, Y., Mohri, H. 1991 A clinical trial of allopurinol (Zyloric) for myocardial protection. *J Thorac Cardiovasc Surg* **101**:713-718

Thompson, J.A., Hess, M.L. 1986 The oxygen free radical system: a fundamental mechanism in the production of myocardial necrosis. *Progress Cardiovasc Diseases* **286**:449-462

van den Berg, J.J.M. 1994 Effects of oxidants and and antioxidants evaluated using parinaric acid as a convenient and sensitive probe. *Redox report* **1**:11-21

Vemuri, R., Yagev, S., Heller, M., Pinson, A.J. 1985 Studies on oxygen and volume restrictions in cultured cardiac cells, I: a model for ischemia and anoxia with a new approach. *In Vitro* **21**:521-525

Zweier, J.L., Broderick, R., Kuppusamy, P., Thompson-Gorman, Lutty, G.A. 1994 Determination of the mechanism of free radical generation in human aortic endothelial cells exposed to anoxia and reoxygenation. *J Biol Chem* **269**: 24156-24162

Subject Index

NATO ASI Series H

Vol. 89: **Molecular and Cellular Mechanisms of H^+ Transport.**
Edited by B. H. Hirst. 504 pages. 1994.

Vol. 90: **Molecular Aspects of Oxidative Drug Metabolizing Enzymes: Their Significance in Environmental Toxicology, Chemical Carcinogenesis and Health.**
Edited by E. Arınç, J. B. Schenkman, E. Hodgson. 623 pages. 1994.

Vol. 91: **Trafficking of Intracellular Membranes: From Molecular Sorting to Membrane Fusion.**
Edited by M. C. Pedroso de Lima, N. Düzgüneş, D. Hoekstra. 371 pages. 1995.

Vol. 92: **Signalling Mechanisms – from Transcription Factors to Oxidative Stress.**
Edited by L. Packer, K. Wirtz. 466 pages. 1995.

Vol. 93: **Modulation of Cellular Responses in Toxicity.**
Edited by C. L. Galli, A. M. Goldberg, M. Marinovich. 379 pages. 1995.

Vol. 94: **Gene Technology.**
Edited by A. R. Zander, W. Ostertag, B. V. Afanasiev, F. Grosveld. 556 pages. 1996.

Vol. 95: **Flow and Image Cytometry.**
Edited by A. Jacquemin-Sablon. 241 pages. 1996.

Vol. 96: **Molecular Dynamics of Biomembranes.**
Edited by J. A. F. Op den Kamp. 424 pages. 1996.

Vol. 97: **Post-transcriptional Control of Gene Expression.**
Edited by O. Resnekov, A. von Gabain. 283 pages. 1996.

Vol. 98: **Lactic Acid Bacteria: Current Advances in Metabolism, Genetics and Applications.**
Edited by T. F. Bozoglu, B. Ray. 412 pages. 1996.

Vol. 99: **Tumor Biology Regulation of Cell Growth, Differentiation and Genetics in Cancer.**
Edited by A. S. Tsiftsoglou, A. C. Sartorelli, D. E. Housman, T. M. Dexter. 342 pages. 1996.

Vol. 100: **Neurotransmitter Release and Uptake.**
Edited by Ş. Pöğün. 344 pages. 1997.

Vol. 101: **Molecular Mechanisms of Signalling and Membrane Transport.**
Edited by K. W. A. Wirtz. 344 pages. 1997.

Vol. 102: **Interacting Protein Domains**
Edited by L. Heilmeyer. 296 pages. 1997.

Vol. 103: **Molecular Microbiology.**
Edited by S. J. W. Busby, C. M. Thomas, N. L. Brown. 334 pages. 1998.

NATO ASI Series H

Vol. 104: **Cellular Integration of Signalling Pathways in Plant Development.**
Edited by F. Lo Schiavo, R. L. Last, G. Morelli, N. V. Raikhel.
330 pages. 1998.

Vol. 105: **Gene Therapy.**
Edited by Kleanthis G. Xanthopoulos.
238 pages. 1998.

Vol. 106: **Lipid and Protein Traffic.**
Pathways and Molecular Mechanisms.
Edited by J. A. F. Op den Kamp.
374 pages. 1998.